Food Engineering, Production and Analysis

Food Engineering, Production and Analysis

Edited by **Hugh Brennan**

SYRAWOOD
PUBLISHING HOUSE

New York

Published by Syrawood Publishing House,
750 Third Avenue, 9th Floor,
New York, NY 10017, USA
www.syrawoodpublishinghouse.com

Food Engineering, Production and Analysis
Edited by Hugh Brennan

International Standard Book Number: 978-1-68286-132-5 (Hardback)

Printed in the United States of America.

Contents

Preface

Food engineering as a field is very diverse and finds applications in agriculture, food processing and associated industries. This book presents researches and studies performed by experts across the globe. It elaborates concepts revolving around food science and technology, food chemistry, properties of various foods, relationship between food and the environment, etc. The extensive content of this book provides the readers with a thorough understanding of the subject. Students, researchers, experts and all associated with food engineering will benefit alike from this book.

This book is a result of research of several months to collate the most relevant data in the field.

When I was approached with the idea of this book and the proposal to edit it, I was overwhelmed. It gave me an opportunity to reach out to all those who share a common interest with me in this field. I had 3 main parameters for editing this text:

1. Accuracy – The data and information provided in this book should be up-to-date and valuable to the readers.
2. Structure – The data must be presented in a structured format for easy understanding and better grasping of the readers.
3. Universal Approach – This book not only targets students but also experts and innovators in the field, thus my aim was to present topics which are of use to all.

Thus, it took me a couple of months to finish the editing of this book.

I would like to make a special mention of my publisher who considered me worthy of this opportunity and also supported me throughout the editing process. I would also like to thank the editing team at the back-end who extended their help whenever required.

Editor

Comparative Evaluation of Diagnostic Tools for Oxidative Deterioration of Polyunsaturated Fatty Acid-Enriched Infant Formulas during Storage

Caroline Siefarth [1,2], **Yvonne Serfert** [3], **Stephan Drusch** [4] **and Andrea Buettner** [1,2,*]

[1] Department of Chemistry and Pharmacy, Emil Fischer Centre, Friedrich-Alexander University of Erlangen-Nürnberg, Schuhstr. 19, Erlangen 91052, Germany; E-Mail: caroline.siefarth@fau.de

[2] Fraunhofer Institute for Process Engineering and Packaging (IVV), Giggenhauser Str. 35, Freising 85354, Germany

[3] Department of Food Technology, University of Kiel, Heinrich-Hecht-Platz 10, Kiel 24118, Germany; E-Mail: yserfert@foodtech.uni-kiel.de

[4] Department of Food Technology and Food Material Science, Institute of Food Technology and Food Chemistry, Technical University of Berlin, Königin-Luise-Str. 22, Berlin 14195, Germany; E-Mail: stephan.drusch@tu-berlin.de

* Author to whom correspondence should be addressed; E-Mail: andrea.buettner@ivv.fraunhofer.de

Abstract: The challenge in the development of infant formulas enriched with polyunsaturated fatty acids (PUFAs) is to meet the consumers' expectations with regard to high nutritional and sensory value. In particular, PUFAs may be prone to fatty acid oxidation that can generate potential rancid, metallic and/or fishy off-flavors. Although such off-flavors pose no health risk, they can nevertheless lead to rejection of products by consumers. Thus, monitoring autoxidation at its early stages is of great importance and finding a suitable analytical tool to perform these evaluations is therefore of high interest in quality monitoring. Two formulations of infant formulas were varied systematically in their mineral composition and their presence of antioxidants to produce 18 model formulas. All models were aged under controlled conditions and their oxidative deterioration was monitored. A quantitative study was performed on seven characteristic odor-active secondary oxidation products in the formulations via two-dimensional high resolution gas chromatography-mass spectrometry/olfactometry (2D-HRGC-MS/O). The sensitivity of the multi-dimensional GC-MS/O analysis was supported by two additional analytical tools for monitoring autoxidation, namely the analysis of lipid hydroperoxides and conjugated dienes. Furthermore, an aroma profile analysis (APA) was performed to reveal the presence and intensities of typical odor qualities generated in the course of fatty acid oxidation. The photometrical analyses of lipid hydroperoxides and conjugated dienes were found to be too insensitive for early indication of the development of sensory defects. By comparison, the 2D-HRGC-MS/O was capable of monitoring peroxidation of PUFAs at low *ppb*-level in its early stages. Thereby, it was possible to screen oxidative variances on the basis of such volatile markers already within eight weeks after production of the products, which is an earlier indication of oxidative deterioration than achievable via conventional methods. In detail, oxidative variances between the formulations revealed that lipid oxidation was low when copper was administered in an encapsulated form and when antioxidants (vitamin E, ascorbyl palmitate) were present.

Keywords: antioxidants; autoxidation; infant formula; mineral composition; two-dimensional GC-MS; off-flavor; PUFA

Abbreviations

2D-HRGC-MS/O: two-dimensional high resolution gas chromatography-mass spectrometry/olfactometry; AA: arachidonic acid; ANOVA: analysis of variance; APA: aroma profile analysis; CLA: conjugated linoleic acid; DHA: docosahexaenoic acid; FID: flame ionization detector; FTIR: fourier transform infrared; GC: gas chromatography; GOS/FOS: galacto oligosaccharides/fructo oligosaccharides; HPLC: high performance liquid chromatography; IDF: international dairy federation; LC: long chain; LOD: limit of detection; LOQ: limit of quantification; LPO: lipid hydroperoxides; MS: mass spectrometry; ND/NA: not detected/not analyzed; n.s./sig.: not significant/significant; (P)UFA: (poly) unsaturated fatty acid; PV: peroxide value; SAFE: solvent assisted flavor evaporation; SD: standard deviation; SIDA: stable isotope dilution assay; UV: ultra violet; VOC: volatile organic compound.

1. Introduction

Polyunsaturated fatty acids (PUFAs) may be easily oxidized to hydroperoxides via light-induced photooxygenation and/or in conjunction with further autoxidation processes. Thereby, radicals are formed in the presence of transition metals or heme proteins by a uni-molecular decomposition of hydroperoxides. Hence, peroxidation is initiated and then accelerates auto-catalytically. Several factors can promote autoxidation processes: fatty acid composition, oxygen partial pressure, concentration and activity of pro-oxidants and anti-oxidants, surface area of the product, and conditions of storage in terms of temperature, light, or water activity [1,2]. Infant formulas are high in PUFAs to match the fatty acid composition of mothers' milk [3]. Thus, these products may be susceptible to fatty acid oxidation. Powdered infant formulas with a low water-activity and a large particle surface area are especially particularly susceptible to autoxidation. Furthermore, infant formulas are increasingly enriched with long chain (LC) *n*-3 and *n*-6 fatty acids [e.g., docosahexaenoic acid (DHA), arachidonic acid (AA), respectively] due to their purported promotion of visual and cognitive development [4].

Lipid oxidation leads to the formation of a broad variety of volatile and non-volatile substances. In several cases, the volatile substances formed are extremely odor-active. Thus, even small concentrations of typical secondary fatty acid oxidation products like aldehydes, ketones or alcohols can lead to rancid, fishy, metallic or cardboard-like off-flavors [5–9]. According to Grosch [10], unsaturated aldehydes with 6, 9 and 10 carbon atoms, respectively, and vinyl ketones with 8 carbon atoms are of special interest due to their high odor intensities and low odor thresholds, respectively [11]. Recently, a large number of odor-active volatile organic compounds (VOCs) were described as being responsible for a fishy off-odor profile in encapsulated fish oil products or formed in human milk during (freeze-) storage [5,12]. The volatile composition of infant formulas has also been previously studied [13,14]. Aldehydes, ketones, short chain fatty acids, sulfides and furanes, which are derived mainly from lipid autoxidation and thermal oxidation, could be identified via headspace sampling and GC-MS analyses. Many of these VOCs match those that were described as key contributors to fishy off-odor profiles, e.g., in stored human milk. Since the most recent study on the volatile profiles of infant formulas [13], comprehensive research has been undertaken and further advancements have been achieved, such as the enrichment of infant formulas with LC-PUFAs rather than replacing the milk fat with plain vegetable oils. As a drawback of such developments, sensory problems in enriched infant formulas might arise due to the increased susceptibility of the LC-PUFA-enriched matrices to lipid autoxidation. To the best of our knowledge, a combined analytical study of non-volatile primary products and odor-active secondary products of lipid autoxidation utilizing different analytical strategies, with additional aroma profile analysis (APA) of varying formulations of LC-PUFA-enriched infant formulas has not been reported in literature before. Accordingly, the aim of the present study was to identify the most sensitive method for monitoring lipid oxidation from its early stages for quality control purposes.

So far, standard tools for monitoring fatty acid oxidation are the determination of lipid hydroperoxides or peroxide value (PV) and conjugated dienes. In these methods, the formation of primary lipid oxidation products is measured either photometrically or iodometrically. A good overview of common methods for analysis of total hydroperoxides is given by Dobarganes and Velasco [15] with special focus on sensitivity and applications. The latter aspect is of great importance, especially when complex food matrices like infant formula emulsions are analyzed for lipid oxidation instead of pure fats and oils. However, an important aspect that must be kept in mind is that these primary oxidation products are just intermediates and prone to decomposition. Thus, another possibility is to measure the volatile, non-volatile and/or polymeric secondary oxidation products that are formed through decomposition reactions of hydroperoxides. With regard to lipid oxidation monitoring, the detection of VOCs and especially individual carbonyl compounds is gaining momentum in addition to the commonly established determination of total carbonyl content. For example, pentanal and hexanal are typically formed during the oxidation of *n*-6 PUFAs, and propanal is known as a characteristic secondary oxidation product of *n*-3 PUFAs [16]. In general, hexanal analysis via headspace GC methods or GC-MS is well established in the food industry as an indicator for lipid oxidation [17,18]. The advantage of GC-MS compared to most common analytical methods used for carbonyl determinations (e.g., spectrophotometry of chromophoric derivatives or combined with high-performance liquid chromatography, HPLC) is its high sensitivity in combination with mass specificity. Furthermore, the accuracy and precision of GC-MS can be further improved when used in combination with stable isotope dilution assays (SIDA) as well as combinatory or multi-dimensional approaches such as GC-GC (heart-cut)-MS [19].

In the present study, three different methods for monitoring lipid oxidation were comparatively applied to 18 LC-PUFA-enriched and spray-dried infant formulas that were formulated to provide a range of combinatory compositions of minerals and antioxidants. A range of odor-active secondary oxidation marker substances were monitored in the formulas by means of 2D-HRGC-MS in combination with SIDA over a storage period of two months. Lipid hydroperoxides were determined in parallel for a storage period of up to one year via the commonly established thiocyanate method, as well as conjugated dienes by conventional UV detection (established protocols, *cf.* Section 2.7).

Quantitative data on odor-active compounds obtained from GC-MS analyses as well as the two standard oxidation parameters were compared with the results of an APA on typical odor qualities of the *n*-3 and *n*-6 fatty acid enriched infant formulas. In addition, a trained sensory panel evaluated if the observed flavor changes were rated as off-flavor in the respective infant formulas.

2. Experimental Section

2.1. Materials

The powdered infant formulas consisted of the following ingredients (supplier in parentheses): skim milk (obtained locally from a milk processing company, Cremilk GmbH, Kappeln, Germany), skim milk powder (Ledor MMP-S UST, Hochdorf Swiss Milk AG, Hochdorf, Switzerland), whey protein concentrate (WPC P45, Prolactal GmbH, Hartberg, Austria), vegetable fats and oils (palm, rapeseed, linseed, coconut, sunflower, high oleic sunflower) (Florin AG, Muttenz, Switzerland), butter (Emmi Schweiz AG, Luzern, Switzerland), maltodextrin (Tereos Syral S.A.S., Marckolsheim, France), lactose (Milei GmbH, Leutkirch, Germany), dietary fibers: galacto-oligosaccharides (GOS, Vivinal® GOS, FrieslandCampina Domo EMEA, Amersfoort, the Netherlands) and fructo-oligosaccharides (FOS, Selectchemie AG, Zürich, Switzerland), soya lecithin (Solae Europe S.A., Le Grand-Saconnex, Switzerland), monoglycerides based on rapeseed and palm oil (Dimodan® PH 100 NS/B Kosher, DuPont™ Danisco®, Danisco Switzerland AG, Switzerland), fish oil (Marinol D-40, Stepan Lipid Nutrition Europe, Koog aan de Zaan, the Netherlands), arachidonic acid oil (Martek Biosciences, DSM Nutritional Products Europe Ltd., Basel, Switzerland), nucleotides (DSM Nutritional Products Europe Ltd., Heerlen, the Netherlands), lactoferrin (Morinaga Milk Industry Co. Ltd., Tokyo, Japan), antioxidants ascorbyl palmitate and DL-α-tocopherol (Ronoxan® A, DSM Nutritional Products Europe Ltd.), vitamins and minerals. A mixture of vitamins and semi-essential nutrients [vitamin C (sodium ascorbate), vitamin E, vitamin A, niacin, vitamin D, pantothenic acid, vitamin K, vitamin B12, biotin, vitamin B1, folic acid, vitamin B6, vitamin B2, choline bitartrate, taurine, inositol, L-carnitine] was purchased from Glanbia Nutritionals Deutschland GmbH (Orsingen-Nenzingen, Germany). Mineral supplements were also purchased as a single mixture (magnesium chloride, manganese sulfate, sodium selenite, zinc sulfate and potassium iodate) from Glanbia Nutritionals Deutschland GmbH. The copper and iron compositions of the infant formulas were varied over 18 different samples. Copper was added either as Cu(II)-sulfate (Glanbia Nutritionals Deutschland GmbH), encapsulated

Cu(II)-sulfate (Vitablend Nederland B.V., Wolvega, the Netherlands) or Cu-lysine-complex. The latter complex was prepared prior to the experiments from Cu(II)-sulfate and L-lysine HCl (Selectchemie AG). Iron was added to the formulas either as Fe(II)-gluconate (Gluconal®, Purac Biochem, Gorinchem, the Netherlands), Fe(III)-pyrophosphate (LipoFer®, Lipofoods, Barcelona, Spain) or encapsulated Fe(II)-sulfate (Vitablend Nederland B.V.).

An overview of the formula composition in relation to the variations of copper and iron is given in Section 2.3.

2.2. Chemicals

GC-MS analysis: The reference compounds with the stated purities were as follows (supplier in parentheses): hexanal 98%, (E,Z)-nona-2,6-dienal 95%, oct-1-en-3-one 50% (Aldrich, Steinheim, Germany), (E,E)-deca-2,4-dienal 85%, (E)-hex-2-enal 97% (Fluka, Steinheim, Germany), (Z)-octa-1,5-dien-3-one 99%, trans-4,5-epoxy-(E)-dec-2-enal 90% (aromaLAB AG, Freising, Germany). The following stable isotope labeled standards were additionally purchased from aromaLAB: [5,5,6,6-^{2}H$_4$]-hexanal, [1,2-^{13}C$_2$]-(E,Z)-nona-2,6-dienal, [1,2-^{2}H$_{2-3}$]-oct-1-en-3-one, [4,5-^{2}H$_2$]-(E,E)-deca-2,4-dienal, [2,3-^{2}H$_2$]-(E)-hex-2-enal, [5,6-^{2}H$_2$]-(Z)-octa-1,5-dien-3-one, [9,9,10,10-^{2}H$_4$]-trans-4,5-epoxy-(E)-dec-2-enal. Dichloromethane p.a. was used as the organic solvent and extracting agent (Th. Geyer GmbH and Co. KG, Renningen, Germany). Anhydrous sodium sulfate was purchased from chemsolute® (Th. Geyer GmbH and Co. KG, Renningen, Germany).

Thiocyanate assay and analysis of conjugated dienes—the following chemicals were used—ammonium thiocyanate (Sigma Aldrich, St. Louis, MO, USA), barium chloride, iron-(II)-sulphate heptahydrate, isooctane (Carl Roth GmbH and Co. KG, Karlsruhe, Germany), 2-propanol (VWR International, Pty Ltd., Murarrie, Australia).

2.3. Experimental Design

Manufactured infant formulas varied in the use of antioxidants (ascorbyl palmitate and DL-α-tocopherol), as well as in copper and iron composition. Either an encapsulated variant of the minerals or a non-encapsulated alternative was used. In the non-encapsulated case, iron variants with different valences (Fe^{2+}, Fe^{3+}) and/or copper variants (Cu^{2+} and a Cu-lysine-complex) were processed (cf. Section 2.1). An overview of the 18 different formulations is given in Table 1.

Table 1. Formulations of the manufactured infant formulas with respect to the presence of antioxidants (ascorbyl palmitate and DL-α-tocopherol) and mineral composition in relation to copper and iron ($n_{formulations} = 18$). X: used ingredient(s).

Formulation	Antioxidants	Cu(II)-Sulfate	Cu-Lysine-Complex	Cu(II)-Sulfate, Encapsulated	Fe(II)-Gluconate	Fe(III)-Pyrophosphate	Fe(II)-Sulfate, Encapsulated
A1	X	X	-	-	X	-	-
A2	X	X	-	-	-	X	-
A3	X	X	-	-	-	-	X
A4	X	-	X	-	X	-	-
A5	X	-	X	-	-	X	-
A6	X	-	X	-	-	-	X
A7	X	-	-	X	X	-	-
A8	X	-	-	X	-	X	-
A9	X	-	-	X	-	-	X
B1	-	X	-	-	X	-	-
B2	-	X	-	-	-	X	-
B3	-	X	-	-	-	-	X
B4	-	-	X	-	X	-	-
B5	-	-	X	-	-	X	-
B6	-	-	X	-	-	-	X
B7	-	-	-	X	X	-	-
B8	-	-	-	X	-	X	-
B9	-	-	-	X	-	-	X

2.4. Manufacturing of Infant Formula

The primary materials for producing infant formulas were mixed in several steps. First of all, fats and oils were heated to 63 °C. Therein, antioxidants and emulsifiers were dissolved, as well as fish oil and arachidonic acid oil. Skim milk was heated to 55 °C and skim milk powder was dissolved in the skim milk. All ingredients were added step-by-step. In a final step, the fat mix was added and the suspension was pre-dispersed using a rotor-stator-blender (Ystral GmbH, Ballrechten-Dottingen, Germany) at 16,000 rpm for 5 min. Subsequently, a two-step homogenization was performed (Panda 2K, Niro Soavi Deutschland, Lübeck, Germany) at 150/50 bar. The formula milk emulsion was spray-dried at a dry substance content of 50% (Mobile Minor, Niro Inc., Copenhagen, Denmark) with two-fluid nozzle at 170 °C/80 °C inlet and outlet temperature, respectively. After spray-drying, lactoferrin was added under dry conditions. The combined formula milk powders were packaged in aluminum bags under 20% vacuum (VM-19/S/CL, Röscher GmbH, Berlin, Germany). The packaged samples were stored unopened at constant temperature of 21 ± 1 °C for up to 12 months. The powdered infant formulas were analyzed for lipid oxidation products immediately after manufacturing, as well as after the following storage periods (months): 1, 2, 3, 6, 9 and 12 at room temperature.

2.5. Sample Preparation

For analyses of lipid oxidation products, the powdered formula milks were reconstituted with tap water as follows: 180 mL of tap water was boiled in an Erlenmeyer flask for 1 min and cooled down to 40 °C. Formula milk powder (27.0 ± 0.1 g) was weighed in a beaker and the tap water was added. The mixture was shaken well for 1 min.

For GC-MS analyses, the reconstituted infant formulas were first subjected to solvent assisted flavor evaporation (SAFE). This distillation technique is known for its fast and gentle isolation of volatiles from food matrices [20]. At room temperature (21 ± 1 °C), 25.0 (±0.1) g formula milk was combined with freshly-distilled dichloromethane at a ratio of 2:1 (milk:solvent). Additionally, 1 mL of labeled internal standards was added to the mixture. Concentrations of seven characteristic odor-active secondary oxidation products (see Section 2.2) were determined in a formula milk reference in preliminary experiments and comparable amounts were added as isotope labeled internal standard to the milk-solvent mixture for SIDA. The combined mixture was stirred for 30 min, after which it was immediately distilled (high vacuum conditions, 50 °C water bath temperature and 55 °C water temperature within the SAFE apparatus). After distillation, additional aliquots of 10 mL of dichloromethane were administered to the residue and distillation was re-performed. The latter step was performed to achieve a complete transfer of the VOCs from the milk matrix residue to the distillate. The organic phase of the distillate was separated and the aqueous phase was twice extracted with 15 mL of dichloromethane. Subsequently, the combined organic phases were dried with anhydrous sodium sulfate, filtered and concentrated to a total volume of 100–200 μL via Vigreux-distillation and micro-distillation at 50 °C [21].

For GC-MS analysis of the powdered infant formulas immediately after production, three independent reconstitutions of the 18 formulations were performed and prepared for GC-injection, as described above. For further GC-MS analyses of lipid oxidation products, three independent reconstitutions were randomly performed for one third of the samples (six formulations).

For analysis of lipid hydroperoxides and conjugated dienes, formula powders were dissolved in distilled water and extracted with isooctane/2-propanol, with 7 mL of the isooctane/2-propanol mixture (1:1, v/v) added to 2.5 mL of the sample. The mixture was stirred for 30 s, centrifuged at a relative centrifugal force of $720\times g$ (Allegra 2IR Centrifuge, Beckmann Coulter, Brea, CA, USA), and the organic phase was used for lipid isolation. An aliquot of the organic phase was transferred to a test tube and the organic solvent was removed under a stream of nitrogen. Analyses were performed in triplicate on two independent reconstitutions of powdered formulas.

2.6. Quantification by Stable Isotope Dilution Assays (SIDA)

Seven selected odorants known as secondary fatty acid oxidation markers (see Section 2.2) were quantified in the infant formulas directly after production, as well as after four and eight weeks of storage, respectively. The selection was based on a previous GC/O screening on about 20 infant formula samples with a fishy off-flavor (results not shown). Compared to a reference, several odor impressions were detected in the off-flavor samples, which could be identified by comparison with reference substances based on the following criteria: odor quality, retention indices on two stationary phases and mass spectra. Identification and quantification were performed using a two-dimensional gas chromatographic system (2D-HRGC-MS/O) which consisted of two helium CP 3800 GCs (Varian Inc., Darmstadt, Germany) in combination with a Saturn 2200 MS (Varian Inc.) and sniffing ports for olfactory detection.

Concentrated distillates (2 μL) containing the target volatiles and labeled internal standards were injected into the GC system using the cold-on-column technique at 40 °C. Injections were performed in triplicate. The 2D-HRGC-MS measurements were executed using the following capillaries: DB-FFAP (30 m × 0.32 mm fused silica capillary, free fatty acid phase FFAP, 0.25 μm; type Chrompack, Varian) in the first oven and DB-5 (30 m × 0.25 mm fused silica capillary DB-5, 1.5 μm; type J and W, Agilent Technologies, Santa Clara, CA, USA) in the second oven. The oven programs were as follows: after 2 min, the oven temperature was raised to 240 °C (DB-FFAP) and 250 °C (DB-5), respectively, at a rate of 10 °C·min^{-1}. The final temperatures were held for 5 min. The flow rate of the helium carrier gas was 2.5 mL·min^{-1}. The effluent was split at the end of the columns using deactivated and uncoated fused silica capillaries (100 cm × 0.2 mm). In the first oven, the effluent was split into a sniffing port and a flame ionization detector (FID). A specific section of the effluent, containing the volatile and its labeled standard (retention time ± 0.2 min) was transferred via a cryo trap to the second oven. At the end of the second oven, the effluent was split into a second sniffing port and the MS. The FID detector was set to 250 °C and the sniffing ports to 300 °C. Mass spectra in the chemical ionization mode (MS/CI) were acquired with methanol as the reagent gas at a flow of 2.5 mL·min^{-1}. Selected ions of the odorants and labeled standards were analyzed in scan mode (m/z 60–249 range) and their intensities were calculated by means of Varian MS-Workstation, MS Data Review (Version 6.9; Service Pack 1, Varian Inc.).

Concentrations were calculated by means of calibration data obtained by measuring defined mixture ratios of respective labeled and unlabeled compounds. Furthermore, MS response factors were determined by measuring defined mixtures of respective labeled and unlabeled compounds. Further details on quantification, e.g., selected ions or calibration factors, are given in Table 2.

Table 2. Odorants, selected ions for quantification by stable isotope dilution assays and details on calibration factors determined on film capillary DB-5. R^2—coefficient of determination.

Odorant	Ion (m/z)	Internal Standard	Ion (m/z)	Calibration Line	R^2
hexanal	83	$[5,5,6,6-^2H_4]$-hexanal	87	$y = 0.4915x - 0.0475$ [1]	0.9853
				$y = 0.6059x - 0.0690$ [2]	0.9958
(E)-hex-2-enal	99	$[2,3-^2H_2]$-(E)-hex-2-enal	101	$y = 0.6555x + 0.0538$ [1]	0.9999
				$y = 0.8132x - 0.1313$ [2]	0.9984
(E,Z)-nona-2,6-dienal	121	$[1,2-^{13}C_2]$-(E,Z)-nona-2,6-dienal	123	$y = 0.8522x + 0.0784$ [1]	0.9996
				$y = 1.0200x - 0.0600$ [2]	0.9938
(Z)-octa-1,5-dien-3-one	125	$[5,6-^2H_2]$-(Z)-octa-1,5-dien-3-one	127	$y = 0.9355x - 0.0323$	0.9998
oct-1-en-3-one	127	$[1,2-^2H_{2-3}]$-oct-1-en-3-one	129–130	$y = 1.2092x - 0.0408$ [1]	0.9970
				$y = 1.6112x - 0.1673$ [2]	0.9989
(E,E)-deca-2,4-dienal	153	$[4,5-^2H_2]$-(E,E)-deca-2,4-dienal	155	$y = 0.7705x - 0.1005$ [1]	0.9952
				$y = 0.8218x - 0.0851$ [2]	0.9988
trans-4,5-epoxy-(E)-dec-2-enal	153	$[9,9,10,10-^2H_4]$-trans-4,5-epoxy-(E)-dec-2-enal	157	$y = 1.4603x - 0.0139$	0.9998

[1] calibration factors used for odorants at low *ppb*-level concentrations; [2] calibration factors used for odorants at high *ppb*-level concentrations.

2.7. Determination of Lipid Hydroperoxides and Conjugated Dienes

The determination of hydroperoxides and conjugated dienes was performed immediately after formula production and additionally at selected times up to 12 months of storage (*cf.* Section 2.4). The sample preparation is described in Section 2.5. After reconstitution of the powders with distilled water, lipids were extracted from the infant formulas using a blend of ethanol, hexane, and ethyl acetate, as described by Satué-Gracia [22].

Hydroperoxides: the International Dairy Federation standard method [23], as described in Drusch *et al.* [24] with slight modifications, was used for the determination of the hydroperoxide content. This photometrical assay is based on the oxidation of Fe(II) to Fe(III) ions that form a colored complex with ammonium thiocyanate. The extinction coefficient was measured at 485 nm.

Conjugated dienes: after dilution of the organic phase of the lipid extractions with 2-propanol, conjugated dienes were determined photometrically at 234 nm. For calculation of the concentration, the results were expressed as millimoles of hydroperoxides per kg fat using a molar coefficient of 26,000 for methyl linoleate hydroperoxides [25].

2.8. Aroma Profile Analysis (APA)

A volume of 20 mL of the respective reconstituted formula milk was filled into sensory glass beakers (140 mL, J. Weck GmbH u. Co. KG, Wehr, Germany) and closed with a lid. Sensory analyses were performed in a sensory panel room at 21 ± 1 °C. Trained panelists ($n = 8$–12, male/female, age 24–45) from the University of Erlangen-Nuremberg (Erlangen, Germany) and Fraunhofer IVV (Freising, Germany) with normal olfactory and gustatory function participated in the APA sessions and exhibited no known illness at the time of examination. Prior to this study, the assessors were recruited in weekly training sessions for the recognition of about 100 selected odor-active compounds according to their odor qualities by means of an in-house developed flavor language.

In a first session the panelists were asked to orthonasally evaluate the samples and the named odor attributes of the formula milks were collected. Attributes that were detected by more than 50% of the panelists were selected for subsequent evaluations. In subsequent sessions, the panelists were asked to score the orthonasally perceived intensities of the selected attributes on a seven-point-scale from 0 (no perception) to 3 (strong perception) in increments of 0.5. The order of presentation of the infant formulas was randomized and no information on the purpose of the experiment or the composition of the samples was given to the panelist. Aroma profile analysis was performed on samples immediately after production, as well as after four, 8 and 24 weeks (six months) of storage. The results for each odor attribute and sample were averaged and plotted in a box-plot diagram.

In addition to the APA, the panelists were asked if they perceived an off-flavor in the respective infant formulas. When an off-flavor was recorded, rating of the overall odor impression of the products was based on a yes/no answer.

To check changes in color over storage period, an instrumental color analysis was carried out using a Chroma Meter CR-300 (Konika Minolta Inc., Marunouchi, Japan) with a DP-301 data processor. Calibration was performed on a white standard (CR-A43, Konika Minolta Inc.). The color was expressed in $L*a*b*$ mode, in which $L*$ represents the lightness value, and $a*$ and $b*$ values the chromaticity coordinates. The samples were analyzed in triplicate.

2.9. Statistical Analyses

For statistical analysis, Student's t-tests (or Welch-tests), Mann-Whitney U-tests and one-way repeated analysis of variances (repeated ANOVA) were carried out (as declared within the respective sections of the manuscript). In the latter case, either the Tukey HSD or the Fisher LSD procedure was used to detect significant differences between specific storage times. Statistical analyses were performed using OriginPro 9G (OriginLab Co., Northampton, MA, USA) and Statistica 10 (StatSoft Europe GmbH, Hamburg, Germany) software, respectively. For all analyses, the level of statistical significance was set at 5%. Outliers were identified as individual values that were not within two standard deviations of the preceding and succeeding storage time data points.

3. Results

3.1. Monitoring of Autoxidation—Quantification of Secondary Lipid Oxidation Products by Means of 2D-HRGC-MS/O

Seven representative compounds of secondary lipid oxidation reactions were chosen among the alkanal, alkenal and alkadienal substance classes, as well as alkenones and alkadienones, which are predominantly formed during autoxidation. According to the literature, these odorants have previously been identified in infant formulas and human breast milk, respectively, since both mothers' milk and infant formulas are rich in PUFAs [5,13]. Furthermore, these compounds were also identified in n-3 fatty acid enriched fish oil capsules that are administered during pregnancy and the lactation period to supplement the mothers' nutritional fatty acid level [26]. Table 3 shows the quantitative data of the selected odorants in infant formulas directly after production (week 0) supplemented with antioxidants (formulations A1–A9) and without supplementation of antioxidants (formulations B1–B9). All compounds were present in the formula milks in the lower $\mu g \cdot kg^{-1}$ (ppb) range, with (E)-hex-2-enal, (Z)-octa-1,5-dien-3-one and (E,Z)-nona-2,6-dienal exhibiting the lowest concentrations below 2 $\mu g \cdot kg^{-1}$. Oct-1-en-3-one and (E,E)-deca-2,4-dienal were detected in the freshly produced formulations at concentrations below 10 $\mu g \cdot kg^{-1}$. $trans$-4,5-Epoxy-(E)-dec-2-enal was determined at concentrations of up to 20 $\mu g \cdot kg^{-1}$ and hexanal at concentrations of between 5 and 50 $\mu g \cdot kg^{-1}$.

Table 3. Quantification of selected odorants in infant formulations directly after production (with and without antioxidants).

Formulation	Concentration (µg·kg⁻¹)													
	Hexanal		(E)-Hex-2-enal		Oct-1-en-3-one		(Z)-Octa-1,5-dien-3-one		(E,Z)-Nona-2,6-dienal		(E,E)-Deca-2,4-dienal		trans-4,5-Epoxy-(E)-dec-2-enal	
	Mean	SD	Mean	SD	Mean	SD	Mean	SD	Mean	SD	Mean	SD	Mean	SD
A1	10.93	±1.85	0.22	±0.14	3.43	±0.49	0.05 [a]	±0.00	0.43 [a]	±0.00	0.99	±0.32	4.22	±0.81
A2	5.73 [a]	±1.24	0.09	±0.01	3.42	±0.25	0.05 [a]	±0.01	0.32 [a]	±0.08	3.07	±0.71	9.66	±0.88
A3	9.87	±2.77	0.20	±0.14	4.02	±0.85	0.06 [a]	±0.06	0.48 [a]	±0.15	2.57	±0.35	6.50	±4.27
A4	14.4	– *	0.13	– *	3.62	– *	<0.01 [a]	– *	<0.01 [a]	– *	4.21	– *	5.49	– *
A5	10.60 [a]	– *	<0.01 [a]	– *	3.23	– *	0.03 [a]	– *	<0.01 [a]	– *	4.60	– *	14.79	– *
A6	10.11 [a]	– *	<0.01 [a]	– *	3.22	– *	0.02 [a]	– *	<0.01 [a]	– *	6.20	– *	15.33	– *
A7	6.23 [a]	±0.73	0.12 [a]	±0.04	2.46	±0.39	<0.01 [a]	±0.01	0.24 [a]	±0.03	3.17	±0.20	5.61	±1.03
A8	8.35 [a]	±4.90	0.04 [a]	±0.03	1.97	±0.22	0.02 [a]	±0.01	0.25 [a]	±0.09	1.92	±0.34	3.44	±0.30
A9	9.68	±3.78	0.17	±0.17	2.61	±0.55	0.01 [a]	±0.01	0.23 [a]	±0.15	2.47	±0.42	6.80	±0.83
B1	35.96	±4.10	0.32	±0.02	1.52	±0.17	0.15	±0.05	1.28	±0.12	3.83	±0.36	9.69	±0.24
B2	36.49	±7.45	0.49	±0.28	1.81	±0.57	0.07 [a]	±0.01	0.60	±0.20	3.50	±2.71	6.53	±0.44
B3	30.81	±8.73	0.21	±0.01	2.14	±0.03	0.01 [a]	±0.01	0.79	±0.06	2.56	±0.35	9.83	±0.87
B4	33.42	±5.34	0.22	±0.04	1.69	±0.12	0.01 [a]	±0.01	0.61	±0.08	4.14	±1.55	10.25	±2.13
B5	37.12	±9.34	0.16	±0.04	1.86	±0.12	0.01 [a]	±0.01	0.64	±0.03	2.98	±0.40	9.73	±1.08
B6	41.11	±7.67	0.37	±0.21	2.16	±0.38	0.03 [a]	±0.01	0.69	±0.19	2.93	±0.26	11.74	±2.13
B7	37.99	±6.71	0.66	– *	2.04	±0.46	0.03 [a]	±0.03	0.72	±0.12	3.88	±0.22	7.50	±1.72
B8	30.43	±10.13	0.42	±0.11	1.13	±0.05	<0.01 [a]	±0.00	0.63	±0.12	2.60	±0.78	6.16	±1.79
B9	33.82	±11.25	0.38	±0.19	1.44	±0.13	<0.01 [a]	±0.01	0.72	±0.19	2.54	±0.23	10.00	±1.44

Reconstitution of powdered formulas was performed in triplicate. * reconstitution of powdered formulas was not performed in triplicate; [a] determinations yielding values below the linearity of the calibration curve (limit of quantification, LOQ).

The detection of low concentrations of the respective odorants in freshly produced infant formulas demonstrates that autoxidation is already initialized during production. For some compounds, differences seemed to arise in the concentrations of A-formulations (with antioxidants) and B-formulations (without antioxidants). To validate the significance of such differences between the formulations with and without antioxidants, a Student's t-test, Welch-test or Mann-Whitney U-test was performed, depending on normal distribution and homogeneity of variances of the data (Table 4). Differences between A-formulations and B-formulations were significant in four out of seven secondary lipid oxidation markers [hexanal, (E)-hex-2-enal, oct-1-en-3-one, (E,Z)-nona-2,6-dienal]. Details on statistical analyses are given in Table 4. The most significant changes were observed for hexanal, as shown in Figure 1. Thus, differences between A- and B-formulations, as shown for hexanal in Figure 1, are detectable by means of multi-dimensional GC-MS even directly after production, highlighting the high sensitivity of this method.

Table 4. Statistical data on comparison of means between formulations with and without antioxidants (week 0). t and $|Z|$ as critical values of the respective statistical test.

Odorant	Comparison of Means		
	$t(16)$, $\lvert Z \rvert$	Significance ($p < 0.05$)	p-Value
hexanal	-17.79 [a]	sig.	5.77×10^{-12}
(E)-hex-2-enal	-4.26 [a]	sig.	5.93×10^{-4}
oct-1-en-3-one	5.58 [a]	sig.	4.12×10^{-5}
(Z)-octa-1,5-dien-3-one	0.26 [c]	n.s.	0.79
(E,Z)-nona-2,6-dienal	-3.54 [c]	sig.	4.01×10^{-4}
(E,E)-deca-2,4-dienal	0.04 [b]	n.s.	0.97
$trans$-4,5-epoxy-(E)-dec-2-enal	-0.67 [b]	n.s.	0.52

[a] Student's t-test; [b] Welch-test; [c] Mann-Whitney U-test.

Figure 1. Overview of hexanal concentrations in formulations with antioxidants (**A**) and without antioxidants (**B**). Reconstitution of powdered formulas was performed in triplicate.

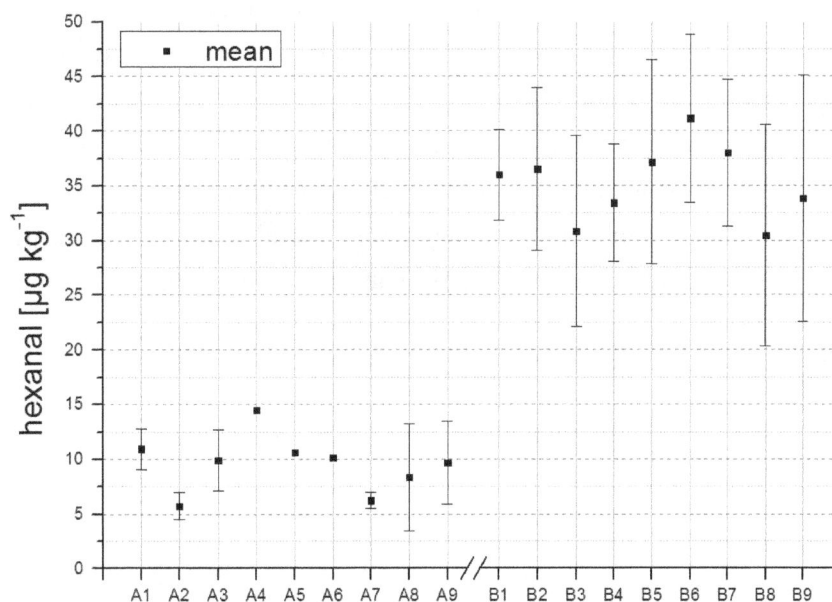

Tables 5 and 6 summarize the results of quantification after four weeks and eight weeks of storage.

After four weeks, the selected odorants were still present in the infant formulas at concentrations in the lower *ppb*-range (below 10 µg·kg^{-1}), with the exception of *trans*-4,5-epoxy-(*E*)-dec-2-enal and hexanal, which showed higher concentrations already directly after production. Nevertheless, the concentrations of the latter two odorants showed an increase compared to week 0 but were still below 100 µg·kg^{-1} after four weeks of storage. After eight weeks of storage, most formulations showed a (further) increase in the formation of secondary lipid oxidation products. Whereas (*E*)-hex-2-enal, (*Z*)-octa-1,5-dien-3-one and (*E,Z*)-nona-2,6-dienal stayed at concentrations below 10 µg·kg^{-1}, the concentration of (*E,E*)-deca-2,4-dienal and oct-1-en-3-one increased in some cases up to 50 to 60 µg·kg^{-1}. The highest increases in concentrations for some formulations were detected for hexanal and *trans*-4,5-epoxy-(*E*)-dec-2-enal, with concentrations of up to 850 µg·kg^{-1}.

These observations on increasing concentrations of secondary lipid oxidation products with storage time confirm the on-going autoxidation processes in the packaged, evacuated infant formulas. Nevertheless, some formulations were obviously quite stable over storage time with respect to the formation of these lipid oxidation products (A1, A4–A6, A7–A9, B7–B8) as shown in Tables 5 and 6, while other formulations (B1–B3, B4–B6, B9) were prone to considerable oxidative effects. The data set of the first eight weeks of storage confirms that A-formulations with antioxidants remained more stable than B-formulations without antioxidants, as demonstrated by the reference data of week 0. Table 7 details the statistical interpretation of the data, showing that differences between A- and B-formulations were significant after four weeks of storage in four out of seven secondary lipid oxidation markers [hexanal, (*E*)-hex-2-enal, (*E,Z*)-nona-2,6-dienal, *trans*-4,5-epoxy-(*E*)-dec-2-enal] and after eight weeks for all odorants under investigation.

Only two formulations without antioxidants were found to be stable: B7 and B8, respectively (Tables 5 and 6). Thus, the mineral combination used in these formulations, which potentially acts as a pro-oxidant, seems to trigger less oxidation than other combinations on basis of the quantified volatile oxidation markers. In these two formulas, encapsulated Cu^{2+} was manufactured together with non-encapsulated $Fe^{2+/3+}$. The equivalent formulas with antioxidants (A7 and A8) also remained very stable over a storage period of eight weeks.

Figure 2 shows the data for a very stable formulation (A7) in comparison to two formulations that were found to be less stable (B4, B6). The plots clearly demonstrate that all chosen oxidation markers were well-suited to reveal differences in the stability of the respective formulations. Statistical analyses (Table 8) demonstrated the following: whereas in most cases a significant increase in the formation of such oxidation markers was observed for the less stable formulations B4 and B6 comparing weeks 0–8, no significant increase (or decrease) was found for the stable formulation A7 during the storage period of eight weeks [only exception: hexanal (weeks 0–8)]. The only exception in the B-formulations was B6 for oct-1-en-3-one (weeks 0–8).

Table 5. Quantification of selected odorants in infant formulations after four and eight weeks of storage with antioxidants.

Storage (Weeks)	Formulation	Concentration (µg·kg⁻¹)						
		Hexanal	(E)-Hex-2-enal	Oct-1-en-3-one	(Z)-Octa-1,5-dien-3-one	(E,Z)-Nona-2,6-dienal	(E,E)-Deca-2,4-dienal	trans-4,5-Epoxy-(E)-dec-2-enal
4	A1	11.40 [a]	0.11	1.83	<0.01 [a]	0.47 [a]	1.52	4.61
8		16.14	0.11	1.69	0.05 [a]	0.47 [a]	5.07	17.65
4	A2	20.97	0.33	5.33	0.18	0.87	5.71	14.94
8		43.54	0.73	6.37	0.46	2.37	3.78	64.84 [a]
4	A3	20.32	0.23	5.22	0.21	0.91	6.56	24.37 [a]
8		70.03	0.64	6.02	0.40	2.54	5.34	79.56 [a]
4	A4	12.97	0.13	2.79	0.05 [a]	0.51 [a]	4.20	6.12
8		15.77	0.20	3.02	0.10 [a]	0.64	3.19	8.62
4	A5	10.90 [a]	0.09	2.11	0.04 [a]	0.36 [a]	3.28	5.21
8		16.70	0.23	5.90	ND	ND	3.24	18.11 [a]
4	A6	14.99	0.18	3.23	0.05 [a]	0.41 [a]	3.95	11.95
8		24.93	0.30	3.61	0.10 [a]	0.81	3.67	27.73 [a]
4	A7	6.99 [a]	0.13	2.01	0.05 [a]	0.31 [a]	2.79	5.87
8		17.56	0.09	1.14	0.02 [a]	0.09 [a]	2.36	6.95
4	A8	3.87 [a]	0.04	1.39	<0.01 [a]	0.25 [a]	2.80	3.67
8		5.72 [a]	0.08	2.21	0.07 [a]	0.52 [a]	7.65	47.55 [a]
4	A9	6.02 [a]	0.09	2.05	0.04 [a]	0.39 [a]	2.53	2.67
8		6.82 [a]	0.06	2.32	0.02 [a]	0.35 [a]	3.23	6.29

[a] Determinations yielding values below as well as above the linearity of the calibration curve (LOQ and upper LOQ, respectively); ND—not detected (below limit of detection, LOD).

Table 6. Quantification of selected odorants in infant formulations after four and eight weeks of storage without antioxidants.

Storage (Weeks)	Formulation	Concentration (µg kg^{-1})						
		Hexanal	(E)-Hex-2-enal	Oct-1-en-3-one	(Z)-Octa-1,5-dien-3-one	(E,Z)-Nona-2,6-dienal	(E,E)-Deca-2,4-dienal	trans-4,5-Epoxy-(E)-dec-2-enal
4	B1	41.34	0.44	2.01	0.25	2.28	5.56	26.09 [a]
8		65.17	0.83	4.71	0.23	2.52	14.16	49.71 [a]
4	B2	35.32	ND	2.62	ND	0.80	4.79	32.77 [a]
8		103.89	1.01	8.38	0.31	2.21	8.69	106.53 [a]
4	B3	53.06	0.69	5.38	0.18	1.69	4.15	8.60
8		259.79 [a]	3.42 [a]	37.76 [a]	0.33	6.35	27.11 [a]	320.51 [a]
4	B4	65.83	0.95	3.45	0.12	1.04	5.41	42.46 [a]
8		240.88 [a]	2.19 [a]	13.62 [a]	0.38	2.20	13.74	97.51 [a]
4	B5	50.24	0.57	6.07	0.16	1.55	5.48	92.97 [a]
8		129.03	0.75	49.31 [a]	0.37	3.02	8.85	92.90 [a]
4	B6	83.80	1.14	5.25	0.17	1.32	6.41	42.76 [a]
8		841.35 [a]	8.72 [a]	31.40 [a]	0.24	4.10	59.16 [a]	310.13 [a]
4	B7	36.56	0.34	2.76	0.09 [a]	0.69	4.44	7.05
8		41.02	0.31	4.36	0.31	2.12	5.02	20.32 [a]
4	B8	28.54	0.28	1.06	0.01 [a]	1.02	2.77	16.59
8		27.92	0.55	3.19	0.06 [a]	0.48 [a]	2.91	11.39
4	B9	29.89	0.29	1.60	0.04 [a]	0.41 [a]	8.85	36.99 [a]
8		58.16	0.25	12.85 [a]	ND	1.94	4.61	53.45 [a]

[a] Determinations yielding values below as well as above the linearity of the calibration curve (LOQ and upper LOQ, respectively); ND—not detected (below limit of detection, LOD).

Table 7. Statistical data on comparison of means between formulations with and without antioxidants (week 4, week 8). t and $|Z|$ as critical values of the respective statistical test.

Odorant	Comparison of Means					
	$t(16)$, $|Z|$		*Significance* ($p < 0.05$)		*p*-Value	
	Week 4	Week 8	Week 4	Week 8	Week 4	Week 8
hexanal	-5.48 [b]	-3.00 [c]	sig.	sig.	2.98×10^{-4}	2.68×10^{-3}
(*E*)-hex-2-enal	-3.77 [b]	-2.91 [c]	sig.	sig.	0.01	3.57×10^{-3}
oct-1-en-3-one	-0.61 [a]	-2.62 [b]	n.s.	sig.	0.55	0.03
(*Z*)-octa-1,5-dien-3-one	-1.52 [a]	-2.41 [b]	n.s.	sig.	0.15	0.04
(*E,Z*)-nona-2,6-dienal	-3.39 [b]	-2.07 [c]	sig.	sig.	0.01	0.04
(*E,E*)-deca-2,4-dienal	-2.08 [b]	-2.30 [c]	n.s.	sig.	0.05	0.02
trans-4,5-epoxy-(*E*)-dec-2-enal	-2.91 [c]	-2.30 [c]	sig.	sig.	3.57×10^{-3}	0.02

[a] Student's *t*-test; [b] Welch-test; [c] Mann-Whitney *U*-test.

Figure 2. Example of a stable formulation (A7) in comparison to two less stable formulations (B4, B6) with respect to lipid autoxidation/formation of secondary lipid oxidation markers. Reconstitution of powdered formulas was performed in triplicate.

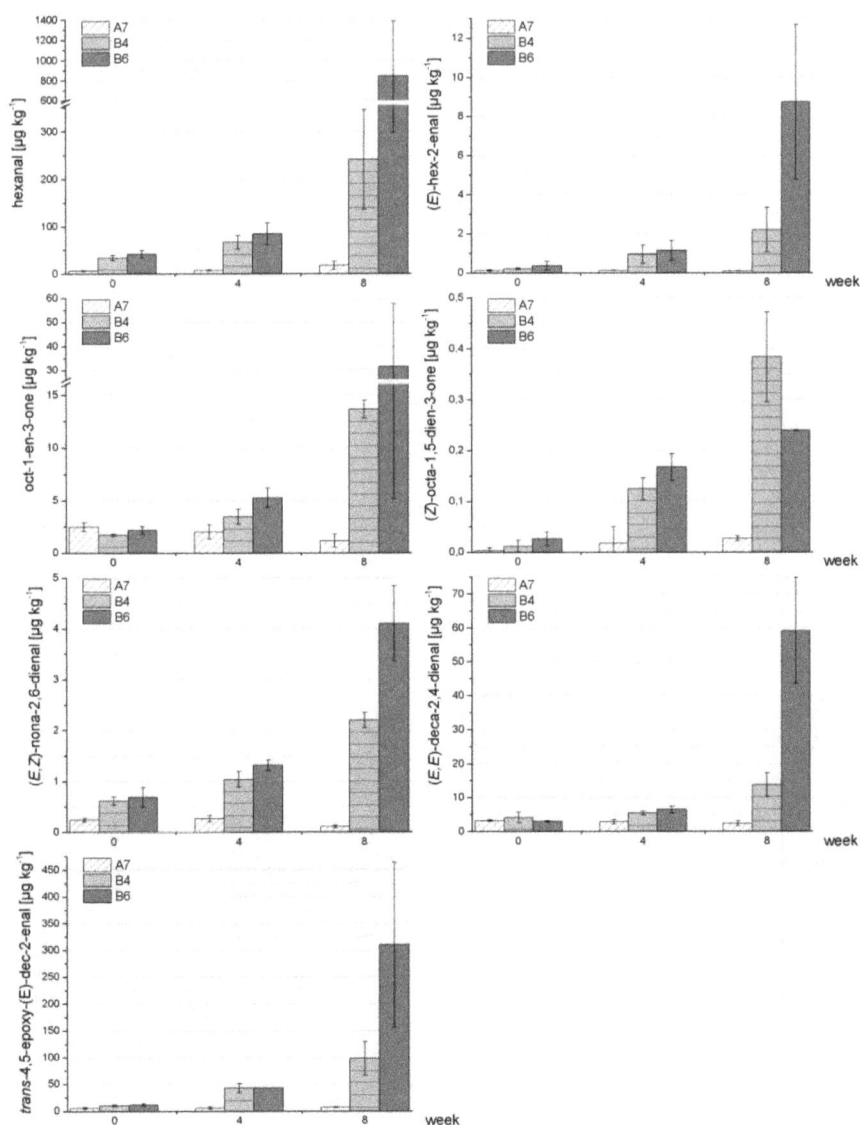

Table 8. Statistical data on one-way repeated ANOVA combined with posthoc-test [1,2] for pairwise comparison of the means: analysis of lipid oxidation via secondary lipid oxidation products over storage time.

Odorant	Formulation	Repeated Measures ANOVA		Post-hoc Test Week 0–4		Post-hoc Test Week 0–8	
		F-Value	p-Value	p-Value	Significance ($p < 0.05$)	p-Value	Significance ($p < 0.05$)
hexanal	A7	5.21	0.08	0.98[1]	n.s.	0.04[2]	sig.
	B4	11.45	0.02	0.78[1]	n.s.	0.01[2]	sig.
	B6	6.51	0.06	0.98[1]	n.s.	0.03[2]	sig.
(E)-hex-2-enal	A7	1.36	0.42*	-	-	-	-
	B4	8.37	0.04	0.21[2]	n.s.	0.02[2]	sig.
	B6	14.50	0.01	0.90[1]	n.s.	0.01[2]	sig.
oct-1-en-3-one	A7	5.39	0.07	0.56[1]	n.s.	0.07[1]	n.s.
	B4	213.20	8.64×10^{-5}	0.10[1]	n.s.	4.41×10^{-5}[2]	sig.
	B6	3.22	0.15*	-	-	-	-
(Z)-octa-1,5-dien-3-one	A7	1.05	0.43*	-	-	-	-
	B4	41.54	2.11×10^{-3}	0.11[1]	n.s.	8.85×10^{-4}[2]	sig.
	B6	140.39	1.97×10^{-4}	8.52×10^{-4}[1]	sig.	8.02×10^{-5}[2]	sig.
(E,Z)-nona-2,6-dienal	A7	8.30	0.11*	-	-	-	-
	B4	91.40	4.59×10^{-4}	0.05[1]	n.s.	4.41×10^{-4}[1]	sig.
	B6	46.51	1.70×10^{-3}	0.32[1]	n.s.	1.82×10^{-3}[1]	sig.
(E,E)-deca-2,4-dienal	A7	4.20	0.10*	-	-	-	-
	B4	12.65	0.02	0.82[1]	n.s.	0.02[1]	sig.
	B6	35.55	2.84×10^{-3}	0.89[1]	n.s.	1.67×10^{-3}[2]	sig.
trans-4,5-epoxy-(E)-dec-2-enal	A7	0.38	0.73*	-	-	-	-
	B4	22.22	0.01	0.15[1]	n.s.	2.74×10^{-3}[2]	sig.
	B6	10.11	0.03	0.69[2]	n.s.	0.01[2]	sig.

* no post-hoc test necessary ($p \geq 0.10$, n.s.); [1] homogeneity of variance: Tukey test; [2] non-homogeneity of variance: Fisher LSD test.

3.2. Monitoring of Autoxidation—Analysis of Lipid Hydroperoxides and Conjugated Dienes

Two of the standard tools for monitoring fatty acid oxidation involve the determination of lipid hydroperoxides (LPO) and of conjugated dienes. As mentioned in Section 1, these methods focus on the formation of primary lipid oxidation products, which can be measured photometrically. Here, both lipid hydroperoxides and conjugated dienes were monitored in all 18 infant formulas over a storage period of 12 months. The results are shown in Tables 9 and 10. LPOs as well as conjugated dienes were detectable in all formulations using the methods described in Section 2.7.

Table 9 shows that evaluation of the process of lipid oxidation via thiocyanate assay by monitoring of the formation of LPOs was very limited directly after production, as well as during the first weeks of storage. LPOs were first quantifiable at concentrations above the LOQ (here: 1 mmol·kg^{-1}) in formulations without antioxidants (B-formulations) after eight weeks of storage, with the exception of formulations B7 and B8. For A-formulations with antioxidants (except A7–A9), LPOs were detectable at concentrations above the LOQ after three months of storage. In general, formulations B7, B8 and A7–A9 exhibited lower concentration levels over a storage time of 12 months (<5 mmol·kg^{-1} for A7–A9, <50 mmol·kg^{-1} for B7 and B8). These findings are in agreement with the results of the 2D-HRGC-MS/O analysis (Section 3.1).

For the formation of conjugated dienes (Table 10), relatively constant concentrations from 10 to 20 mmol·kg^{-1} were detected over the first three months of storage. An initial increase in concentration was found for B-formulations after six months and A-formulations after nine months with concentrations above 20 mmol·kg^{-1}. After nine months of storage only the formulations A4, A7–A9, B7 and B8 showed conjugated diene concentrations below 20 mmol·kg^{-1}. These results again confirm that both stable and less stable formulas could be found among all 18 formulations investigated. Again, formulations A7–A9, B7 and B8 were the most stable formulations, indicating that encapsulated Cu^{2+} seems to inhibit lipid autoxidation independent from the presence of (non-)encapsulated iron.

In general, it seems that the process of oxidative deterioration can be evaluated by means of the formation of LPOs and conjugated dienes as primary markers of fatty acid oxidation. Nevertheless, comparably long storage intervals were necessary in the present study (when samples are stored under vacuum) to confirm differences in the stability of the formulas. To confirm the latter statement, an ANOVA was carried out on the obtained data. Figure 3 displays the formation of LPOs and conjugated dienes over nine months of storage. As in Section 3.1, the formulations A7, B4 and B6 were chosen as representative samples for stable (A7) and less stable (B4, B6) formulations. Details on the respective statistical data for these formulations are shown in Table 11 for the formation of LPOs and Table 12 for the formation of conjugated dienes. In the case of LPOs, storage data was compared to month 1 since all data of month 0 was below the limit of quantification (LOQ). Due to outliers (*cf.* Section 2.9), month 1 was not included in the one-way repeated ANOVA on conjugated dienes.

Table 9. Analysis of lipid hydroperoxides in infant formulations with and without antioxidants via thiocyanate assay.

Formulation	Concentration (mmol·kg^{-1})													
	Month 0		Month 1		Month 2		Month 3		Month 6		Month 9		Month 12	
	Mean	SD	Mean	SD	Mean	SD	Mean	SD	Mean	SD	Mean	SD	Mean	SD
A1	ND	-	<1.0	-	<1.0	-	1.3	±0.2	17.8	±0.3	49.9	±1.6	NA	-
A2	ND	-	ND	-	1.6	±0.2	11.1	±0.6	42.5	±1.4	>100.0	-	NA	-
A3	ND	-	ND	-	1.2	±0.5	7.6	±1.1	38.4	±2.4	>100.0	-	NA	-
A4	<1.0	-	ND	-	<1.0	-	1.6	±0.1	11.2	±0.3	32.2	±2.4	NA	-
A5	<1.0	-	ND	-	<1.0	-	11.1	±1.0	20.0	±1.3	54.4	±1.9	NA	-
A6	<1.0	-	ND	-	<1.0	-	2.3	±0.1	15.0	±0.8	44.3	±0.8	NA	-
A7	ND	-	ND	-	<1.0	-	<1.0	-	<1.0	-	<1.0	-	2.9	±0.9
A8	<1.0	-	ND	-	<1.0	-	<1.0	-	<1.0	-	<1.0	-	2.6	±0.6
A9	<1.0	-	ND	-	<1.0	-	<1.0	-	<1.0	-	1.2	±0.4	1.2	±0.8
B1	<1.0	-	<1.0	-	1.4	±0.3	3.6	±0.3	18.2	±0.4	45.6	±1.6	NA	-
B2	ND	-	<1.0	-	2.2	±0.1	5.0	±0.3	20.9	±0.8	54.5	±1.9	NA	-
B3	ND	-	<1.0	-	4.1	±0.2	1.9	±0.4	40.4	±1.9	>100.0	-	NA	-
B4	ND	-	<1.0	-	2.6	±0.1	3.1	±0.5	32.1	±3.0	>100.0	-	NA	-
B5	ND	-	<1.0	-	3.3	±0.2	8.2	±1.0	32.6	±1.3	>100.0	-	NA	-
B6	ND	-	<1.0	-	4.5	±0.2	13.2	±0.3	46.6	±1.4	>100.0	-	NA	-
B7	ND	-	<1.0	-	<1.0	-	2.5	±0.2	7.3	±0.4	17.5	±0.9	48.0	±2.2
B8	<1.0	-	<1.0	-	<1.0	-	1.1	±0.0	4.0	±0.3	15.1	±1.7	40.3	±1.7
B9	ND	-	<1.0	-	1.6	±0.1	2.8	±0.1	12.5	±0.2	40.8	±2.9	NA	-

Analyses were performed in triplicate on two independent reconstitutions of powdered formulas. ND—not detected (below LOD); NA—not analyzed (since upper LOQ was exceeded at previous point of investigation).

Table 10. Analysis of conjugated dienes in infant formulations with and without antioxidants via UV absorption.

Formulation	Concentration (mmol·kg⁻¹)													
	Month 0		Month 1		Month 2		Month 3		Month 6		Month 9		Month 12	
	Mean	SD	Mean	SD	Mean	SD	Mean	SD	Mean	SD	Mean	SD	Mean	SD
A1	12.9	±0.5	13.3	±1.0	14.4	±1.5	10.8 *	±1.6	15.8	±0.5	21.9	±1.5	NA	-
A2	11.0	±0.2	13.7	±0.4	11.9 *	±0.4	16.5	±0.4	19.0	±0.4	30.9	±0.7	NA	-
A3	11.0	±0.1	13.4	±0.8	11.6 *	±0.2	15.7	±0.3	18.2	±0.5	29.9	±0.8	NA	-
A4	11.1	±0.2	11.8	±0.7	10.2 *	±0.4	11.0	±0.2	14.6	±0.2	19.0	±1.0	NA	-
A5	11.3	±0.0	10.4	±1.2	10.8	±0.5	15.7	±0.2	16.5	±0.4	24.5	±1.0	NA	-
A6	12.5	±1.2	13.1	±0.5	10.6	±0.2	11.3	±0.4	15.7	±0.3	22.8	±0.4	NA	-
A7	10.9	±0.1	14.4 *	±0.8	10.7	±0.3	11.8	±0.4	11.6	±0.1	12.1	±0.8	11.7	±0.2
A8	11.2	±0.2	13.5 *	±0.6	10.9	±0.4	11.5	±0.5	11.8	±0.3	13.1	±0.7	11.7	±0.3
A9	10.9	±0.5	11.5	±0.8	10.4	±0.2	10.7	±0.2	11.1	±0.1	13.0	±0.2	11.3 *	±0.6
B1	13.7	±0.2	14.4	±0.5	14.4	±2.0	14.7	±0.6	16.9	±0.5	23.1	±0.7	NA	-
B2	13.9	±0.1	14.3	±0.1	13.5	±0.2	14.0	±0.4	18.7	±0.9	25.6	±0.6	NA	-
B3	12.5	±0.6	14.4	±0.5	14.4	±0.2	13.8 *	±0.2	22.1	±0.7	31.2	±0.8	NA	-
B4	14.5	±0.4	12.7 *	±0.5	14.4	±0.3	11.5 *	±0.2	20.5	±0.5	29.0	±0.4	NA	-
B5	15.1	±1.8	11.1 *	±5.1	14.6	±0.4	15.6	±0.6	21.2	±0.2	29.0	±0.9	NA	-
B6	14.4	±0.0	12.8 *	±0.5	14.5	±0.9	16.3	±0.7	23.6	±0.8	33.5	±0.7	NA	-
B7	12.5	±0.3	13.7	±0.2	14.0	±0.4	11.6 *	±0.1	14.4	±0.5	16.8	±1.0	22.2	±0.5
B8	11.6	±1.0	12.9	±0.6	14.0	±0.4	10.3 *	±0.7	13.9	±0.3	16.5	±0.7	21.8	±0.3
B9	14.0	±0.2	12.9 *	±0.4	14.1	±0.4	13.5	±0.2	16.2	±1.2	23.2	±1.2	NA	-

Analyses were performed in triplicate on two independent reconstitutions of powdered formulas. NA—not analyzed (*cf.* Table 9); * outlier (*cf.* Section 2.9).

Figure 3. Example of a stable formulation (A7) compared to two less stable formulations (B4, B6) with respect to lipid autoxidation/formation of primary lipid oxidation markers. Analyses were performed in triplicate on two independent reconstitutions of powdered formulas. ND—not detected (below LOD); < below LOQ; > upper LOQ exceeded.

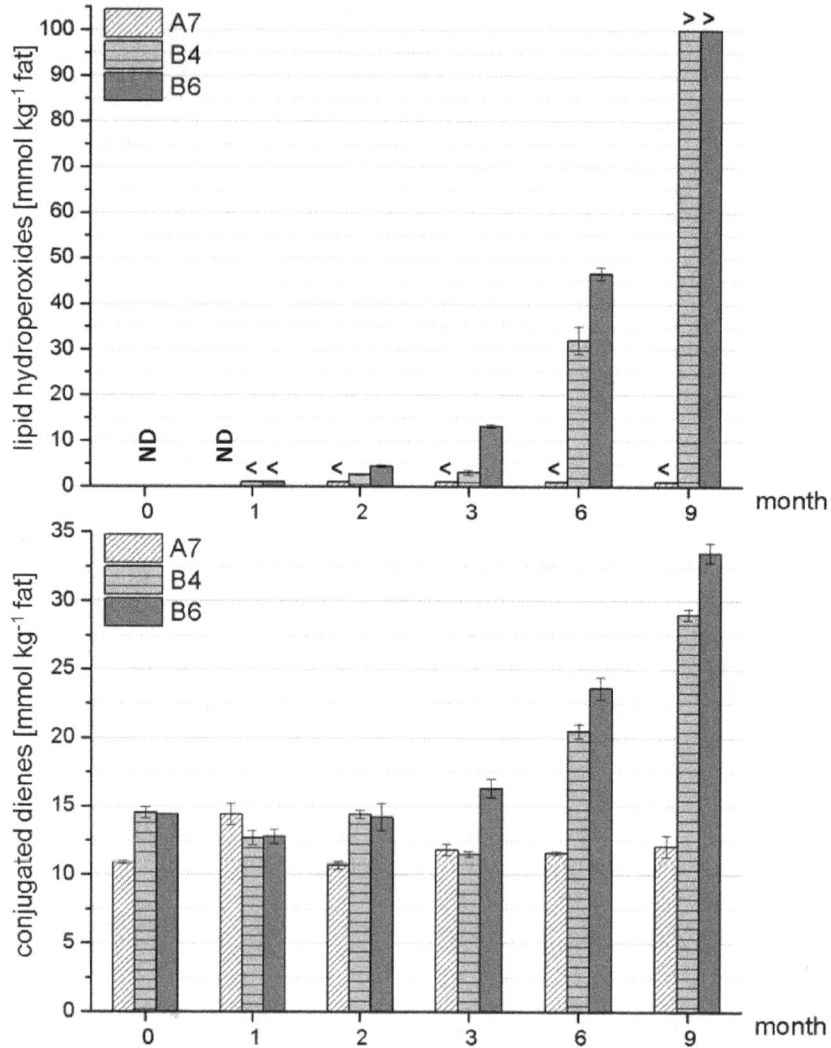

Table 11. Statistical data on one-way repeated ANOVA combined with a post-hoc test [1,2] for pairwise comparison of the means: analysis of lipid oxidation via lipid hydroperoxides over storage period.

Odorant	Formulation	Repeated Measures ANOVA		Post-hoc Test Month 1–2		Post-hoc Test Month 1–3		Post-hoc Test Month 1–6		Post-hoc Test Month 1–9	
		F-Value	p-Value	p-Value	Significance ($p < 0.05$)	p-Value	Significance ($p < 0.05$)	p-value	Significance ($p < 0.05$)	p-Value	Significance ($p < 0.05$)
Lipid hydroperoxides	A7	- *	- *	-	-	-	-	-	-	-	-
	B4	2142.26	5.14×10^{-26}	0.99^1	n.s.	0.56^2	n.s.	$1.85 \times 10^{-6\,2}$	sig.	0.00^2	sig.
	B6	712.14	2.97×10^{-21}	1.00^1	n.s.	0.32^2	n.s.	$1.10 \times 10^{-3\,2}$	sig.	0.00^2	sig.

[1] homogeneity of variance: Tukey test; [2] non-homogeneity of variance: Fisher LSD test; * insufficient data for ANOVA (month 1: below LOD).

Table 12. Statistical data on one-way repeated ANOVA combined with a post-hoc test [1,2] for pairwise comparison of the means: analysis of lipid oxidation via conjugated dienes over storage period.

Odorant	Formulation	Repeated Measures ANOVA		Post-hoc Test Month 0–2		Post-hoc Test Month 0–3		Post-hoc Test Month 0–6		Post-hoc Test Month 0–9	
		F-Value	p-Value	p-Value	Significance ($p < 0.05$)	p-Value	Significance ($p < 0.05$)	p-Value	Significance ($p < 0.05$)	p-Value	Significance ($p < 0.05$)
Conjugated dienes	A7	15.53	6.30×10^{-6}	0.98^1	n.s.	$1.90 \times 10^{-3\,2}$	sig.	0.02^1	sig.	$9.87 \times 10^{-5\,2}$	sig.
	B4	2357.51	1.98×10^{-26}	0.99^1	n.s.	$2.10 \times 10^{-7\,1,*}$	sig. *	0.00^1	sig.	0.00^1	sig.
	B6	788.05	1.09×10^{-21}	0.66^2	n.s.	$1.63 \times 10^{-4\,2}$	sig.	$1.78 \times 10^{-15\,2}$	sig.	0.00^2	sig.

[1] homogeneity of variance: Tukey test; [2] non-homogeneity of variance: Fisher LSD test; * outlier.

The findings reported in Tables 11 and 12 confirm that it was possible to detect significant differences in the stability of the formulas via the photometrical detection of LPOs and conjugated dienes. However, statistical significance of the data was achieved only with storage period intervals of at least six months for LPOs and three to six months for conjugated dienes. Hence, neither method was sufficiently sensitive in the current case to identify and monitor the emergence of oxidative lipid deterioration processes in its very early stages.

To complete the data analyses performed in Section 3.1, A-formulations were compared with B-formulations to reveal if the measurements of primary products of lipid oxidation via the presented methods were successful with regard to detection of significant differences between formulations with and without antioxidants. In the case of the thiocyanate assays (formation of LPOs), no significances could be found over storage period [$t(16)$ = −0.19373, 1.03466, −1.18395, $p < 0.05$ (n.s.) (3, 6 and 9 months)]. Data from week 0 to 8 were not investigated with regard to significances, since concentrations in most of the samples were below the LOQ of the method.

With respect to conjugated dienes, a significant difference was found between A-formulations and B-formulations directly after production, as well as after a storage time of eight weeks and six months [$t(16)$ = −4.608 (sig.), −0.818 (n.s.), −6.532 (sig., Welch), −0.668 (n.s.), −2.454 (sig.), −1.460 (n.s.), $p < 0.05$ (0, 1, 2, 3, 6 and 9 months)]. The occurrence of these differences between A- and B-formulations was not constant over the storage period, since differences between week 4, month 3 and month 9 were not significant.

3.3. Sensory Characterization of Infant Formulas during Storage—Aroma Profile Analysis

Samples of each formulation of reconstituted milk were evaluated by APA. The following odor attributes were scored via orthonasal evaluation with regard to their intensities: cooked milk-like, fatty, green-grassy, metallic, fishy, oily, blood-like. Table 13 shows the mean values of the perceived odor intensities as rated by the panel. All attributes were generally perceived in all formulations but were ranked with considerably divergent intensities in the different formulations at different storage periods. The attributes fatty and cooked milk-like were more related to the original flavor profile of the fresh infant formulas, as can be deduced from Table 13. For most formulas, the latter two attributes were ranked in the freshly produced samples as being clearly perceivable (odor intensities between 1.0 and 2.0). Also the attribute metallic was ranked with intensities of between 1 and 2 for most formulations. Nevertheless, all other odor attributes were evaluated in the freshly produced formulations with very low intensity (below 1.0), with only very few exceptions.

Table 13. Intensity rating of specific orthonasally perceived odor attributes in aroma profile analysis (APA) of reconstituted infant formulations with and without antioxidants. Mean values of evaluations by trained panelists ($n_{panelists} = 8\text{–}12$).

Storage (Week)	Formulation	With Antioxidants								Formulation	Without Antioxidants							
		Fishy	Metallic	Oily	Green, Grassy	Blood-Like	Fatty	Cooked Milk-Like	Off-Flavor		Fishy	Metallic	Oily	Green, Grassy	Blood-Like	Fatty	Cooked Milk-Like	Off-Flavor
0	A1	0.90	1.40	0.65	0.95	0.80	1.50	1.35	-	B1	1.06	2.28	0.89	1.06	0.44	1.39	1.33	*
4		0.61	1.33	0.78	1.22	0.44	2.00	1.83	-		1.28	1.78	0.72	1.00	1.00	1.17	0.72	*
8		0.78	1.33	0.50	1.00	0.72	1.22	1.56	*		1.20	2.00	0.95	1.00	0.90	1.25	1.05	*
24		1.17	2.17	1.75	1.42	1.42	1.42	1.08	*		0.96	2.08	1.08	1.13	1.21	1.38	1.25	-
0	A2	0.69	1.56	0.63	1.00	1.06	1.13	0.81	-	B2	0.33	0.61	0.39	0.89	0.11	1.44	2.11	-
4		1.32	1.82	1.09	0.91	1.09	1.27	1.45	*		0.89	1.28	0.61	0.50	0.72	0.94	1.22	-
8		1.22	1.83	1.28	1.11	1.33	1.56	0.72	*		1.30	2.40	1.00	0.95	1.15	1.30	1.20	*
24		0.68	2.23	1.05	1.09	1.45	0.82	1.05	*		0.86	1.95	1.00	1.00	1.27	0.91	1.18	*
0	A3	1.19	1.94	1.06	0.81	1.63	0.94	0.38	*	B3	0.91	1.50	0.91	0.82	1.00	0.86	1.27	-
4		1.18	1.86	0.91	1.14	0.95	1.23	1.18	*		0.70	1.64	0.80	0.95	1.09	1.09	1.50	*
8		1.39	2.11	1.06	0.72	1.39	1.17	1.00	*		1.27	2.09	1.05	0.77	1.55	1.05	0.86	*
24		0.95	1.82	1.14	0.86	1.14	1.05	0.77	-		1.36	2.14	1.36	1.09	1.27	1.14	0.68	-
0	A4	0.44	0.88	0.25	1.00	0.50	0.75	1.44	-	B4	0.60	0.90	0.55	0.55	0.60	1.05	1.55	*
4		0.44	0.90	0.39	0.61	0.28	1.17	1.94	-		1.11	1.94	1.11	0.56	1.00	1.17	1.28	*
8		1.35	1.65	1.30	0.75	1.25	1.20	1.25	*		0.83	2.28	0.50	0.89	1.28	1.00	1.50	*
24		0.77	1.91	0.82	0.73	1.27	0.95	1.18	*		1.21	2.04	1.63	1.21	1.42	1.54	0.54	*
0	A5	0.86	2.05	1.09	1.05	1.00	0.91	1.23	-	B5	0.45	0.95	0.35	0.70	0.55	0.85	1.05	-
4		0.60	1.73	0.55	0.40	1.10	0.95	1.14	-		0.72	1.72	0.72	0.94	1.17	1.06	1.06	-
8		0.94	1.50	1.21	0.85	1.50	0.90	1.50	*		1.00	2.44	0.67	1.17	1.83	0.67	0.89	*
24		0.79	2.25	1.63	1.50	1.50	1.50	0.96	*		1.54	2.13	2.04	1.42	1.58	1.29	0.71	*

* more than 2/3 of the panelists agreed on the presence of an off-flavor.

Table 13. Cont.

Intensity of Odor Quality (Scale 0–3, Mean Values)

Storage (Week)	With Antioxidants									Without Antioxidants								
	Formulation	Fishy	Metallic	Oily	Green, Grassy	Blood-Like	Fatty	Cooked Milk-Like	Off-Flavor	Formulation	Fishy	Metallic	Oily	Green, Grassy	Blood-Like	Fatty	Cooked Milk-Like	Off-Flavor
0	A6	0.69	1.38	0.56	1.00	0.88	0.44	0.50	-	B6	0.65	0.95	0.40	0.15	0.55	0.70	1.10	-
4		0.67	1.67	0.56	0.83	0.83	1.00	1.39	*		0.72	1.50	1.00	0.50	1.11	1.22	1.17	*
8		1.15	1.50	1.05	0.65	1.30	1.20	0.85	*		1.22	2.44	1.17	0.83	1.94	1.00	1.06	*
24		0.64	2.14	1.00	1.05	1.18	1.05	1.23	*		1.83	2.25	1.88	1.21	1.46	1.04	0.92	*
0	A7	0.44	1.31	0.31	0.88	0.81	0.88	1.38	-	B7	0.91	1.18	0.85	0.82	0.68	1.14	1.32	-
4		0.77	1.00	0.68	0.82	0.64	1.23	1.68	-		0.50	0.70	0.44	0.40	0.80	1.14	1.82	-
8		0.61	1.28	0.67	0.56	0.89	1.33	1.67	-		0.68	1.64	0.91	0.59	1.05	0.95	1.45	*
24		0.29	1.25	0.54	0.83	1.25	1.50	1.67	-		1.46	1.75	1.63	1.25	1.42	1.33	1.04	*
0	A8	0.56	1.31	0.81	0.88	1.00	0.69	0.75	-	B8	0.73	1.18	0.95	0.68	0.82	0.86	1.05	-
4		0.06	0.78	0.06	0.72	0.06	0.67	1.22	-		0.11	0.83	0.17	0.20	0.67	0.82	1.41	-
8		0.83	1.22	0.56	0.56	0.94	1.00	1.00	-		0.50	1.27	0.64	0.36	0.82	1.00	1.55	-
24		0.58	1.29	0.58	0.79	1.38	1.63	1.75	-		0.41	1.14	0.50	0.45	0.77	1.27	1.55	-
0	A9	0.56	0.63	0.56	0.81	0.69	1.56	1.50	-	B9	0.75	1.40	0.80	0.85	0.90	1.05	1.10	-
4		1.00	2.17	0.72	1.06	0.89	0.83	1.67	*		1.17	2.00	0.72	0.56	1.00	1.00	1.06	*
8		1.65	1.75	1.40	0.75	1.35	1.25	1.20	*		0.94	2.00	0.67	0.78	1.00	0.56	1.22	*
24		0.71	1.38	0.25	0.71	1.00	1.42	2.04	-		1.67	2.33	1.92	0.96	1.83	1.17	0.88	*

* more than 2/3 of the panelists agreed on the presence of an off-flavor.

The sensory profile of the infant formulas changed with storage period and attributes like metallic, green-grassy, blood-like, oily and fishy became increasingly perceivable. Based on off-flavors determined via APA, only a few formulations remained stable over the entire storage period. It was shown in Sections 3.1 and 3.2 that formulation A7 (among others) was very stable in terms of the formation of primary and secondary lipid oxidation markers. Formulations B4 and B6 were less stable, as displayed in Figures 2 and 3. To demonstrate changes in the sensory profile over storage period, these three formulations are discussed here again as representative samples to show the results of the APA in more detail. Thus, the data is represented as box-plot diagrams for five of the seven odor attributes (Figure 4). Instead of the arithmetic mean, median values are plotted due to their robustness towards outliers.

In the case of formulation A7, all off-odor related odor attributes listed in Figure 4 were rated as being very low in odor intensity, with median values between not perceivable (0) and weakly perceivable (1) in most cases. Only the odor attribute cooked milk-like (Table 13) was rated as clearly perceivable, with intensity values greater than 1.5 from a storage period of four weeks upwards. Accordingly, the results from the quantification experiments revealing A7 as a very stable formulation were confirmed by APA. For formulations B4 and B6, odor intensities were also rated as being very low for week 0 samples. The APA of week 4 revealed increasing median values for the metallic and blood-like attributes, with high ranking for both parameters also in week 8 and 24 (*cf.* Figure 4). For the odor qualities fishy, oily and green-grassy, the same trend towards a steadily increasing ranking with storage period was observed, especially when focusing on the 25%–75% quartiles of the box-plots. These results are in good correlation with the quantification results from Sections 3.1 and 3.2, respectively.

Nevertheless, even if a number of off-flavor related odor attributes were ranked as being well perceivable and, in general, correlated well with the quantitative data, an APA does not directly reveal if the overall odor profile is still acceptable or not. Furthermore, the overall flavor profile may change from one dominating attribute to another (e.g., metallic to fishy) and thus, specific attributes may decrease over time even if an overall off-flavor is arising or intensifying, as can be seen in Figure 4 in the case of the attribute blood-like in formulations B4 and B6 after six months (week 24). In these samples, the odor profile changed such that attributes like fishy and oily were perceived as being stronger (presumably because these attributes are more closely related to an off-flavor). Thus, the sensory panel was additionally asked to evaluate if the sample was related to an overall off-flavor or not (indicated with an asterisk in Figure 4 when more than 2/3 of the panelists agreed on the presence of an off-flavor).

Figure 4. Box-plots of sensory evaluation of a stable formulation (A7) and two less stable formulations (B4, B6) over storage period (weeks). *Median*: •; *whiskers*: ±minimum-maximum ratings (without outliers); *box*: percentiles 25%–75%. * off-flavor (perceived by more than 2/3 of the panelists).

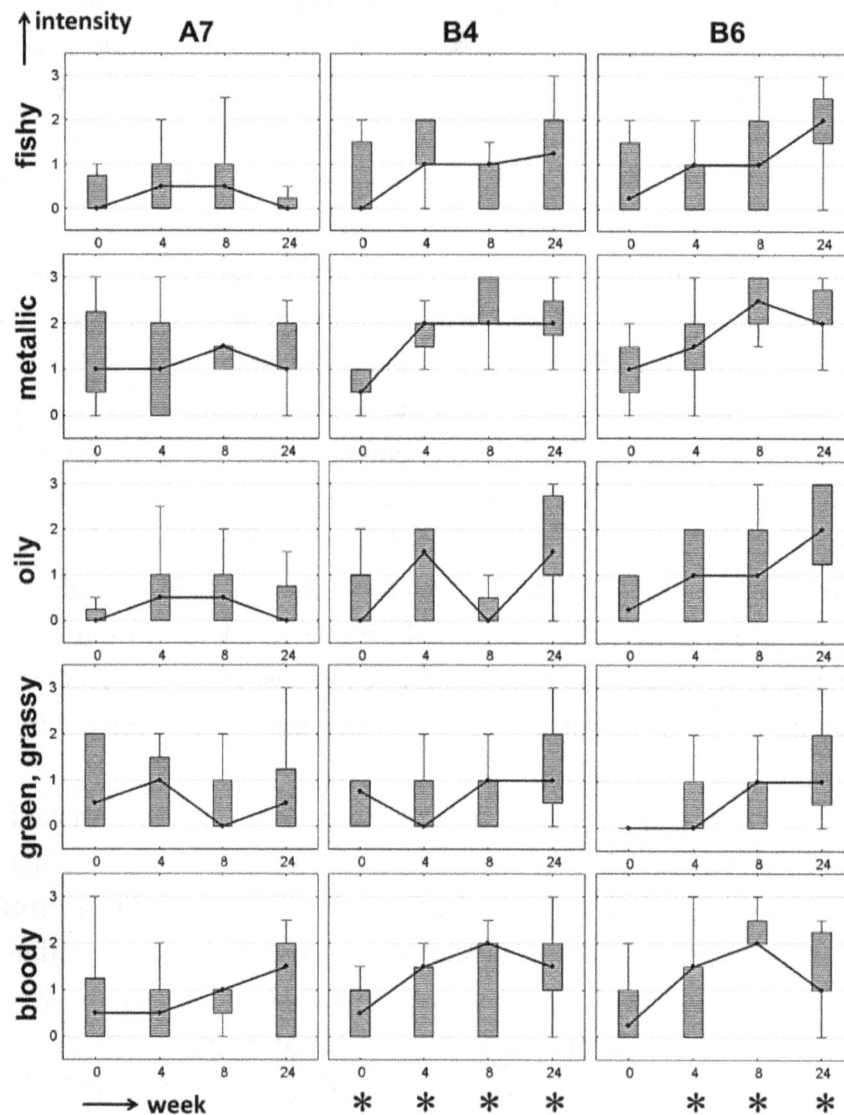

For a better overview, Table 13 also includes a column indicating where an off-flavor was reported in the formulations. As can be seen from Table 13, neither formulation A7, A8 nor B8 were rated as having an off-flavor throughout the storage period. These observations are consistent with the results of the APA and the analyses reported in Sections 3.1 and 3.2 on the stability of the different infant formulas in terms of oxidative deterioration. Nevertheless, comparison of the off-flavor rankings, the APA results, and the results of the quantification experiments are not compatible in some minor cases. One example shall be given for the odor attribute metallic and the quantitative data obtained for *trans*-4,5-epoxy-(*E*)-dec-2-enal, an odor-active compound with a metallic flavor [5]. For most formulations over the storage period there is a good correlation of sensory and quantitative experiments, except for formulations A5 and B5 after four weeks of storage. In the latter example the panelists did not agree on the presence of an off-flavor, even if the concentration of *trans*-4,5-epoxy-(*E*)-dec-2-enal was detected in formulation B5 at a concentration that was

18 times higher than in formulation A5. At the same time, the panelists performing the APA rated the odor attribute *metallic* as being clearly perceivable and comparable in both samples (APA rating 1.73 *vs.* 1.72 for formulations B5 and A5, respectively). This is a very good example to demonstrate that an off-flavor [here: metallic] is perceivable and the concentrations of the volatiles responsible for this off-odor [here: *trans*-4,5-epoxy-(*E*)-dec-2-enal] are elevated, but the overall flavor profile is a mixture resulting from the concentration ratio of alkanals, alkenals, alkadienals, alkenones and alkadienones. At a certain (unknown) point, when concentration ratios change due to on-going autoxidation, the overall flavor profile will also change, presumably from metallic towards fishy, oily or blood-like, respectively.

In this context, there is another interesting finding from the data of sensory evaluation and off-flavor ranking: a metallic off-flavor seems to be more accepted by the panelists in the case of infant formulas. However, this acceptance changes to rejection if the concentration ratio of the volatiles changes into a fishy, oily or blood-like flavor note, as can be seen for all formulations: as soon as the intensity rating for fishy, oily or blood-like exceeded being well-perceivable (APA rating 1.5 or higher), the panelists agreed on the presence of an off-flavor, indicating that also consumers would reject these products.

Thus, it can be summarized that powdered infant formulas were found to develop fishy, oily or blood-like flavor notes that arise from a mixture of saturated and unsaturated aldehydes and ketones and impart an off-flavor to those products. A few formulations—namely A7, A8 and B8—were found to be very stable over the whole storage period with respect to quantitative experiments as well as sensory evaluations. Thus, it can be assumed that a combination of encapsulated Cu^{2+} and non-encapsulated Fe^{3+} is indeed a favorable combination to manufacture stable infant formulas with respect to their odor profile.

In addition to APA, an instrumental color analysis was carried out over storage time to check any browning developments due to interactions between lipid oxidation products with proteins or polymerization of lipid oxidation products. No significant color changes could be detected by means of a paired samples *t*-test (week 0 *vs.* week 8), which showed the following results: L^* $t(17) = 1.368$, $p = 0.19$ (n.s.), a^* $t(17) = -0.718$, $p = 0.48$ (n.s.) and b^* $t(17) = -1.807$, $p = 0.09$ (n.s.), respectively.

Overall, it was found that there was a very good correlation between the sensory profile, the off-flavor ranking, and the concentration of primary and secondary oxidation markers. Furthermore, the results of a multi-dimensional GC-MS analysis on odor-active aldehydes and ketones (*cf.* Section 3.1) demonstrated that it was possible to clearly distinguish formulations with and without antioxidants directly after production, even before a sensory panel could detect any obvious differences in the flavor profiles of the formulations. This finding is of enormous importance for quality monitoring and to hasten and optimize product developments of PUFA-enriched infant formulas, as well as of PUFA-enriched foods in general.

4. Discussion

Primary and secondary lipid oxidation markers were used as diagnostic tools in the process of oxidative deterioration of PUFA-enriched infant formulas during storage. The monitoring of volatiles from secondary fatty acid oxidation via GC-MS is described in the literature as a common operation tool to characterize fatty acid oxidation in foods [27,28]. The stability of the lipid fraction in infant formulas is frequently investigated via GC-MS. So far, the monitoring of the formation of saturated aldehydes such as propanal, pentanal, and hexanal has been the focus of many studies [16,18]. Nevertheless, the results of the present study show that it was possible to confirm further secondary oxidation markers besides the common alkanals: alkenals, alkadienals, alkenones and alkadienones, which were present at even lower concentrations but still with high odor potencies and described previously for UFA/conjugated linoleic acid (CLA)-enriched butter [27]. All volatiles investigated are extremely odor-active and thus of great sensorial importance for PUFA-enriched infant formulas and an early detection of on-going autoxidation processes. Moreover, it was possible by two-dimensional GC-MS measurements to clearly distinguish stable and less stable formulations directly after manufacturing and thus, even before a sensory panel could detect clear differences in the respective flavor profiles. Spitzer and Buettner [29] carried out a quantitative study on lipid autoxidation in human milk. All seven odorants investigated in the present study were similarly part of the investigation of human milk. Concentrations in human milk after six months storage at -19 °C were to a great extent comparable with the concentrations described in the present study for spray-dried infant formula milks. Both this and the aforementioned study were based on multi-dimensional GC-MS/O analyses: thus, this analytical approach has proved to be a very sensitive tool to determine lipid oxidation in milk matrices quickly after production—in the current case within the first weeks. These first weeks are of the utmost interest for the food industry in terms of quality control as well as to speed up product improvements. The results of quantification further confirmed that alkyl radicals (RO•) and peroxyl radicals (ROO•) were formed during production, leading to the formation of hydroperoxides (ROOH) in the presence of oxygen by a free radical chain mechanism [30,31]. Once these hydroperoxides are established, lipid oxidation may even proceed further under exclusion of oxygen, leading to increasing concentrations of secondary lipid oxidation products, as shown in Section 3.1. These findings are confirmed by the investigations of Chávez-Servín et al. [32] on packaged infant formulas.

When regarding previous studies on volatile oxidative degradation products, hexanal is reported to be one of the most common markers. Static headspace GC measurements have detected hexanal at concentrations in the low mg·kg^{-1} range (ppm); dynamic headspace sampling, purge-and-trap sampling, and classical liquid injection GC-MS measurements were found to be more sensitive (*ppb*-range) in terms of hexanal values [13,27,32,33]. Nevertheless, it should be noted that the present study—using a multi-dimensional approach—was able to even detect volatiles in the very low *ppb*-range [<0.1 µg·kg^{-1} for concentrations of (E)-hex-2-enal]. This is of great importance for detecting sensory changes in PUFA-enriched infant formulas from on-going autoxidation processes due to the very low odor thresholds of such compounds.

The formation of primary lipid oxidation markers via the photometrical methods presented in the current study also confirmed on-going autoxidation processes in the packaged, evacuated infant formulas even if longer time intervals were necessary to obtain significant results. Conjugated dienes provide a valuable tool to draw conclusions on the "thermal history" of oils as recently reported by Cihelkova et al. [34]. Geometrical isomerization of double bonds of fatty acids may occur during deodorization and refining of oils. Since hydroperoxides are degraded during these process steps and volatile compounds are removed, conjugated dienes are a reliable parameter for quality evaluation of the oils used [35]. A comprehensive review of the relevant literature reveals that the determination of peroxide value (PV) or hydroperoxides in infant formulas is generally not reported in the context of very early detection of oxidation processes. For instance, Chávez-Servín et al. [18,32] described a storage period of 70 days for infant formulas after the packaging was opened and 18 months in the case of packaged infant formulas for detection of these primary oxidation markers. Only Manglano et al. [36] found significant increases in PV and hydroperoxides during the first five months of storage of commercial vacuum-packed infant formulas. Therefore, the findings of the present study on sensitivity and consistency of methods for determination of primary oxidation products reflect the previous reports in the literature quite well. These methods have been described before as standard tools in industrial routine control for monitoring fatty acid oxidation in fats and oils as well as milk matrices [37,38] and are used in standard protocols for the determination of the hydroperoxide content in anhydrous milk fat [23]. Therefore, depending on the matrix under investigation, it might be useful to monitor both primary and secondary lipid oxidation products in samples where the oxidative status is unknown.

Having a closer look at the results of this study, it was shown that A-formulations with antioxidants remained more stable than B-formulations without antioxidants, which is in accordance with previous reports [30]. Again, statistical analyses revealed differences in the sensitivity of the different methods used, with 2D-HRGC-MS/O measurements of secondary lipid oxidation products providing the most reliable results directly after production and during the first weeks of storage. To the best of our knowledge, no other studies involving an experimental design in production of infant formulas with and without antioxidants and with additional monitoring of their oxidative as well as sensory stability have been reported in the literature.

Another outcome of this study is that an important technological aspect could be elaborated using the combinatory approach of analytical analyses and comparative evaluation of sensitivity: formulations 7 and 8 with encapsulated Cu(II)-salts and non-encapsulated Fe(II) and Fe(III)-salts appeared to be more stable than other mineral combinations of copper and iron. In general, transition metals such as copper and iron are described as key promotors of lipid oxidation in the aqueous phase of oil-in-water emulsions [30]. As in the present study, Manglano et al. [36] performed a systematic investigation on infant formulas, albeit just in relation to different Fe(II)-salts. All formulas were supplemented with antioxidants, thereby only differing in tocopherol source. An outcome of the study of Manglano et al. [36] was that it was not possible to detect significant differences within the infant formulas with respect to lipid oxidation. These results are quite consistent with the findings of the present study, indicating a greater influence of copper on fatty acid oxidation together with the presence of antioxidants than of iron.

Sensory evaluations were performed in the present study since the challenge in the development of n-3 and n-6 fatty acid enriched infant formulas is to meet the consumers' expectations by avoiding aversive off-flavors. As already described by Spitzer et al. [5] for human milk, the formation of the odorants investigated in the present study are obviously closely related to the development of an off-flavor in PUFA-enriched milk samples. The corresponding off-flavor in human milk has previously been described by Spitzer et al. [5] as fishy, metallic, rancid and fatty; the same flavor attributes that were also detected in the present study. For infant formulas, there are hardly any reports in the literature regarding the formation of off-flavors. García-Martínez [39] determined the rancidity of an infant formula without any PUFA-enrichment, judged by a panel consisting of six untrained panelists. Chávez-Servín *et al.* [32] performed sensory evaluations by means of Duo-Trio and Pairs comparison tests where attributes such as "better smell", "better flavor" or "more rancid flavor" were used. Others, like Hausner *et al.* [13] did not perform APA but suggested a correlation between the detected volatile compounds and typical (off-) flavor notes in heated milk from citations in the literature. To close this gap, the present study implemented description of the aroma profiles of 18 different infant formulas over storage period and evaluation of the samples with regard to a subjective off-flavor rating by the panelists. Overall, these findings fit in the majority of cases to the results obtained from the analyses of primary and secondary products of lipid oxidation. At this point it should be highlighted again that the two-dimensional GC-MS analysis on secondary oxidation products is even more sensitive than the sensory panel in distinguishing stable and less stable formulations directly after production. This finding is of huge importance for the industry in the context of quality monitoring and product development of PUFA-enriched milk-based products.

In addition, it was shown that specific sensory attributes from APA continuously increased but, in some cases, also decreased during storage (specifically when one dominating sensory impression changed to another). It was found that the overall flavor profile is dependent on a mixture of volatiles and resulting from specific concentration ratios of alkanals, alkenals, alkadienals, alkenones and alkadienones. At a certain point, when concentration ratios change due to an on-going autoxidation, the overall flavor profile will also change, e.g., from metallic towards a fishy, oily or a blood-like flavor. This phenomenon of the development of a fishy off-flavor has been described before for dry spinach [6]. Masanetz *et al.* [6] found that a shift in the concentration of (*Z*)-octa-1,5-dien-3-one and methional was responsible for a more dominant hay-like or fishy odor of dry spinach. Moreover, Venkateshwarlu *et al.* [7] investigated the influence of different alkenals, alkadienals, alkenones on the development of fishy and metallic off-flavors in pasteurized milk samples. A synergistic effect was found for (*E,Z*)-nona-2,6-dienal and (*Z*)-hept-4-enal with regard to a developing fishy off-flavor. In another study, Venkateshwarlu *et al.* [8] investigated fish oil-enriched milk and assumed that fishy and metallic off-flavors are the result of a combination of potent odorants that were identified in their study. The latter findings confirmed a high correlation of flavor changes in foods dependent on mixing ratios of specific volatiles.

Another aspect with high relevance to the present study on PUFA-enriched infant formulas is that Venkateshwarlu *et al.* [8] reported that GC/O analysis revealed no volatile compound(s) with a fishy flavor note. This finding is also in agreement with the GC/O results of the present study (results not shown).

With regard to the sensory evaluations and respective off-flavor rankings of the present study, it may be concluded that a dominant metallic flavor was found to be more acceptable for the panelists in the case of infant formulas than other off-odors that were generated over the course of product storage. This acceptance changed to a more aversive rating if the concentration ratio of the volatiles changed in a way that a fishy, oily or blood-like flavor note arose. To the best of our knowledge, no other studies have been reported that deal with the acceptance of panelists or consumers towards different (off-) flavors arising in PUFA-enriched infant formulas.

Thus, it can be concluded that it is not a mono-dimensional process to correlate specific volatile compounds with the formation of off-flavors. To understand the complexity of oxidative deterioration and off-flavor formation, the necessity of performing comprehensive sensorial investigations together with a detailed analytical elucidation of the underlying molecular processes could therefore be demonstrated with the present study, paving the way for future investigations in this important field of research. This study highlights the huge potential for technological improvements of infant formulas based on the applied combinatory approach.

5. Conclusions

According to McClements and Decker [30], progress in the development of fatty foods like infant formulas is strongly dependent on improved diagnostic methods to control and monitor their oxidative stability. The aim of this study was to monitor fatty acid oxidation in PUFA-enriched infant formulas in its early stages to support product development in terms of formula optimization. Thus, three different methods of monitoring lipid oxidation were applied in a comparative approach. All diagnostic tools of this study were found to be suitable with regard to confirmation of autoxidation processes under the present study conditions. Nevertheless, different statistical analyses revealed differences in the sensitivity of the different methods used, with respect to the results directly after production and during the first weeks of storage: 2D-HRGC-MS/O analysis of odorous marker substances was the most sensitive tool for early detection and monitoring of autoxidation in the milk matrices. This technique was even more sensitive than a sensory panel in differentiating stable and less stable formulations directly after production. Methods determining lipid hydroperoxides and conjugated dienes were suitable to indicate oxidative changes, albeit after sensory changes became apparent.

In addition, the composition of all infant formulas was varied with regard to antioxidants and the minerals iron and copper in the frame of an experimental design. By combining the data from chemo-analytical investigations with the results from sensory examinations it was shown that lipid oxidation was low when copper was administered in an encapsulated form and when anti-oxidants (vitamin E, ascorbyl palmitate) were additionally present. Accordingly, such findings offer the possibility to predict oxidative processes and to establish optimization strategies for development of infant formulas.

Conflicts of Interest

The authors declare no conflict of interest.

References

1. Loncin, M.; Bimbenet, J.J.; Lenges, J. Influence of the activity of water on the spoilage of foodstuffs. *Int. J. Food Sci. Technol.* **1968**, *3*, 131–142.
2. Belitz, H.-D.; Grosch, W.; Schieberle, P. *Lehrbuch der Lebensmittelchemie*, 6th ed.; Springer: Berlin, Heidelberg, Germany, 2008; pp. 190–229.
3. Innis, S.M. Human milk: Maternal dietary lipids and infant development. *Proc. Nutr. Soc.* **2007**, *66*, 397–404.

4. Koletzko, B.; Lien, E.; Agostoni, C.; Boehles, H.; Campoy, C.; Cetin, I.; Decsi, T.; Dudenhausen, J.W.; Dupont, C.; Forsyth, S.; *et al.* The roles of long-chain polyunsaturated fatty acids in pregnancy, lactation and infancy: Review of current knowledge and consensus recommendations. *J. Perinat. Med.* **2008**, *36*, 5–14.

5. Spitzer, J.; Buettner, A. Characterization of aroma changes in human milk during storage at −19 °C. *Food Chem.* **2010**, *120*, 240–246.

6. Masanetz, C.; Guth, H.; Grosch, W. Fishy and hay-like off flavours of dry spinach. *Z. Lebensm. Unters. Forsch. A* **1998**, *206*, 108–113.

7. Venkateshwarlu, G.; Let, M.B.; Meyer, A.S.; Jacobsen, C. Modeling the sensory impact of defined combinations of volatile lipid oxidation products on fishy and metallic off-flavors. *J. Agric. Food Chem.* **2004**, *52*, 1635–1641.

8. Venkateshwarlu, G.; Let, M.B.; Meyer, A.S.; Jacobsen, C. Chemical and olfactometric characterization of volatile flavor compounds in a fish oil enriched milk emulsion. *J. Agric. Food Chem.* **2004**, *52*, 311–317.

9. Let, M.B.; Jacobsen, C.; Meyer, A.S. Effects of fish oil type, lipid antioxidants and presence of rapeseed oil on oxidative flavour stability of fish oil enriched milk. *Eur. J. Lipid Sci. Technol.* **2004**, *106*, 170–182.

10. Grosch, W. Aromastoffe aus der lipidperoxidation. *Eur. J. Lipid Sci. Technol.* **1989**, *91*, 1–6.

11. Boerger, D.; Buettner, A.; Schieberle, P. Structure/odour Relationships in Homologous Series of Aromaactive Allylalcohols and Allylketones. In *Flavour Research at the Dawn of the Twenty-First Century, Proceedings of the 10th Weurman Flavour Research Symposium*; le Quere, J.L., Etievant, P.X., Eds.; Lavoisier: London, UK, 2003; pp. 281–286.

12. Sandgruber, S.; Buettner, A. Comparative human-sensory evaluation and quantitative comparison of odour-active oxidation markers of encapsulated fish oil products used for supplementation during pregnancy and the breastfeeding period. *Food Chem.* **2012**, *133*, 458–466.

13. Hausner, H.; Philipsen, M.; Skov, T.; Petersen, M.; Bredie, W. Characterization of the volatile composition and variations between infant formulas and mother's milk. *Chemosens. Percept.* **2009**, *2*, 79–93.

14. Van Ruth, S.M.; Floris, V.; Fayoux, S. Characterisation of the volatile profiles of infant formulas by proton transfer reaction-mass spectrometry and gas chromatography-mass spectrometry. *Food Chem.* **2006**, *98*, 343–350.

15. Dobarganes, M.C.; Velasco, J. Analysis of lipid hydroperoxides. *Eur. J. Lipid Sci. Technol.* **2002**, *104*, 420–428.

16. Romeu-Nadal, M.; Castellote, A.I.; López-Sabater, M.C. Headspace gas chromatographic method for determining volatile compounds in infant formulas. *J. Chromatogr. A* **2004**, *1046*, 235–239.

17. Ulberth, F.; Roubicek, D. Monitoring of oxidative deterioration of milk powder by headspace gas-chromatography. *Int. Dairy J.* **1995**, *5*, 523–531.

18. Chávez-Servín, J.L.; Castellote, A.I.; López-Sabater, M.C. Volatile compounds and fatty acid profiles in commercial milk-based infant formulae by static headspace gas chromatography: Evolution after opening the packet. *Food Chem.* **2008**, *107*, 558–569.

19. Bruenner, B.A.; Jones, A.D.; German, J.B. Simultaneous determination of multiple aldehydes in biological tissues and fluids using gas chromatography/stable isotope dilution mass spectrometry. *Anal. Biochem.* **1996**, *241*, 212–219.

20. Engel, W.; Bahr, W.; Schieberle, P. Solvent assisted flavour evaporation—A new and versatile technique for the careful and direct isolation of aroma compounds from complex food matrices. *Eur. Res. Technol.* **1999**, *209*, 237–241.

21. Bemelmans, J.M.H. Review of Isolation and Concentration Techniques. In *Progress in Flavour Research*; Land, D.G., Nursten, H.E., Eds.; Applied Science Publisher: London, UK, 1979; pp. 79–88.

22. Satué-Gracia, T.; Frankel, E.N.; Rangavajhyala, N.; German, J.B. Lactoferrin in infant formulas: Effect on oxidation. *J. Agric. Food Chem.* **2000**, *48*, 4984–4990.

23. IDF (International Diabetes Federation). *Indernational IDF Standard 74A:1991: Anhydrous Milkfat: Determination of Peroxide Value*; IDF: Brussels, Belgium, 1991.

24. Drusch, S.; Serfert, Y.; Scampicchio, M.; Schmidt-Hansberg, B.; Schwarz, K. Impact of physicochemical characteristics on the oxidative stability of fish oil microencapsulated by spray-drying. *J. Agric. Food Chem.* **2007**, *55*, 11044–11051.

25. Chan, H.S.; Levett, G. Autoxidation of methyl linoleate. Separation and analysis of isomeric mixtures of methyl linoleate hydroperoxides and methyl hydroxylinoleates. *Lipids* **1977**, *12*, 99–104.

26. Sandgruber, S.; Much, D.; Amann-Gassner, U.; Hauner, H.; Buettner, A. Sensory and molecular characterisation of the protective effect of storage at −80 °C on the odour profiles of human milk. *Food Chem.* **2012**, *130*, 236–242.

27. Mallia, S.; Escher, F.; Dubois, S.B.; Schieberle, P.; Schlichtherle-Cerny, H. Characterization and quantification of odor-active compounds in unsaturated fatty acid/conjugated linoleic acid (UFA/CLA)-enriched butter and in conventional butter during storage and induced oxidation. *J. Agric. Food Chem.* **2009**, *57*, 7464–7472.

28. Serfert, Y.; Drusch, S.; Schwarz, K. Sensory odour profiling and lipid oxidation status of fish oil and microencapsulated fish oil. *Food Chem.* **2010**, *123*, 968–975.

29. Spitzer, J.; Buettner, A. Monitoring aroma changes during human milk storage at −19 °C by quantification experiments. *Food Res. Int.* **2013**, *51*, 250–256.

30. McClements, D.J.; Decker, E.A. Lipid oxidation in oil-in-water emulsions: Impact of molecular environment on chemical reactions in heterogeneous food systems. *J. Food Sci.* **2000**, *65*, 1270–1282.

31. Frankel, E.N. Lipid oxidation. *Prog. Lipid Res.* **1980**, *19*, 1–22.

32. Chávez-Servín, J.L.; Castellote, A.I.; Martín, M.; Chifré, R.; López-Sabater, M.C. Stability during storage of LC-PUFA-supplemented infant formula containing single cell oil or egg yolk. *Food Chem.* **2009**, *113*, 484–492.

33. Park, P.S.W.; Goins, R.E. Determination of volatile lipid oxidation products by dynamic headspace-capillary gas chromatographic analysis with application to milk-based nutritional products. *J. Agric. Food Chem.* **1992**, *40*, 1581–1585.

34. Cihelkova, K.; Schieber, A.; Lopes-Lutz, D.; Hradkova, I.; Kyselka, J.; Filip, V. Quantitative and qualitative analysis of high molecular compounds in vegetable oils formed under high temperatures in the absence of oxygen. *Eur. Food Res. Technol.* **2013**, *237*, 71–81.

35. Shahidi, F.; Zhong, Y. Lipid Oxidation: Measurement Methods. In *Bailey's Industrial Oil and Fat Products*, 6th ed.; Shahidi, F., Ed.; John Wiley & Sons Inc.: Hoboken, NJ, USA, 2005; pp. 357–385.

36. Manglano, P.; Lagarda, M.J.; Silvestre, M.D.; Vidal, C.; Clemente, G.; Farré, R. Stability of the lipid fraction of milk-based infant formulas during storage. *Eur. J. Lipid Sci. Technol.* **2005**, *107*, 815–823.

37. Gallaher, J.J.; Hollender, R.; Peterson, D.G.; Roberts, R.F.; Coupland, J.N. Effect of composition and antioxidants on the oxidative stability of fluid milk supplemented with an algae oil emulsion. *Int. Dairy J.* **2005**, *15*, 333–341.

38. Drusch, S.; Serfert, Y.; van den Heuvel, A.; Schwarz, K. Physicochemical characterization and oxidative stability of fish oil encapsulated in an amorphous matrix containing trehalose. *Food Res. Int.* **2006**, *39*, 807–815.

39. García-Martínez, M.D.C.; Rodríguez-Alcalá, L.M.; Marmesat, S.; Alonso, L.; Fontecha, J.; Márquez-Ruiz, G. Lipid stability in powdered infant formula stored at ambient temperatures. *Int. J. Food Sci. Technol.* **2010**, *45*, 2337–2344.

2

Fermented Brown Rice Flour as Functional Food Ingredient

**Muna Ilowefah, Chiemela Chinma, Jamilah Bakar, Hasanah M. Ghazali,
Kharidah Muhammad * and Mohammad Makeri**

UPM-BERNAS Research Laboratory, Faculty of Food Science and Technology, Universiti Putra
Malaysia, 43400 UPM, Serdang, Selangor, Malaysia; E-Mails: mona.milad2005@gmail.com (M.I.);
Chinmachiemela@yahoo.com (C.C.); jamilah@putra.upm.edu.my (J.B.);
hasanah@putra.upm.edu.my (H.M.G.); makeri50@yahoo.com (M.M.)

* Author to whom correspondence should be addressed; E-Mail: kharidah@putra.upm.edu.my

---

Abstract: As fermentation could reduce the negative effects of bran on final cereal products, the utilization of whole-cereal flour is recommended, such as brown rice flour as a functional food ingredient. Therefore, this study aimed to investigate the effect of fermented brown rice flour on white rice flour, white rice batter and its steamed bread qualities. Brown rice batter was fermented using commercial baker's yeast (Eagle brand) according to the optimum conditions for moderate acidity (pH 5.5) to obtain fermented brown rice flour (FBRF). The FBRF was added to white rice flour at 0%, 10%, 20%, 30%, 40% and 50% levels to prepare steamed rice bread. Based on the sensory evaluation test, steamed rice bread containing 40% FBRF had the highest overall acceptability score. Thus, pasting properties of the composite rice flour, rheological properties of its batter, volume and texture properties of its steamed bread were determined. The results showed that peak viscosity of the rice flour containing 40% FBRF was significantly increased, whereas its breakdown, final viscosity and setback significantly decreased. Viscous, elastic and complex moduli of the batter having 40% FBRF were also significantly reduced. However, volume, specific volume, chewiness, resilience and cohesiveness of its steamed bread were significantly increased, while hardness and springiness significantly reduced in comparison to the control. These results established the effectiveness of yeast fermentation in reducing the detrimental effects of bran on the sensory properties of steamed white rice bread and encourage the usage of brown rice flour to enhance the quality of rice products.

Keywords: fermentation; brown rice flour; pasting properties; rheological properties; bread volume; texture properties

1. Introduction

The past two decades have seen a rapid increase in consumer demand for healthy foods, which has prompted recent research to find methods for production of healthy and functional foods. The usage of whole grain cereal instead of milled cereals is one such trend for production of healthy and functional foods, as consumption of whole grain foods has shown a reduction in the risk of several diseases, such as cardiovascular diseases, obesity, diabetes and some types of cancers [1,2]. One of the most significant components in the whole-cereal grains that play a significant part in its health properties is dietary fiber and phenolics, which are mainly concentrated in the outer layers of the cereal grain [1]. Production of whole grain foods is a complicated task for the food industry due to the mechanical negative effects of the bran on protein network formation and consequently on the sensory properties of the end product [3]. Accordingly, the most common forms of cereals composition are milled products, such as white wheat flour or white rice.

Rice is a unique crop due to its colorless, soft taste, low sodium levels, easy digestible carbohydrates and hypoallergenic properties. Therefore, its flour is an attractive food material to be used for making gluten free foods [4]. Steamed bread is a traditional product made from white rice and is known as Apam in Malaysia. It is very popular in this part of the world and consumed at breakfast. It is formulated with white rice flour, sugar, salt, water and yeast. This kind of product is an example of gluten free foods.

Gluten free materials such as white rice flour do not have the required characteristics for production of leavened foods, since their proteins have no ability to develop a viscoelastic network such as gluten. In addition, lack of nutritional value has existed as a health problem of gluten free products specifically produced from white rice flour. Consequently, in recent years, there has been an increasing interest in using different food materials such as, gums, hydrocolloids and starches to enable developing a similar gluten network [5]. Moreover, whole grain flour such as millet, brown rice and sorghum were used to enrich gluten free foods [6,7]. Accordingly, using whole grain cereals flour could be promising for the development of healthy and acceptable gluten free foods.

Modification of whole-cereal flour prior to its usage by using simple food processing, such as fermentation, could eliminate the negative effects of the bran. Pre-fermentation of the whole-flour might increase fiber solubility due to enzyme reactions on the cell wall structure. The addition of pre-fermented wheat bran to wheat dough caused an increase in the bread volume of high fiber wheat bread [3]. The addition of yeast-fermented peeled bran to wheat bread increased its volume by 10%–15% and softened the crumb structure by 25%–35% compared to its unfermented bran [8]. Also, the addition of pre-fermented flour positively affected the texture, shelf life, aroma and nutritional value of gluten free foods, most probably because of metabolic activities of the microbes [9]. The positive effects associated with fermentation are the partial degradation of fiber and softening of bran particles [3]. The acidification rate of the dough is also an important property to achieve appropriate crumb structure and higher bread volume, since it affects the enzyme activities [10].

Recently, fermentation became a trend for production of healthy foods from whole grain cereals. Industrial application of the biotechnology of fermentation for the production of gluten free baked products is a promising innovation in the health foods industry. Thus, the objectives of the current study is to investigate the effect of fermented brown rice flour on white rice flour, white rice batter and its steamed bread (Apam) characteristics, which could encourage the usage of whole rice grains.

2. Materials and Methods

2.1. Materials

Brown rice grains (MR219) and commercial baker's yeast (Eagle CY 1266, China) were bought from a local supermarket in Selangor, Malaysia. Brown rice grains were ground using a Cyclotech™ (1093) grinder (FOSS, Sweden) and a 500 μm sieve to obtain brown rice flour (BRF), which was used for the fermentation process. Brown rice grains were milled using a rice miller (Satake TM 05C, Australia) to obtain white rice grains that were ground to flour in a Cyclotech™ (1093) grinder (FOSS, Sweden) and sieved to produce white rice flour (WRF). Flour samples were packaged in polyethylene bags and stored at 4 °C.

2.2. Fermentation Process

Fermentation conditions (time, temperature and yeast concentration) were optimized to achieve moderate acidity (pH 5.5) of fermented brown rice batter using Minitab 14 software (data not shown). Fermented brown rice batter at the optimum conditions was dried in an air oven drier at 50 °C for 3 h, ground to flour sieved in a 500 μm sieve and verified as fermented brown rice flour (FBRF).

2.3. Steamed Bread Making Process

The formula of 100 g of WRF which contains 2% sugar, 2% salt, 3% baker's yeast and 93% volume water based on the flour weight was used to prepare steamed white rice bread (SWRB). The SWRB formula of 60% white rice flour substituted with 40% FBRF achieved the highest bread volume and the overall acceptability of the tested sensory properties among the other ratios that included 0%, 10%, 20%, 30%, and 50% FBRF (data not shown) (Figure 1). To prepare rice bread; instant yeast was dissolved in a solution of water and sugar and conditioned at 32 °C for 10 min. Dry ingredients, such as WRF, salt and FBRF were mixed together, and then all the ingredients were mixed manually in a beaker for 3 min. White rice batter without FBRF was considered as the control. Finally, rice batters were positioned in cups and fermented in a fermenting box at 32 °C for 30 min. After fermentation, the samples were steamed for 15 min, cooled at room temperature (25 °C) for 1 h before measurements. Bread samples were made in five replicates.

Figure 1. Digital images of steamed white rice bread (SWRB) and steamed white rice breads substituted with fermented brown rice flour (FBRF).

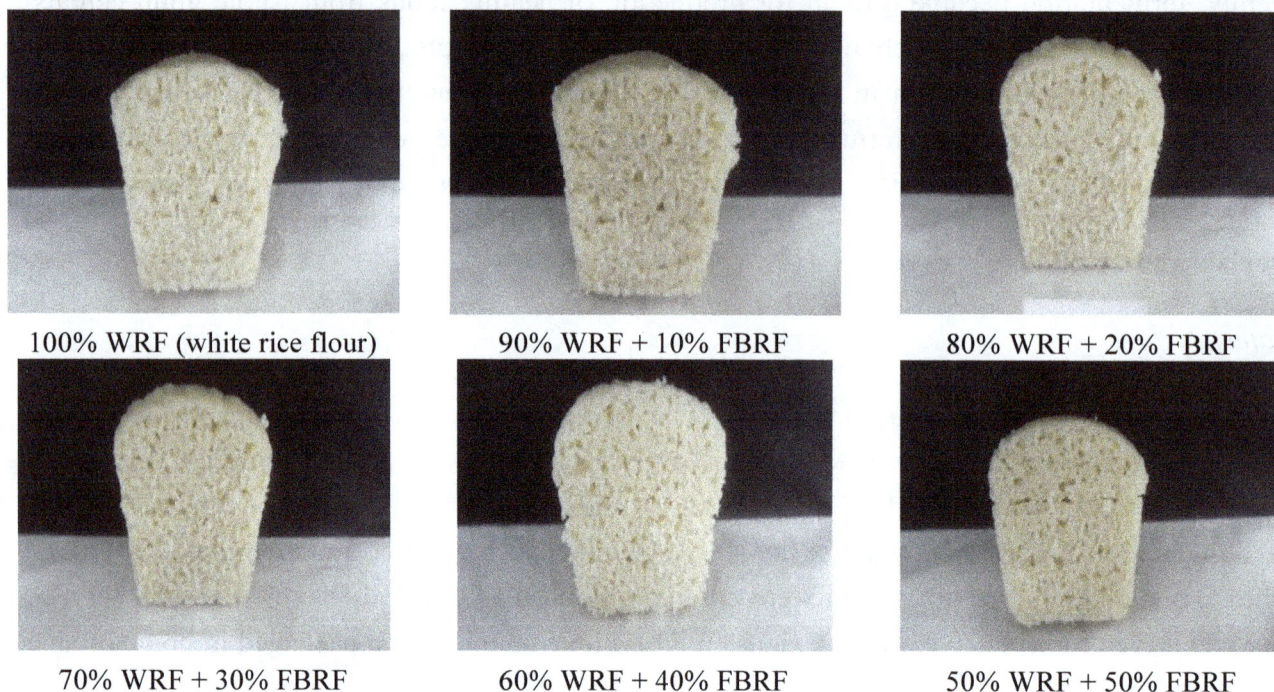

100% WRF (white rice flour)	90% WRF + 10% FBRF	80% WRF + 20% FBRF
70% WRF + 30% FBRF	60% WRF + 40% FBRF	50% WRF + 50% FBRF

2.4. Determination of Flour Pasting Properties

Pasting properties of WRF and WRF substituted with 40% FBRF was determined using a rapid viscos analyser (Newport Scientific Pty Ltd., New South Wales NSW, Australia) following the method of AACC 61-02 [11]. Peak viscosity, breakdown, final viscosity and setback of the samples were recorded by the Thermocline software.

2.5. Determination of Dynamic Rheological Properties of White Rice Batter

The dynamic rheological test was performed using a HAAKE Rheowin 600 rheometer, (Thermo, Germany) at 32 °C using parallel plate geometry (35 mm diameter and 1 mm gap). Batter samples were prepared as mentioned earlier, but without adding the yeast to avoid interference of air cells formation. After placing the batter samples between plates; batter samples were allowed to rest for 2 min to relax the residual stress. The linear viscoelastic part was measured by performing stress sweep experiment at 1 Hz frequency prior to determination. Based on the linear viscoelastic region, frequency sweep test was carried out at 1% strain between 0.1 and 10 Hz. Storage modulus (G′), loss modulus (G″), complex modulus (G*) and tan delta (δ) values were obtained from the Rheowin software (3.3). All the rheological tests were done in five replicates and their average values were stated in the results.

2.6. Measurement of Steamed Bread Volume

The volume of steamed breads was measured after 1 h of steaming according to the seed displacement method described by Hallén, *et al.* [12] using sago pearl. The loaf was inserted in a container with known volume (V_1) and filled with sago pearl. After that, the volume of sago pearl used was recorded (V_2). Bread volume (V) was calculated based on the following formula:

$$V \text{ (mL)} = V_1 - V_2$$

Specific volume was determined by dividing the volume of the bread by its weight (cm^3/g).

2.7. Texture Profile Analysis

The properties of the crumb texture of the breads were measured by using a Texture Analyzer (TA-XT2, UK) equipped with P/75 mm plate, 30 N load cell and trigger force of 5 g. After cooling at room temperature for 1 h, bread samples were prepared with approximately 2.5 × 2.5 × 2.5 cm dimensions taken from the crumb and compressed to 40% of its thickness [13], at a post and pre-speed of 2 mm/s and the interval time between first and second compression was 5 s. The texture properties, which include hardness, cohesiveness, chewiness, springiness and resilience were recorded using the Exponent software (32) (Stable Micro System, UK). All the samples were analyzed in five replicates.

2.8. Statistical Analyses

One way analysis of variance (ANOVA) and Tukey's multiple range tests with a confidence interval of 95% were used to report the significant differences between the obtained results.

3. Results and Discussion

3.1. Pasting Properties of White Rice Flour (WRF) and White Rice Flour Supplemented with 40% FBRF

Figure 2 shows the pasting properties of WRF and WRF with 40% FBRF. It can be observed that with addition of FBRF the peak viscosity significantly increased from 142.21 to 154.13 RVU, however, breakdown, final viscosity and setback were significantly reduced. The breakdown is associated with the ability of starch granules to be more resistant to being broken during heating and shearing. The results of this study indicated that the breakdown value was significantly reduced with the addition of FBRF. This result may be explained by the fact that, with an increase in protein content, some rice proteins could protect starch granules from being broken and increase pasting viscosity [14], as FBRF had a higher protein content after fermentation (data not shown). Also, the increase of peak viscosity and the decrease in breakdown values might indicate higher resistance to deformation and higher stability of the paste during baking. Moreover, the findings of the current study are consistent with those of Wang, et al. [15] who reported that final viscosity of rice paste decreased at low pH (4.10) and, in this study, the acidity of the rice batter with FBRF was higher than that of the control (data not shown). Another important finding was that the retrogradation phenomena could be significantly reduced in the final product, where setback—which is an indication of retrogradation phenomena—declined after addition of FBRF.

Figure 2. Pasting properties of white rice flour (WRF) and white rice flour with 40% FBRF.

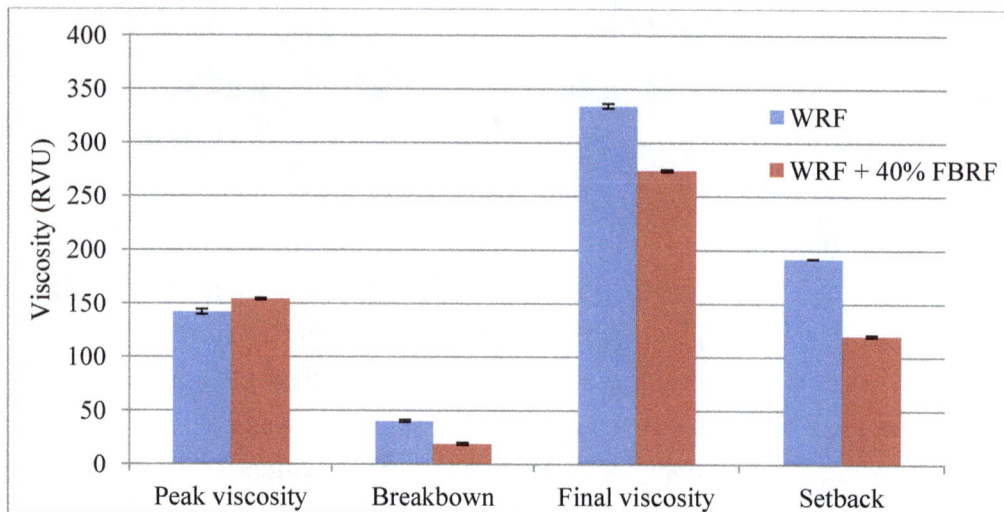

3.2. Viscoelastic Properties of White Rice Batter (WRB) and White Rice Batter Supplemented with 40% FBRF

The viscoelastic properties of the WRB and WRB containing 40% FBRF were studied using a dynamic oscillatory test. The mechanical spectra of the WRB sample indicated that both elastic (G') and viscos module (G'') values were higher than that of WRB with FBRF at all the tested frequency ranges (Figure 3). In addition, G' and G'' of WRB containing FBRF, independent of the frequency, compared to the reference sample. This suggested that the structure of WRB containing FBRF became softer and stronger than the control. It is reported that addition of acid to wheat dough reduced its extensibility and increased its resistance to extension [16]. It seems possible that these results are due to the rise in the positively charged proteins during acid treatment, where the pH of the batter containing FBRF was lower than that of the control. Nevertheless, acidity is not the only cause, since Rieder, *et al.* [16] reported that wheat dough containing fermented oat bran did not possess low pH, but exhibited higher resistance to extension and low extensibility. That means other factors associated with pre-fermented flour addition affected the dough rheology and thus the bread quality. The observed change in the viscoelastic properties could be attributed to the softening effect on the insoluble fiber due to the addition of FBRF. Rieder, *et al.* [16] also indicated that the addition of fermented barley flour to composite wheat bread degraded the fiber structure as indicated by a reduction in β-glucan molecular weight. In the current study, complex module decreased with addition of FBRF. However, the phase angle was not changed at low and high frequency, but decreased at 2.5 to 5.5 Hz (Figure 3). Another study reported that with addition of acetic and lactic acids to wheat dough, the results showed an obvious reduction in the complex modulus and the phase angle values [17].

Figure 3. Viscoelastic properties of white rice batter and white rice batter with 40% FBRF.

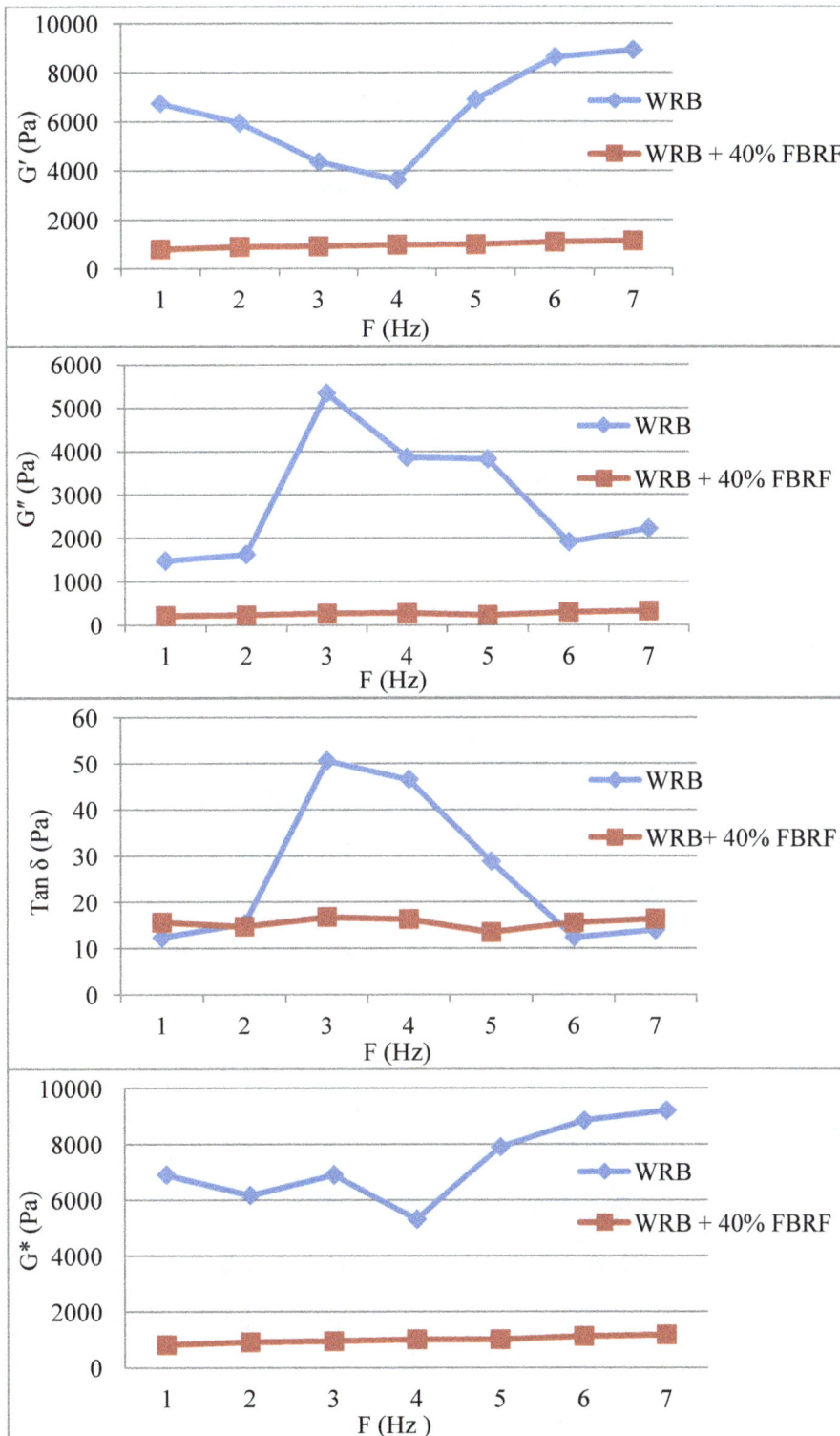

3.3. Steamed Bread Volume

Figure 4 shows that SWRB containing 40% FBRF gained higher volume (80 cm^3) and specific volume (2.46 cm^3/g) in comparison to the reference sample (70 cm^3 and 2.06 cm^3/g, respectively). This finding supports previous research, where the addition of fermented barley flour to whole barley dough improved the overall quality of its bread. Its bread volume was significantly higher than that of the control [16]. Dough acidification is an important issue, since it affects bread volume, crumb softness and retrogradation phenomena. The reduction in the pH during dough fermentation activates some enzymes such as α-amylase and protease. During addition of pre-fermented flour to wheat dough the acidity performs on the gluten network, which could improve extensibility and softness of the dough that helped to retain higher amount of CO_2 produced during fermentation, consequently increase the loaf volume [10]. However, high rate of acidity might increase hydrolyzation of the protein network; resulted in less elastic and softer dough that lead in a reduction in the bread volume, and elevate staling rate and bread firmness, as indicated by the addition of sourdough with high acidity to wheat dough [18]. The moderate decrease in the pH (5–6) of the dough because of microbial fermentation positively influences its structure specifically in high fiber breads, where addition of cereal fiber causes detrimental effects on the dough and bread structure. Also, the increase in bread volume of SWRB containing FBRF could be linked with the reduction in the loss and storage moduli of its batter and the reduction in breakdown of starch granules, which make the structure of the batter softer, stronger and more stable during steaming.

Figure 4. Volume and specific volume of steamed white rice bread (SWRB) and steamed white rice bread with 40% FBRF.

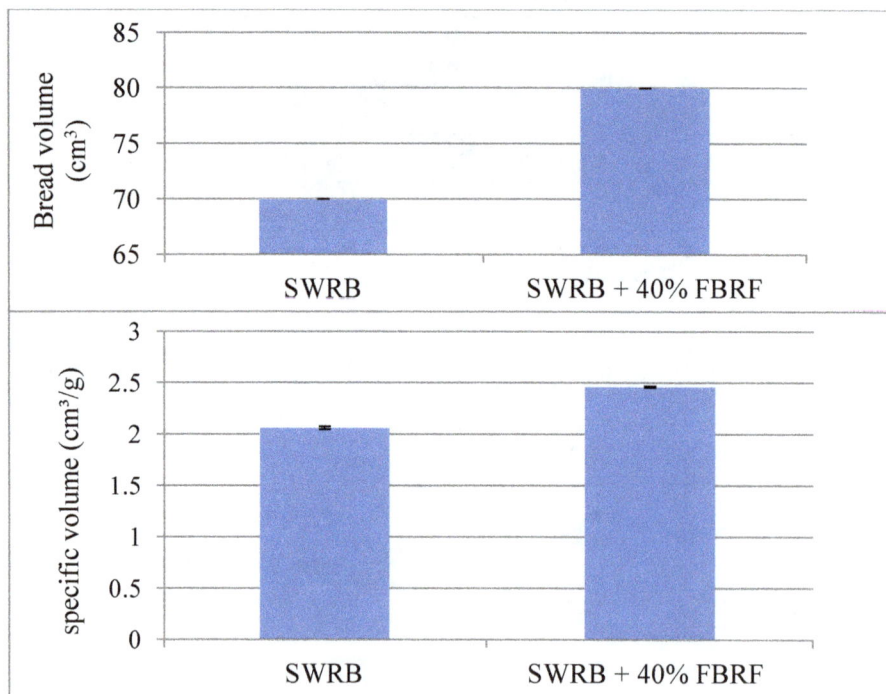

3.4. Steamed Bread Texture

Figure 5 shows the texture properties, which include hardness, springiness, cohesiveness, chewiness and resilience of the SWRB and SWRB supplemented by 40% FBRF. Hardness is the peak force during the first compression, and springiness or elasticity is defined as the height which the food recovered between the end of the first bit and beginning of the second bit. On the other hand, cohesiveness is the ratio of the positive force area during the second compression cycle to that during the first compression and adhesiveness is the negative force area for the first test and it is considered necessary to move the plunger away from the sample. Chewiness is assessed from the results of multiple analyses of gumminess and springiness [19]. The results of the current study indicated that the addition of FBRF softened the bread by recording less hardness value (6398.61 g) in comparison to the control (6948.13 g). The SWRB containing FBRF had higher values of chewiness, cohesiveness and resilience but a lower springiness value that indicated FBRF significantly improved the texture properties of SWRB. Rizzello, *et al.* [20] reported that pre-fermented fine and coarse bran fractions improved texture properties of the leavened baked product and the coarse fraction was therefore more effective. Fermented wheat germ is also observed to increase the bread volume of wheat bread and reduce its hardness rate [21]. Also, freeze-dried pre-fermented wheat germ had a positive influence on the texture and sensory characteristics of white wheat bread. The values of hardness and resilience of breads with fermented wheat germ were lower than those of the control, which means bread containing fermented germ was softer than its counterparts [21]. According to Salmenkallio-Marttila, *et al.* [3] addition of pre-fermented wheat bran to wheat bread supplemented with bran improved the crumb texture properties; specifically the elasticity. Also, it improved the bread volume and its shelf life. It is reported that adding pre-fermented bran with yeast and lactic acid bacteria improved the CO_2 retention during dough proofing, and as a result increased the bread volume and the crumb softness [3,22].

Figure 5. Texture properties of steamed white rice bread (SWRB) and steamed white rice bread with 40% FBRF.

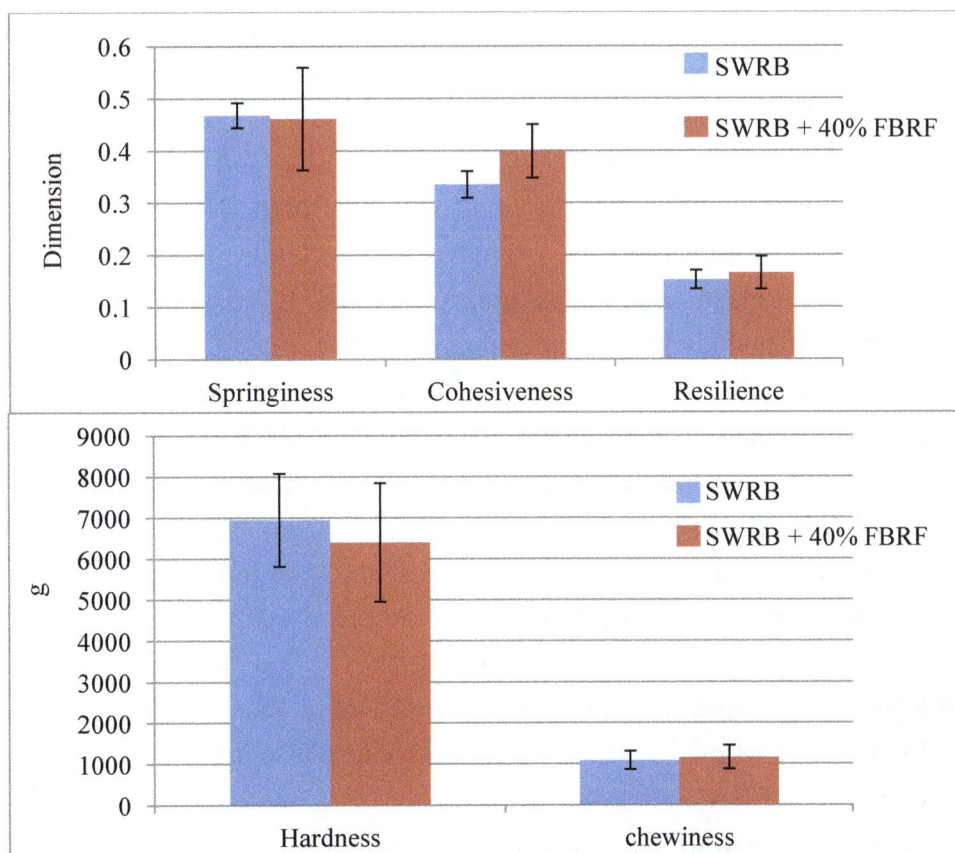

4. Conclusions

Addition of fermented brown rice flour to steamed white rice bread significantly improved rheological and textural properties as well as volume of the bread as indicated in this study. Currently, investigation of the effect of fermented brown rice flour on the nutritional value of steamed white rice bread is under way.

Acknowledgments

The authors are grateful to Universiti Putra Malaysia and Padiberas Nasional Berhad (BERNAS) for the support and facilities provided in accomplishing this research.

Conflicts of Interest

The authors declare no conflict of interest.

References

1. Slavin, J. Whole grains and human health. *Nutr. Res. Rev.* **2004**, *17*, 99–110.
2. Topping, D.J. Cereal complex carbohydrates and their contribution to human health. *Cereal Sci.* **2007**, *46*, 220–229.
3. Salmenkallio-Marttila, M.; Katina, K.; Autio, K. Effects of bran fermentation on quality and microstructure of high-fiber wheat bread. *Cereal Chem.* **2001**, *78*, 429–435.
4. Gujral, H.S.; Guardiola, I.; Carbonell, J.V.; Rosell, C.M. Effect of cyclodextrinase on dough rheology and bread quality from rice flour. *J. Agric. Food Chem.* **2003**, *51*, 3814–3818.
5. Marcoa, C.; Rosell, C.M. Effect of different protein isolates and transglutaminase on rice flour properties. *J. Food Eng.* **2008**, *84*, 132–139.
6. Alvarez-Jubete, L.; Arendt, E.K.; Gallagher, E. Nutritive value and chemical composition of pseudocereals as gluten-free ingredients. *Int. J. Food Sci. Nutr.* **2010**, *21*, 106–113.
7. Repo-Carrasco-Valencia, R.; Hellström, J.K.; Pihlava, J.M.; Mattila, P.H. Flavonoids and other phenolic compounds in Andean indigenous grains: Quinoa (*Chenopodium quinoa*), kañiwa (*Chenopodium pallidicaule*) and kiwicha (*Amaranthus caudatus*). *Food Chem.* **2010**, *120*, 128–133.
8. Katina, K.; Juvonen, R.; Laitila, A.; Flander, L.; Nordlund, E.; Kariluoto, S.; Poutanen, K. Fermented wheat bran as a functional ingredient in baking. *Cereal Chem.* **2012**, *89*, 126–134.
9. De Vuyst, L.; Vancanneyt, M. Biodiversity and identification of sourdough lactic acid bacteria. *Food Microbiol.* **2007**, *24*, 120–127.
10. Clarke, C.I.; Schober, T.J.; Arendt, E.K. Effect of single strain and traditional mixed strain starter cultures on rheological properties of wheat dough and on bread quality. *Cereal Chem.* **2002**, *79*, 640–647.
11. AACC Method 61–02: Determination of the Pasting Properties of Rice with Rapid Visco-Analyser. In *Approved Methods of Analysis*, 9th ed.; American Association of Cereal Chemists International: St. Paul, MN, USA, 1995.
12. Hallén, E.; İbanoğlu, Ş.; Ainsworth, P.J. Effect of fermented/germinated cowpea flour addition on the rheological and baking properties of wheat flour. *Food Eng.* **2004**, *63*, 177–184.
13. Schober, T.J.; Messerschmidt, M.; Bean, S.R.; Park, S.H.; Arendt, E.K. Gluten-free bread from sorghum: Quality differences among hybrids. *Cereal Chem.* **2005**, *82*, 394–404.

14. Zhu, L.J.; Liu, Q.Q.; Sang, Y.; Gu, M.H.; Shi, Y.C. Underlying reasons for waxy rice flours having different pasting properties. *Food Chem.* **2010**, *120*, 94–100.

15. Wang, H.H.; Sun, D.W.; Zeng, Q.; Lu, Y. Effect of pH, corn starch and phosphates on the pasting properties of rice flour. *J. Food Eng.* **2000**, *46*, 133–138.

16. Rieder, A.; Holtekjølen, A.K.; Sahlstrøm, S.; Moldestad, A. Effect of barley and oat flour types and sourdoughs on dough rheology and bread quality of composite wheat bread. *J. Cereal Sci.* **2012**, *55*, 44–52.

17. Wehrle, K.; Grau, H.E.; Arendt, K. Effects of lactic acid, acetic acid, and table salt on fundamental rheological properties of wheat dough. *Cereal Chem.* **1997**, *74*, 739–744.

18. Gocmen, D.; Gurbuz, O.; Kumral, A.Y.; Dagdelen, A.F.; Sahin, I. The effects of wheat sourdough on glutenin patterns, dough rheology and bread properties. *Eur. Food Res. Technol.* **2007**, *225*, 821–830.

19. Bourne, M.C. Texture profile analysis. *Food Technol.* **1978**, *32*, 62–66.

20. Rizzello, C.G.; Coda, R.; Mazzacane, F.; Minervini, D.; Gobbetti, M. Micronized by-products from debranned durum wheat and sourdough fermentation enhanced the nutritional, textural and sensory features of bread. *Food Res. Int.* **2012**, *46*, 304–313.

21. Rizzello, C.G.; Nionelli, L.; Coda, R.; Di Cagno, R.; Gobbetti, M. Use of sourdough fermented wheat germ for enhancing the nutritional, texture and sensory characteristics of the white bread. *Eur. Food Res. Technol.* **2010**, *230*, 645–654.

22. Katina, K.; Arendt, E.; Liukkonen, K.H.; Autio, K.; Flander, L.; Poutanen, K. Potential of sourdough for healthier cereal products. *Trends Food Sci. Technol.* **2005**, *16*, 104–112.

3

^{1}H NMR Spectroscopy and Multivariate Analysis of Monovarietal EVOOs as a Tool for Modulating Coratina-Based Blends

Laura Del Coco, Sandra Angelica De Pascali and Francesco Paolo Fanizzi *

Department of Biological and Environmental Sciences and Technologies (Di.S.Te.B.A.),
University of Salento, Prov.le Lecce-Monteroni, Lecce 73100, Italy;
E-Mails: laura.delcoco@unisalento.it (L.D.C.); sandra.depascali@unisalento.it (S.A.D.P.)

* Author to whom correspondence should be addressed; E-Mail: fp.fanizzi@unisalento.it

Abstract: Coratina cultivar-based olives are very common among 100% Italian extra virgin olive oils (EVOOs). Often, the very spicy character of this cultivar, mostly due to the high polyphenols concentration, requires blending with other "sweetener" oils. In this work, monovarietal EVOO samples from the Coratina cultivar (Apulia, Italy) were investigated and compared with monovarietal EVOO from native or recently introduced Apulian (Italy) cultivars (Ogliarola Garganica, Ogliarola Barese, Cima di Mola, Peranzana, Picholine), from Calabria (Italy) (Carolea and Rossanese) and from other Mediterranean countries, such as Spain (Picual) and Greece (Kalamata and Koroneiki) by ^{1}H NMR spectroscopy and multivariate analysis (principal component analysis (PCA)). In this regard, NMR signals could allow a first qualitative evaluation of the chemical composition of EVOO and, in particular, of its minor component content (phenols and aldehydes), an intrinsic behavior of EVOO taste, related to the cultivar and geographical origins. Moreover, this study offers an opportunity to address blended EVOOs tastes by using oils from a specific region or country of origin.

Keywords: olive oil; ^{1}H NMR spectroscopy; monovarietal EVOO; PCA

1. Introduction

Extra virgin olive oil (EVOO) is undoubtedly an essential ingredient in the Mediterranean diet, and its beneficial effects on human health, such as the reduction of coronary heart disease risk factors, the prevention of several types of cancer and the modification of immune and inflammatory responses, are well known [1,2]. These benefits are mainly due to both the elevated oleic acid content and the antioxidant properties of its minor components [3], such as phytosterols, carotenoids, tocopherols and hydrophilic phenols. EVOO contains at least 30 phenolic compounds. The major phenolic compounds are oleuropein derivatives containing hydroxytyrosol, which are strong antioxidants and radical scavengers. Moreover, phenolic compounds are related to the sensory and nutritional qualities of EVOOs [4], which play an important role in the olive oil blending. In this regard, several investigations have been devoted to understand the correlations of phenolic compounds with the organoleptic properties (flavor, astringency and hardness) of foods [5] and to assess their content in olive oil. Recently, an experimental investigation was performed on blend and monocultivar EVOOs to estimate the perceived bitterness intensity by correlating sensory and chemical analysis (total phenol content spectrophotometrically measured) [6]. Other analytical methods, such as paper, thin-layer and column chromatography, as well as UV spectroscopy were applied to polyphenol analysis. Substantial developments have been realized by using high-resolution gas chromatography (GC) and high-performance liquid chromatography (HPLC) [4,7]. However, during the last decade, proton nuclear magnetic resonance spectroscopy (^{1}H NMR) has been successfully employed in olive oil analysis [8]. The usefulness of ^{1}H NMR spectroscopy has been increasingly known for its sample preparation ease, quickness and reasonable sensitivity to a wide range of compounds in a single measurement. Difficulties arise in relation to very low concentrated molecules detection and handling of all the information obtained from the spectra of multicomponent mixtures, such as olive oil [5]. In this regard, the multi-signal suppression sequence represents an interesting approach, since it exploits NMR signals not only of major lipid components (triglycerides), but also of minor compounds, without requiring any sample pretreatment, such as extraction and/or purification steps [9]. On the other hand, computer-based multivariate analyses software packages nowadays allow for the easy treatment of complex datasets coming from NMR spectra. In particular, ^{1}H NMR spectroscopy coupled with chemometric studies has been used for the identification of EVOOs related to specific production areas and/or olive cultivars [10,11]. The absolute concentration and relative proportion of minor components are typical of each oil and may be used for production area and/or potential adulteration identification purposes. The fine composition of olive oil and, therefore, its sensory characteristics are influenced by several factors, such as climate, soil conditions, agricultural practices beside the specific cultivar used for its production [12,13]. In this work, we investigated by ^{1}H NMR spectroscopy and multivariate analysis on monovarietal EVOO samples from different native or recently introduced Apulian (Italy) cultivars (Coratina, Ogliarola Garganica, Ogliarola Barese, Cima di Mola, Peranzana, Picholine), from Calabria (Italy) (Carolea and Rossanese) and from other Mediterranean countries, such as Spain (Picual) and Greece (Kalamata and Koroneiki). A pattern recognition method based on multivariate statistical analyses has been performed, by applying spectral bucketing rather than the integration of NMR signals. This approach provides first a global profiling overview of the data and allows one, thereafter, to focus only on the discriminant variables (buckets and then related metabolites signals) responsible for class differences, extracting the systematic variation simultaneously for all samples. Furthermore, variable selection may be used also in pattern recognition, in concert with domain knowledge, to select only biologically meaningful bucket regions of datasets for classification or dimensionality reduction [14]. The aim of this work was to offer an opportunity to address blended EVOO tastes, correlating the chemical composition and, in particular, the minor components (phenols and aldehydes) of monocultivar EVOOs with their use as potential taste smoothers and/or enhancers in the commercial blending preparation, using oils from a specific region or country of origin compliant with recent regional [15] or EU regulation [16]. Since EVOO is based on a growing business, where blends of different varietals of olive oils represent a high percentage of the market, the phenol and aldehyde NMR signals could allow the first qualitative evaluation of the characteristics for different monovarietal EVOO samples. This approach was also oriented towards the study of differences in polyphenol content among olive oil cultivars, their geographical origins and the suggestions for the blending of EVOOs for marketing purposes. Indeed, due to the strong taste, the EVOOs from the Coratina cultivar are less appreciated by some consumers [17]. Although the polyphenolic content is related to the bitterness and astringency of foods, it confers more oxidative stability and contributes to the sensory and nutraceutical quality of the oil. In this regard, we attempted to compare the EVOO Coratina samples with the other taste smoothing monocultivar EVOOs from Italy and/or other countries.

2. Experimental Section

2.1. Sample Collection

Monocultivar authentic EVOO samples were obtained during the harvesting period of 2012–2013, from Italy (Apulia and Calabria), Spain and Greece. In particular, monocultivar EVOO samples were: Coratina (10 samples from Bari, Apulia), Ogliarola Barese and Cima di Mola (10 samples for each cultivar, from Bari, Apulia), Ogliarola Garganica (10 samples from Gargano, Foggia, Apulia), Picholine (10 samples from north of Bari, Apulia) Peranzana (3 samples from San Severo, Foggia, Apulia), Carolea and Rossanese (8 and 3, respectively, from Calabria), Picual (3 samples from Spain) and Kalamata and Koroneiki (3 samples for each cultivar, from Greece) (Table 1). EVOOs were supplied from the company "Certified Origins Italia Srl" (Località Il Madonnino, Grosseto, Tuscany, Italy).

Table 1. Cultivar, the number of samples and the area of origin of the monocultivar authentic extra virgin olive oils (EVOOs) collected.

Cultivar (*cv.*)	Number of samples	Geographical origin
Coratina	10	Bari (Apulia, Italy)
Ogliarola Garganica	10	Gargano, Foggia (Apulia, Italy)
Ogliarola Barese	10	Bari (Apulia, Italy)
Cima di Mola	10	Bari (Apulia, Italy)
Picholine	10	Bari (Apulia, Italy)
Peranzana	3	San Severo, Foggia (Apulia, Italy)
Carolea	8	Calabria
Rossanese	3	Calabria
Picual	3	Spain
Kalamata	3	Greece
Koroneiki	3	Greece

2.2. Chemicals

All chemical reagents for analysis were of analytical grade. Chloroform-d (CDCl$_3$, 99.8% atom D) and tetramethylsilane (TMS; 0.03% v/v) were purchased from Armar Chemicals (Switzerland).

2.3. NMR Spectroscopy

For NMR sample preparation, ~140 mg of olive oil were dissolved in deuterated chloroform (CDCl$_3$ with TMS as the internal standard) adjusting the mass ratio of olive oil:CDCl$_3$ to 13.5%:86.5%. Six hundred microliters of the prepared mixture were transferred into a 5-mm NMR tube. NMR spectra were recorded on a Bruker Avance III spectrometer (Bruker, Karlsruhe, Germany), operating at 400.13 MHz for ^{1}H observation and a temperature of 300 K, equipped with a BBO (Broadband Observe) 5-mm direct detection probe incorporating a *z*-axis gradient coil. NMR spectra were acquired using Topspin 2.1 (Bruker). Automated tuning and matching, locking and shimming using the standard Bruker routines, ATMA (automatic tuning and matching in automatic mode), LOCK (frequency-field lock to offset the effect of the natural drift of the NMR's magnetic field B_0) and TopShim, were used to optimize the NMR conditions. Experiments were run in automation mode after loading individual samples on a Bruker Automatic Sample Changer interfaced with the software, IconNMR (Bruker). For each sample, after a 5-min waiting period for temperature equilibration, three ^{1}H NMR experiments were performed: standard one-dimensional ^{1}H ZG NMR experiment; one-dimensional ^{1}H NOESYGPPS NMR pulse sequence; JRESGPPSGF with suppression of the strong lipid signals (20 frequencies), in order to enhance signals of the minor components present in EVOOs. Spectra were obtained by the following conditions: zg pulse program (for ^{1}H ZGNMR), 64 K time domain, spectral width of 20.5555 ppm (8223.685 Hz), p1 (F1 channel—90° ^{1}H transmitter pulse) 12.63 µs, pl1 −1.00 db (decibel), 16 repetitions; noesygpps1d.comp2 pulse program (for ^{1}H NOESYGPPS NMR) 32 K time domain, spectral width 20.5555 ppm (8223.685 Hz), p1 12.63 µs, pl1 −1.00 db, 32 repetitions; jresgppsqf pulse program 8 K time domain, spectral width 16.7082 ppm (6684.492 Hz), pl1 −1 db, 4 repetitions for 40 experiments.

2.4. NMR Data Reduction and Preprocessing

NMR data were processed using Topspin 2.1 (Bruker) and visually inspected using Amix 3.9.13 (Bruker, Biospin). ^{1}H NMR spectra were obtained by the Fourier transformation (FT) of the FID (free induction decay), applying an exponential multiplication with a line broadening factor of 0.3 Hz. The resulting ^{1}H NMR spectra were manually phased and baseline corrected using the Bruker Topspin software. Chemical shifts were reported with respect to the TMS signal set at 0 ppm. ^{1}H NMR spectra were segmented in rectangular buckets of a fixed 0.04-ppm width and integrated using the Bruker Amix software. Bucketing of the ^{1}H NOESYGPPS NMR spectra was obtained within the range of 9.76–6.52 ppm. The spectral region between 7.60 and 6.92 ppm was discarded, because of the peak due to the residual protic chloroform signal at 7.24 ppm. The remaining buckets (64) were then normalized to the total area to minimize small differences and/or acquisition conditions among samples and, subsequently, mean-centered. The description of statistical analyses refers to unscaled data. The data table generated with all the spectra was submitted to multivariate data analysis.

2.5. Multivariate Statistical Analysis

Principal component analysis (PCA) was applied to NMR spectroscopic data. The multivariate statistical analysis and graphics were obtained using Simca-P version 13.0.2 (Umetrics, Sweden). PCA, an unsupervised pattern recognition method, was performed to examine the intrinsic variation in the dataset. The variables used for chemometric analyses were the buckets, which represent the portion of the NMR spectrum of phenolic components. Three separate PCA analyses describe the different performance of Coratina EVOOs with respect to monocultivar oils from various countries: Italy, Spain and Greece. Since the data is coming from different geographical scales (regional, inter-regional and national) and the limits defined by the principal axes were different, separate analyses were conducted. PCA was used to obtain a general overview of the NMR data and to describe the natural variability in the olive oils originating from specific cultivars. The R^2 and Q^2 are the two specific parameters considered for the description of the soundness of the models referring to the considered components. The former (R^2) explains the total variation in the data, whereas the latter (Q^2) is an internal cross-validation parameter, which indicates the predictability of the model. The predictive power of the model is estimated by determining how accurately we can internally predict the data, randomly removing some of them and verifying their correct classification. Therefore, a valid model consists of a good compromise between these two related parameters, R^2 and Q^2, indicating the explained variation and prediction goodness, respectively.

3. Results and Discussion—NMR Spectroscopy and Multivariate Statistical Analyses

The study of multi-suppression ^{1}H NMR and the 2D J-resolved NMR spectra of monocultivar EVOO samples revealed a completely different profile in polyphenols contents. On the basis of the literature data [18,19], the identification of the main polyphenols was performed. As expected, monocultivar Coratina samples were characterized by intense NMR signals in the range at δ = 6.50–7.10 ppm, which could be assigned to phenyl alcohol moieties (tyrosol and hydroxytyrosol) of oleuropein and ligstroside and their aglycone. NMR signals at δ = 9.63 ppm were attributed to aldehydic forms of oleuropein and ligstroside; NMR signals at $\delta = 9.52$ ppm and in the range of $\delta = 9.25–9.19$ ppm were attributed to the dialdehydic forms of oleuropein and ligstroside [19]. It is well known that EVOO samples from *cv.* Coratina confer a bitter and astringent flavor to the oil. The bitter taste of olives is principally caused from oleuropein glucoside and its aglycone [18,20]. Then, oils obtained from olive fruits rich in polyphenols, e.g., cultivar Coratina, are expected to be more bitter and pungent. Recently Lauri *et al.* [18], in order to assess which chemical components are responsible for a given sensory descriptor, have looked for all possible correlations between the NMR signals (in the polyphenol spectral regions) and the analyzed sensory descriptors using OPLS (Orthogonal Partial least squares projections to latent structures) models. In particular, they explored the analytical potentiality of the NMR spectroscopy as a "magnetic tongue" in the analysis of EVOO, with particular attention to the quantitative measure of minor compounds related to the sensory description. Particularly, the phenol and aldehyde NMR signals allowed the first prediction of the sensory characteristics of EVOO.

A further level of investigation was performed using the unsupervised exploratory statistical technique (PCA) on a bucket table made of 64 NMR buckets, from the polyphenol spectral region of 9.76–6.52 ppm. The bucket width was 0.04 ppm, and the spectra were all scaled to the total intensity before proceeding with the bucketing. PCA allowed the remapping the original dataset in a new multivariate coordinate space, where the dimensions are ordered by the decreasing explained variance in the data. The principal components were displayed as a set of scores (t), which highlight clustering or outliers, and a set of loadings (p), which highlight the influence of input variables on t. In all the models studied, PCA showed that by the first two components, a very useful model of the data was built. A first PCA was performed on EVOO samples of Coratina (Italy), Picual (Spain), Kalamata and Koroneiki (Greece), to obtain a general overview of the data grouping when Coratina was compared to cultivars of other countries (Figure 1). Thus, the first dimensions (principal components, t1/t2) highlight in a straightforward manner the spectra that were significantly different from each other and, ultimately, samples that showed distinct biochemical composition. Of the original 64 variables per spectrum, two components were enough to describe 98% of the variance of the entire NMR dataset, giving $R^2 = 0.98$ and $Q^2 = 0.943$. A good clustering of Coratina samples was observed along t1, whereas the Greek EVOO samples (Kalamata and Koroneiki, which were almost overlapped) were well separated from the Spanish EVOOs on the second component, t2. By analysis of the loading plot, Coratina samples showed a higher content of NMR signals at $\delta = 6.78$, 9.22 and 9.62 ppm, which could be assigned to aldehydic and dialdehydic forms of oleuropein and ligstroside. These NMR signals in the recent work of Lauri *et al.* [18] have been correlated with bitter, pungent and artichoke tastes. The loading plot of t2 showed that the NMR Greek samples exhibited a higher concentration of signals at $\delta = 9.74$, 9.50, 9.26 and 6.86 ppm. The signal at 9.74 ppm could be attributed to hexanal [21], molecule, associated with a sweet taste [22].

Figure 1. (A) PCA (t1/t2) score plot for monovarietal Coratina, Picual, Kalamata and Koroneiki samples (two components give $R^2 = 0.98$ and $Q^2 = 0.943$); (B) expansions of the ^{1}H NMR spectra of monovarietal Coratina, Picual, Kalamata and Koroneiki samples used for the bucketing.

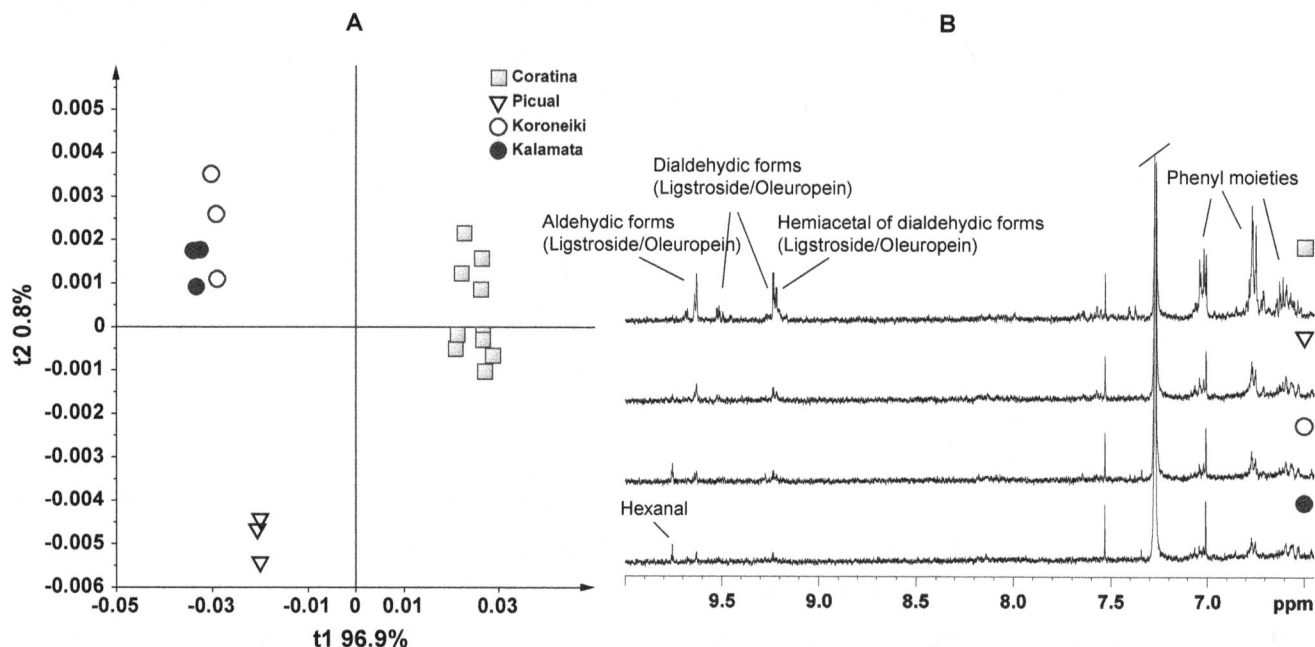

A second PCA was performed on the Coratina and EVOO samples from Calabria (Figure 2). Also in this case, two components were able to describe 97% of the variance of the entire NMR dataset, giving $R^2 = 0.97$ and $Q^2 = 0.95$. Interestingly, there is a complete separation of the Coratina samples along t1 with respect to the Rossanese and Carolea, which, in turn, were split along t1. By examining the loadings of the original bucket variables, Coratina samples were characterized by variables with positive loadings on t1. These loadings, responsible for the marked clustering of Coratina, were again associated with the aldehydic and dialdehydic forms of oleuropein and ligstroside. Moreover, the Rossanese samples separated from the Carolea on the second component (t2), due to the positive loadings at 6.58 and 6.54 ppm, which could be assigned to phenyl alcohols moieties (tyrosol and hydroxytyrosol) of oleuropein and ligstroside aglycones [23]. The lack of aldehydic signals is usually related to a sweet taste.

Figure 2. (**A**) PCA (t1/t2) score plot for monovarietal Coratina, Rossanese and Carolea samples (two components give $R^2 = 0.97$ and $Q^2 = 0.95$); (**B**) expansions region of the ^{1}H NMR spectra of monovarietal Coratina, Rossanese and Carolea samples used for the bucketing.

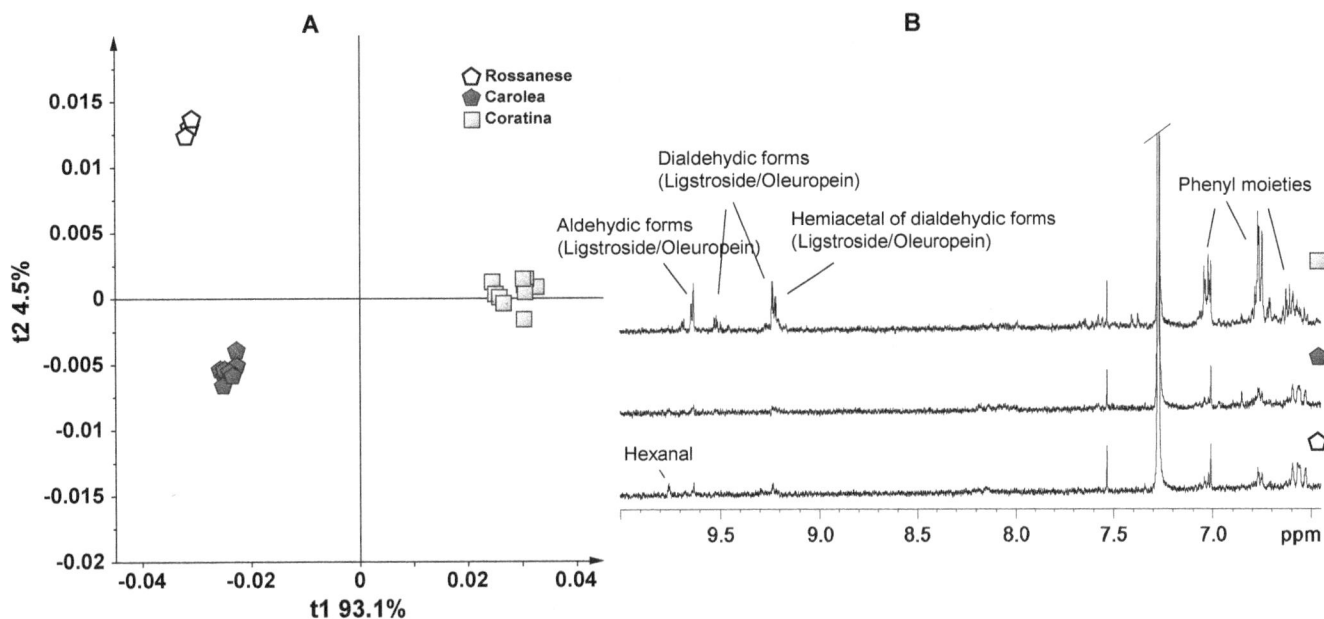

The last scatter plot was obtained by the PCA on all the EVOOs produced from Apulian cultivars (native: Coratina, Ogliarola Garganica, Ogliarola Barese, Cima di Mola; or recently introduced: Picholine and Peranzana) (Figure 3). This model built with six cultivars had a good descriptive ability with four components (t1 explains 87.9%, t2 3.7%, t3 2%, t4 1.1% of the total variance), which describe 95% of the variance with $R^2 = 0.95$ and $Q^2 = 0.89$. Looking at the score plot t1/t2, a good clustering of Coratina samples was observed. In particular, they were positioned at negative values of both t1 and t2, separating on t1 from the other native cultivar and along t2 with respect to the introduced cultivar. On the other hand, a certain degree of overlap was observed between two native cultivars, Ogliarola Barese and Cima di Mola, while Ogliarola Garganica was located at negative values of t1, but positive ones of t2. Finally, the recently introduced Picholine and Peranzana cultivars were clustered at positive values of both t1 and t2. Also in this case, a certain degree of variation in loadings levels was observed by the analysis of the loading plots. Along t1, the Coratina EVOOs showed again a higher content of molecules at $\delta = 6.78$, 9.22 and 9.62 ppm, assigned to aldehydic and dialdehydic forms of oleuropein and ligstroside. The loading plot along t2 showed a clear separation of Coratina from both Picholine and Peranzana samples, due to a higher concentration for Coratina samples of buckets in the range from 6.74 to 6.78 ppm, from 9.18 to 9.22 ppm and at 9.50 ppm, which could be assigned to other aldehydic form of oleuropein [19]. Indeed, by comparison of the ^{1}H NMR spectra of aromatic region of all native Apulian cultivars, a gradual decrement of aldehydic species from Coratina to Cima di Mola was observed. Interestingly, the contribution plot comparison (data not shown) of Ogliarola Barese and Ogliarola Garganica samples, which separated along t2 component, showed that the Ogliarola Barese EVOOs exhibited a higher concentration of molecules in the range of

7.94–8.10, 9.18, 9.26 and 9.50 ppm with respect to Ogliarola Garganica. In general, all these signals could be related to the rosemary flavor [18].

It should be noted that Coratina, Ogliarola Barese and Cima di Mola were the basic cultivars used for the Protected Designation of Origin (PDO) "Terra di Bari", the most important PDO in the regional context and the second in Italy [24]. In all the studied cases, Coratina was the cultivar with the highest content of polyphenols, which could be related to the characteristic bitter and pungent taste. Among all the other cultivar considered, only the Picholine and Peranzana, cultivated in Apulia, showed a reasonable content in polyphenols.

Figure 3. (**A**) PCA (t1/t2) score plot for monovarietal Apulian cultivar samples (native: Coratina, Ogliarola (Ogl.) Garganica, Ogliarola (Ogl.) Barese, Cima di Mola; recently introduced: Picholine and Peranzana). $R^2 = 0.95$ and $Q^2 = 0.89$; (**B**) expansions of the ^{1}H NMR spectra of monovarietal Apulian cultivars samples used for the bucketing.

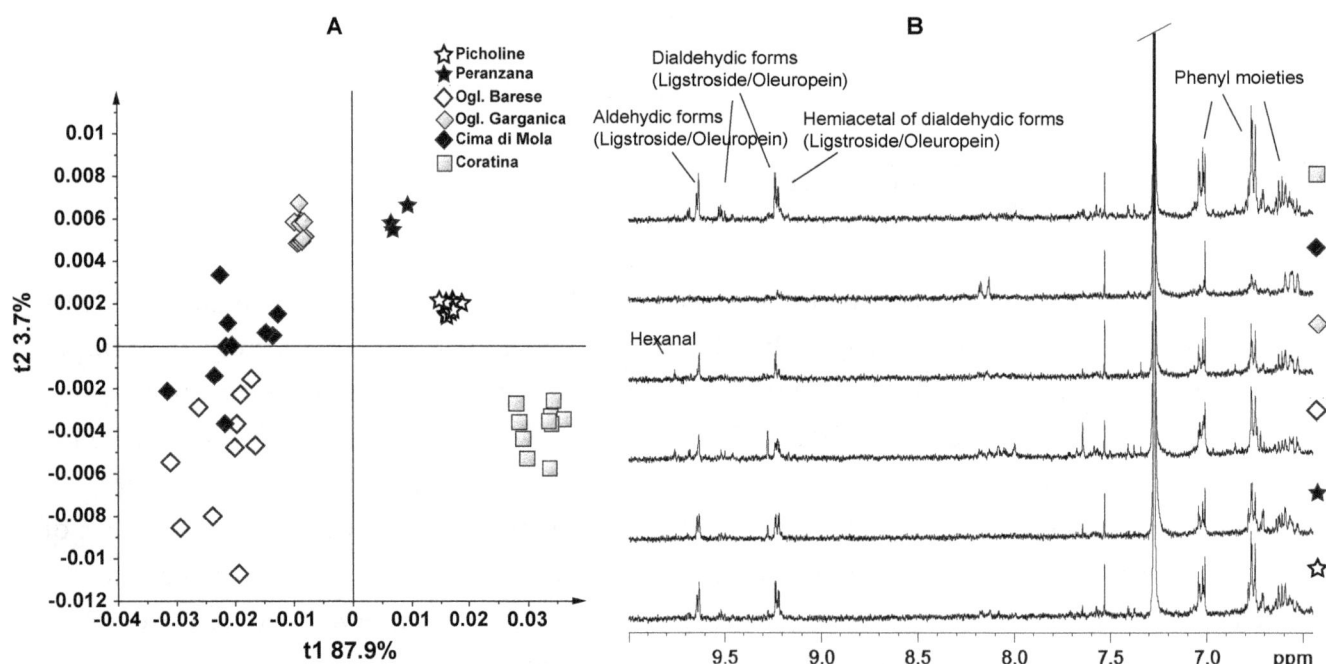

4. Conclusions

This study provides an initial evaluation of how natural variability in olive oil might affect blends originating from specific cultivars. Due to the importance of olive oil marketing for Italy and for other Mediterranean countries, it is very important to preserve its authentic characteristics. Since 90% of the entire oil production comes from Southern Italian Regions (Sicily, Calabria and Apulia), there is a real need to define the characteristics of local olive oil production. In particular, it should be noted that Coratina is the most popular olive cultivar of the Apulia region, among the leading olive oil producers in Italy, accounting for almost 40% of the total country production. As the Coratina cultivar has higher bitterness and very strong pungency, as declared by sensory analysis (panel test), the comparison of its molecular profile with other cultivars could be used to recognize the potential taste smoothers and/or enhancers in the commercial blending preparation. In this regard, NMR spectroscopy coupled with statistical multivariate analysis could be potentially applied to support and buttress taste analysis. All the studied samples were monocultivar EVOOs obtained from various Mediterranean countries (Greece, Spain) and Italian regions (Calabria, Apulia). The chemical profiles of the EVOO samples were analyzed by 1D and 2D ^{1}H NMR spectroscopy, and assignment hypotheses were performed for NMR signals of the minor components of EVOOs. Different polyphenols were detected, especially in Coratina samples, even though no separation and/or isolation procedure of the phenolic fraction was conducted. According to the related loading plots, Coratina was the cultivar with the highest content of polyphenols, which could be related to the characteristic bitter and pungent taste. Among all the other cultivar considered, only the Picholine and Peranzana, cultivated in Apulia, showed a reasonable content of polyphenols. These signals were nearly absent in the oils commonly used to "smooth" the Coratina taste. Our intention was an explorative attempt to address the focused blending of monocultivar EVOOs being performed. The aim of the present work was therefore to give an indication of the possible taste smoother cultivar to be used rather than a specific blend production receipt. On the other hand, an example of a specific comparison for a blend with its monocultivar constituent EVOOs has been already recently reported [9]. This study also offers an opportunity to modify blended EVOO tastes by using oils from a specific region or country of origin compliant with recent regional [15] or EU regulation [16].

Acknowledgments

This work was supported by the grant project, the PON (Programma Operativo Nazionale) 254/Ricerca. Potenziamento del "Centro Ricerche per la Salute dell'Uomo e dell'Ambiente" Code PONa3_00334.

Thanks to the company "Certified Origins Italia Srl" (Località Il Madonnino, Grosseto, Tuscany, Italy) for supplying the samples.

Conflicts of Interest

The authors declare no conflict of interest.

References

1. Willett, W.C.; Sacks, F.; Trichopoulou, A.; Drescher, G.; Ferro-Luzi, A.; Helsing, E.; Trichopoulos, D. Mediterranean diet pyramid a cultural model for healthy eating. *Am. J. Clin. Nutr.* **1995**, *61*, 1402S–1406S.

2. Lipworth, L.; Martinez, M.E.; Angell, J.; Hsien, C.C.; Trichopoulos, D. Olive oil and human cancer, an assessment of the evidence. *Prev. Med.* **1997**, *26*, 81–190.

3. Owen, R.W.; Giacosa, A.; Hull, W.E.; Haubner, R.; Wurtele, G.; Spiegelhalder, B.; Bartsch, H. Olive-oil consumption and health: The possible role of antioxidants. *Lancet Oncol.* **2000**, *1*, 107–112.

4. Tsimidou, M. Polyphenols and quality of virgin olive oil in retrospect. *Ital. J. Food Sci.* **1998**, *10*, 99–116.

5. Bendini, A.; Cerretani, L.; Carrasco-Pancorbo, A.; Gómez-Caravaca, A.M.; Segura-Carretero, A.; Fernández-Gutiérrez, A.; Lercker, G. Phenolic molecules in virgin olive oils: A survey of their sensory properties, health effects, antioxidant activity and analytical methods. an overview of the last decade. *Molecules* **2007**, *12*, 1679–1719.

6. Favati, F.; Condelli, N.; Galgano, F.; Caruso, M.C. Extra virgin olive oil bitterness evaluation by sensory and chemical analyses. *Food Chem.* **2013**, *139*, 949–954.

7. Morales, M.T.; Tsimidou, M. The role of volatile compounds and polyphenols in olive oil sensory quality. In *Handbook of Olive Oil: Analysis and Properties*; Aparicio, R., Harwood, J., Eds.; Kluwer Academic/Plenum Publishers: New York, NY, USA, 1999; Chapter 12, pp. 393–455.

8. Vlahov, G. Application of NMR to the study of olive oils. *Prog. Nucl. Magn. Reson. Spectrosc.* **1999**, *35*, 341–357.

9. Del Coco, L.; de Pascali, S.A.; Fanizzi, F.P. NMR-metabolomic study on monocultivar and blend salento EVOOs including some from secular olive trees. *Food Nutr. Sci.* **2014**, *5*, 89–95.

10. Oms-Oliu, G.; Odriozola-Serrano, I.; Martín-Belloso, O. Metabolomics for assessing safety and quality of plant-derived food. *Food Res. Int.* **2013**, *54*, 1172–1183.

11. Aghemo, C.; Albertino, A.; Gobetto, R.; Lussiana, C.; de Maria, A.; Isocrono, D. Piedmont olive oils: Compositional characterization and discrimination from oils from other regions. *Eur. J. Lipid Sci. Tech.* **2012**, *114*, 1409–1416.

12. Del Coco, L.; Schena, F.P.; Fanizzi, F.P. 1H Nuclear magnetic resonance study of olive oils commercially available as italian products in the United States of America. *Nutrients* **2012**, *4*, 343–355.

13. Del Coco, L.; Perri, E.; Cesari, G.; Muzzalupo, I.; Zelasco, S.; Simeone, V.; Schena, F.P.; Fanizzi, F.P. NMR-based metabolomic approach for EVOO from secular olive trees of Apulia region. *Eur. J. Lipid Sci. Technol.* **2013**, *115*, 1043 1052.

14. Worley, B.; Powers, R. Multivariate analysis in metabolomics. *Curr. Metab.* **2013**, *1*, 92–107.

15. REGOLAMENTO (CE) N. 207/2009 DEL CONSIGLIO. Available online: http://eurlex.europa.eu/LexUriServ/LexUriServ.do?uri=OJ:L:2009:078:0001:0042:it:PDF (accessed on 31 January 2014).

16. The Commission of the European Communities. COMMISSION REGULATION (EC) No 182/2009 of 6 March 2009 Amending Regulation (EC) No 1019/2002 on Marketing Standard for Olive Oil. *Off. J. Eur. Union* **2009**, *L63*, 6–8. Available online: http://eur-lex.europa.eu/LexUriServ/LexUriServ.do?uri=OJ:L:2009:063:0006:0008:EN:PDF (accessed on 31 January 2014).

17. McEwan, J. Consumer attitudes and olive oil acceptance: The potential consumer. *Grasas Aceites* **1994**, *45*, 9–15.

18. Lauri, I.; Pagano, B.; Malmendal, A.; Sacchi, R.; Novellino, E.; Randazzo, A. Application of "magnetic tongue" to the sensory evaluation of extra virgin olive oil. *Food Chem.* **2013**, *140*, 692–699.

19. Christophoridou, S.; Dais, P.; Tseng, L.H.; Spraul, M. Separation and identification of phenolic compounds in olive oil by coupling high-performance liquid chromatography with postcolumn solid-phase extraction to nuclear magnetic resonance spectroscopy (LC-SPE-NMR). *J. Agric. Food Chem.* **2005**, *53*, 4667–4679.

20. Aparicio, R.; Luna, G. Characterisation of monovarietal virgin olive oils. *Eur. J. Lipid Sci. Technol.* **2002**, *104*, 614–627.

21. Mannina, L.; Patumi, M.; Proietti, N.; Bassi, D.; Segre, A.L. Geographical characterization of Italian extra virgin olive oils using high-field ^{1}H NMR spectroscopy. *J. Agric. Food Chem.* **2001**, *9*, 2687–2696.

22. Morales, M.T.; Alonso, M.V.; Rios, J.J.; Aparicio, R. Virgin olive oil roma: Relationship between volatile compounds and sensory attributesby chemometrics. *J. Agric. Food Chem.* **1995**, *43*, 2925–2931.

23. Christophoridou, S.; Dais, P. Detection and quantification of phenol compounds in olive oils by high resolution ^{1}H nuclear magnetic resonance spectroscopy. *Anal. Chim. Acta* **2009**, *633*, 283–292.

24. Production Specification. Available online: http://www.extra-virgin-olive-oil.it/dop-terra-di-bari.asp (accessed on 31 January 2014).

4

Eggs and Poultry Purchase, Storage, and Preparation Practices of Consumers in Selected Asian Countries

Kadri Koppel [1,*], Suntaree Suwonsichon [2,†], Uma Chitra [3,†], Jeehyun Lee [4,†], and Edgar Chambers IV [1]

[1] The Sensory Analysis Center, Kansas State University, 1310 Research Park Drive, Manhattan, KS 66502, USA; E-Mail: eciv@ksu.edu

[2] Kasetsart University Sensory and Consumer Research Center, Department of Product Development, Kasetsart University, 50 Paholyothin Road, Jatujak, Bangkok 10900, Thailand; E-Mail: fagisrsu@ku.ac.th

[3] Department of Nutrition and Dietetics, Kasturba Gandhi College for Women, Secunderabad 500026, India; E-Mail: umachitra7@gmail.com

[4] Department of Food Science and Nutrition, College of Human Ecology, Pusan National University, 30 Jangjeon-Dong, Geumjeoung-Ku, Busan 609 735, Korea; E-Mail: jeehyunlee@pusan.ac.kr

[†] These authors contributed equally to this work.

[*] Author to whom correspondence should be addressed; E-Mail: kadri@ksu.edu

Abstract: The objective of this study was to begin characterizing purchase, storage, handling, and preparation of poultry products and eggs by selected consumers in three Asian countries: India, Korea, and Thailand. Approximately 100 consumers in each location were recruited to participate in this study. The consumers were surveyed about eggs and poultry purchase behavior characteristics, such as temperatures and locations, storage behavior, such as storage locations in the refrigerator or freezer, preparation behavior, such as washing eggs and poultry before cooking, and handling behavior, such as using cutting boards during cooking. The results indicated differences in purchase and storage practices of raw eggs. Most Korean consumers purchased refrigerated eggs and stored the eggs in the refrigerator, while Indian and Thai consumers bought eggs that were stored at room temperature, but would refrigerate the eggs at home. Approximately half of the consumers in each country froze raw meat, poultry, or seafood. Food preparation practices showed potential for cross-contamination during cooking, such as using the same cutting board for different kinds of foods or not washing hands with soap and water. The results presented in this pilot study may lead to development of educational messages and raising consumer awareness of food safety practices in Asian countries.

Keywords: food safety; consumer; eggs; poultry

1. Introduction

One key aspect of food security is food safety; sufficient and available food quantities alone do not guarantee a healthy and nourished population. Foodborne diseases are widespread in the world and consumer education on food safety topics is of high importance [1]. World Health Organization (WHO) has introduced the Five Keys Program [2] that stresses importance of cleanliness, avoiding cross-contamination, cooking and storing at proper temperatures, and using safe water to avoid foodborne diseases. Some of the key food ingredients that are related to salmonellosis outbreaks, if not handled correctly, are eggs and poultry. The main determinants of contamination while handling raw meat and poultry are related to washing practice of contaminated hands, cutting boards, and knives during and after cooking [3]. Egg processing in Asia is the largest in the world, however, in India, egg products consumption is low and production has been reduced. In Korea and Thailand, egg products are being manufactured and quantities are predicted to increase [4]. Furthermore, poultry production and consumption in Asia has been increasing in recent years [5,6]. For example, in India alone, chicken consumption has increased from 1.8 to 2.2 kg/person/year. This seemingly low increase in average consumption becomes critical considering that about 1.2 trillion people live in India.

Consumer behavior regarding eggs and poultry usage has been studied in several countries. For example consumers in Belgium [7], New Zealand [8], the U.S. [9,10], China [11], Brazil [12], Turkey [13], and Greece [14] have been studied regarding their food safety practices while handling eggs and poultry. Food safety concerns in India have been studied with a focus on street food safety and quality [15] and consumer behavior [16,17]. Similarly, in Korea, food safety knowledge, attitudes, or practices were studied among at risk population, such as middle school students [18], the elderly [19,20], and people who prepare food for others, such as kitchen employees at school [21], senior welfare center employees [22], house wives [23], and culinary and hospitality major college students [24,25]. Few studies have done so in Thailand [26]. No studies were found that would compare food safety practices across several Asian countries.

Some studies have looked at food safety issues around street foods that are common in Asian countries. For example, Sudershan *et al.* [15] studied street foods prepared with poultry in the Hyderabad region of India. They found that several of the foods were contaminated with *Staphylococcus aureus* or/and *Bacillus cereus*. Furthermore, local people did not consider street foods safe and thought that foods prepared at home were more hygienic [17]. Those authors reported that safety of home-prepared meals is ensured by cleanliness of the cook, surroundings, and the food. However in Italy, researchers surveyed food safety practices among consumers to develop food safety education program, based on the belief that the home was a source of foodborne disease [27].

According to Henley *et al.* [9], there may be some culturally unique behaviors related to food handling and preparation among Asian cultures. Examples of those include washing raw poultry with hot water or with acidic solutions and purchasing live poultry. A study comparing food safety issues of consumers from Seoul and Shanghai concluded that culture has an influence on perception and behavior [28]. The benefits of a study that compares consumers from several Asian countries include increased knowledge regarding possible culturally specific behaviors, cross-contamination, spoilage, and resulting food-borne illnesses and a uniform approach to consumers that allows data comparison. The objective of this pilot scale study was to begin to characterize consumers' purchase, storage, handling, and preparation of poultry products and eggs in three different Asian countries: India, Korea, and Thailand.

2. Experimental Section

2.1. Questionnaire and Data Collection

The questionnaire was based on questionnaires used in previous research [29], was developed in English, and pretested on food safety experts to ensure collection of relevant information and to ensure uniform understanding of the questions. The questionnaire also was pretested in the U.S. with a group of consumers before translation and with local staff, "in country", after translation. The questionnaire was translated into Korean, and Thai, and then back-translated into English by native speakers to ensure the questionnaire was comparable across countries. In India, the questionnaire was given in English, as all consumers were fluent in the English language. Data was collected in the beginning of 2013 in India and Thailand, and in the middle of 2013 in Korea. The information was gathered via an online survey (Korea) or at a central location where the consumers filled in paper questionnaires (India and Thailand).

The questionnaire collected information on a wide range of home storage and handling practices. Only information regarding 11 multiple-choice questions regarding consumers' eggs and poultry related purchase, storage, preparation, and refrigeration practices are presented.

Demographic information regarding consumer gender, age (<35; ≥35), household income (equivalent as determined by the country to "Low": $<\$25,000$; "Medium": $\$25,000-\$50,000$; "High": $>\$50,000$), and education level (less than college; some college courses or higher) was collected in the end of the questionnaire. Demographic information regarding the consumers who participated in the study is shown in Table 1.

2.2. Consumers

Consumers were tested in Bangkok, Thailand; Hyderabad, India; and Busan and Suncheon, Korea; countries where established food and nutrition researchers were available and willing to conduct the test. In addition, consumers in these countries present a broad diversity of Asian consumers. Consumers were selected from those who were already in databases maintained by researchers, from advertisements in newspapers and newsletters, and word-of-mouth referrals and, thus, are a convenience sample rather than a nationally representative sample. The number and location of the consumers cannot represent the entire population of a country, but serves as a "first look" of practices in those countries.

All participating consumers were pre-screened according to the following conditions: age >18; the consumer was the primary food shopper in the household or shared food shopping responsibility with someone else; the consumer was the food preparer and knew about food storage in the household or was one of several people who cooked and knew about food storage in household; the consumer had a refrigerator in their home. A total of 115 consumers participated in India, 100 in Thailand, and 101 in Sunchoen, Korea. Only consumers who were food shoppers and food preparers were included in this survey, and this limits the consumers available for sampling. In India, all consumers, and in Korea and Thailand most surveyed consumers were female (Table 1) because food preparation in those countries typically is a task to be conducted by females in the household. In addition, because refrigeration is a key aspect of suggested egg and poultry storage, only those consumers with access to a refrigerator were selected. In India, in particular, this decreased potential participation, but consumers without a refrigerator, by default, could not store poultry or eggs as recommended.

Table 1. Demographic characteristics of the surveyed consumers.

Demographic segmentation, %	India ($n = 115$)	Korea ($n = 101$)	Thailand ($n = 100$)
Gender			
Male		15.0	27.0
Female	100.0	86.0	73.0
Age			
<35	28.7	35.0	13.0
>35	71.3	66.0	87.0
Education			
Less than college	21.7	39.0	23.0
Some college courses or more	78.3	62.0	77.0
Income			
Low	46.0	24.0	3.0
Medium	27.0	33.0	36.0
High	27.0	44.0	61.0

2.3. Data Analysis

Chi-square tests were performed for the relationships between various sociodemographic variables. The analysis was conducted using Excel function CHITEST (Microsoft Excel 2010, Microsoft, Redmond, WA, USA).

3. Results and Discussion

3.1. Purchase and Storage Practices

Most surveyed consumers in India, Thailand, and Korea purchased eggs from the store or supermarket (Table 2). Less than 10% of respondents in each country purchased eggs directly from farmers or raised their own chicken. In addition 32% of Thai consumers purchased eggs from the market. In India, respondents who were older than 35 years were significantly more likely to buy eggs from the market than younger consumers ($p = 0.03$). This might imply a heightened risk of pathogens on eggshells as a washing step often is not conducted for eggs sold at markets. Total viable counts of microorganisms, from eggshell, have been found to be significantly higher on unwashed eggshell compared to washed eggshell [30].

Table 2. Consumers' egg purchase location. Where do you usually buy eggs?

Demographic segmentation, %	India (n = 115)			Korea (n = 101)			Thailand (n = 100)		
	Farmer/raise chicken	Store	Market	Farmer/raise chicken	Store	Market	Farmer/raise chicken	Store	Market
Total	2.6	75.7	21.7	6.9	74.2	8.9	5.0	63.0	32.0
Gender									
Male					13.9	1.0	1.0	18.0	8.0
Female	2.6	75.7	21.7	6.9	70.3	7.9	4.0	45.0	24.0
Age									
<35	**0.9**	**21.7**	**6.1**	3.0	26.7	5.0		9.0	4.0
>35	**1.7**	**53.9**	**15.7**	4.0	57.4	4.0	5.0	54.0	28.0
Education									
Less than college		14.8	7.0	**4.0**	**27.7**	**6.9**	3.0	15.0	2.0
Some college courses or more	2.6	60.9	14.8	**3.0**	**56.4**	**2.0**	2.0	48.0	27.0
Income									
Low	0.9	33.9	11.3	**2.0**	**14.9**	**6.9**		2.0	1.0
Medium	1.7	19.1	6.1	**1.0**	**31.7**	**0.0**	5.0	22.0	9.0
High		22.6	4.3	**4.0**	**37.6**	**2.0**		39.0	22.0

Note: values shown in bold are statistically significantly different among those sociodemographic portions according to chi-square test for ($p < 0.05$).

At the time of purchase the eggs were usually refrigerated in Korea, most likely because the Korean Ministry of Food and Drug Safety regulates the distribution of eggs and requires them to be at 0–15 °C [31]. However, in Thailand and India eggs were at room temperature when purchased (Table 3). In addition 16% of consumers in India, 10% of consumers in Korea, and 16% of consumers in Thailand stored raw eggs at room temperature (Table 4). These practices could be related to inclusion of lacto-ovo vegetarians in the study, whose households have only small refrigerators for the most perishable products, but also could subject some Indian, Thai, and Korean consumers to a higher risk for foodborne diseases. Salmonellosis is a leading cause of foodborne diseases throughout the world [32]. A previous study has shown that bacteria causing salmonellosis can multiply within eggs and reach high concentrations when stored at room temperature [33]; this could be exacerbated in warmer climates. In addition, it was reported that the Haugh unit, which indicates freshness of egg albumen, was negatively influenced by warmer temperature [30]. According to modeling studies [34] the risk of foodborne illness is reduced if the shelf life is less than seven days.

Table 3. Percentage of eggs refrigerated at time of purchase. When you usually buy eggs are they refrigerated or at room temperature?

Demographic segmentation, %	India (n = 115)		Korea (n = 101)		Thailand (n = 100)	
	Refrigerated	Room temperature	Refrigerated	Room temperature	Refrigerated	Room temperature
Total		100.0	78.2	22.8	36.0	64.0
Gender						
Male			12.9	3.0	8.0	19.0
Female		100.0	65.3	19.8	28.0	45.0
Age						
<35		28.7	28.7	5.9	**9.0**	**4.0**
>35		71.3	48.5	16.8	**27.0**	**60.0**
Education						
Less than college		21.7	31.7	6.9	6.0	17.0
Some college courses or more		78.3	45.5	15.8	30.0	47.0
Income						
Low		46.1	17.8	5.9	1.0	2.0
Medium		27.0	26.7	5.9	14.0	22.0
High		26.9	32.7	10.9	21.0	40.0

Note: values shown in bold are statistically significantly different among those sociodemographic portions according to chi-square test for ($p < 0.05$).

Table 4. Consumers' storage behavior of raw eggs in the shell. Where would you store the food items?

Demographic segmentation, %	India (n = 115)			Korea (n = 94) *		Thailand (n = 100) *	
	Refrigerator	Room temperature	Do not know	Refrigerator	Room temperature	Refrigerator	Room temperature
Total	81.7	16.5	1.7	89.4	10.6	74.0	16.0
Gender							
Male				16.0		22.0	5.0
Female	81.7	16.5	1.7	73.4	10.6	62.0	11.0
Age							
<35	25.2	3.5		**36.2**	**1.1**	12.0	1.0
>35	56.5	13.0	1.7	**53.2**	**9.6**	72.0	15.0
Education							
Less than college	16.5	4.3	0.9	**27.7**	**7.4**	18.0	5.0
Some college courses or more	65.2	12.2	0.9	**61.7**	**3.2**	66.0	11.0
Income							
Low	42.6	7.8	1.7	17.0	3.2	3.0	
Medium	21.7	5.2		29.8	4.3	30.0	6.0
High	23.5	3.5		42.6	3.2	51.0	10.0

Notes: values shown in bold are statistically significantly different among those sociodemographic portions according to chi-square test for ($p < 0.05$). * No consumers in Korea and Thailand answered "Do not know" option.

When it comes to storing cooked eggs in the shell, behavior became more diverse. For example 37% of surveyed Indian consumers stored cooked eggs in the shell at room temperature and 25% of the Indian consumers didn't know how to store cooked eggs in the shell (Table 5). Furthermore, 34% of surveyed Korean and 42% of Thai consumers stored cooked eggs in the shell at room temperature. Korean consumers older than 35 were significantly more likely to store cooked eggs in the refrigerator than younger consumers, more of whom did so at room temperature ($p = 0.004$).

Table 5. Consumers' storage behavior of cooked eggs in the shell. Where would you store the food items?

Demographic segmentation, %	India (n = 115)			Korea (n = 91)			Thailand (n = 99)		
	Refrigerator	Room temperature	Do not know	Refrigerator	Room temperature	Do not know	Refrigerator	Room temperature	Do not know
Total	37.4	37.4	25.2	64.8	34.1	1.1	55.6	42.4	2.0
Gender									
Male				8.8	7.7		14.1	12.1	
Female	37.4	37.4	25.2	56.0	26.4	1.1	41.5	30.3	2.0
Age									
<35	10.4	11.3	7.0	**17.6**	**20.9**		8.1	4.0	
>35	27.0	26.1	18.3	**46.2**	**13.2**	**1.1**	47.5	38.4	2.0
Education									
Less than college	**4.3**	**13.0**	**4.3**	22.0	11.0	1.1	11.1	10.1	1.0
Some college courses or more	**33.0**	**24.3**	**20.9**	42.9	23.1		44.5	32.3	1.0
Income									
Low	18.3	20.0	7.8	12.1	5.5	1.1	1.0	2.0	
Medium	9.6	9.6	7.8	24.2	9.9		21.2	13.1	2.0
High	9.6	7.8	9.6	28.6	18.7		33.4	27.3	

Note: values shown in bold are statistically significantly different among those sociodemographic portions according to chi-square test for ($p < 0.05$).

A total of 52%–73% of surveyed consumers in all locations stored raw meat in the freezer (Table 6). Other consumers stored raw meat on the top or middle shelf of the refrigerator (16% of Indian, 17% of Korean, and 21% of surveyed Thai consumers), or elsewhere in the refrigerator. Similar findings were reported by Godwin and Coppings [29] for US consumers' raw meat storage practices. The top and middle shelf and the door of the refrigerator are more likely to have elevated temperatures and storing temperature-sensitive foods is not recommended in these locations [25]. In addition these locations are more likely to contaminate foods on lower shelves of the refrigerator [35]. The potential for poultry juice dripping onto items on lower shelves is a problem that seemed to exist in all countries studied. Thus far, none of the studies have looked at the type of refrigerator shelving (meshed or glass, for example) to determine cross-contamination likelihood. Raw foods often are a source of bacterial pathogens. Vindigni *et al.* [26] conducted a study in Bangkok, Thailand, to determine bacterial pathogens on raw food samples. They found *Enterococcus* spp. on 94%, and *Salmonella* spp. on 61%, of the samples.

Table 6. Consumers' raw meat, seafood, and poultry refrigeration and freezing practices. Where would you store the raw meat in the refrigerator? Available options included: Top shelf (1), middle shelf (2), bottom shelf (3), drawer (4), door (5), wherever there is room (6), on the counter or in the cabinet (7), in the freezer (8).

Demographic segmentation, %	India (n = 115)								Korea (n = 101)							Thailand (n = 100)					
	1	2	3	4	5	6	7	8	1	2	3	4	6	7	8	1	2	3	4	6	8
Total	7.0	9.6	1.7	2.6	0.9	2.6	3.5	72.2	9.9	7.9	7.0	8.9	11.9	2.0	52.4	17.0	4.0	2.0	3.0	1.0	73.0
Gender																					
Male									1.0		2.0	1.0	4.0		6.9	5.0	3.0			1.0	18.0
Female	7.0	9.6	1.7	2.6	0.9	2.6	3.5	72.2	8.9	7.9	5.0	7.9	7.9	2.0	45.5	12.0	1.0	2.0	3.0		55.0
Age																					
<35	**1.7**	**3.5**	**0.9**	**0.9**	**0.9**	**0.9**	**0.9**	**19.1**	2.0	5.9	4.0	2.0	5.0		15.8	2.0	1.0	2.0	1.0	1.0	8.0
>35	**5.2**	**6.1**	**0.9**	**1.7**		**1.7**	**2.6**	**53.0**	7.9	2.0	3.0	6.9	6.9	2.0	36.6	15.0	3.0	2.0	2.0		65.0
Education																					
Less than college	**2.6**	**3.5**	**1.0**	**0.9**		**0.9**	**2.6**	**12.2**	3.0	2.0	1.0	1.0	4.0	2.0	25.7	**2.0**	**2.0**	**1.0**		**1.0**	**19.0**
Some college courses or more	**4.3**	**6.1**	**1.7**	**1.7**	**0.9**	**1.7**	**0.9**	**60.0**	6.9	5.9	5.9	7.9	7.9		26.7	**17.0**	**2.0**	**1.0**	**3.0**		**54.0**
Income																					
Low	6.1	5.2	0.9	1.7	0.9	0.9	3.5	27.0	**3.0**	**1.0**		**1.0**	**2.0**	**2.0**	**14.9**						3.0
Medium	0.9	3.5	0.9					20.9	**4.0**	**1.0**	**3.0**	**2.0**	**7.9**		**14.9**			2.0		1.0	28.0
High		0.9		0.9		0.9	0.9	24.3	**3.0**	**5.9**	**4.0**	**5.9**	**2.0**		**22.8**	12.0	4.0		3.0		48.0

Notes: options 7 and 8 in Thailand and 5 in Korea were not checked by any consumers; values shown in bold are statistically significantly different among those sociodemographic portions according to chi-square test for (p < 0.05).

Although surveyed consumers stored raw meat on the refrigerator shelf, only 22%–30% placed something underneath the meat, probably to avoid spilling or dripping of meat juices (Table 7). This behavior trend was similar regardless of gender, age, education level, or income of the consumers in all the countries studied. These juice-catching utensils included plates ($n = 22$ in Korea, $n = 32$ in India, $n = 6$ in Thailand) and paper toweling or meshed bowls, with an additional plate or plastic bags (data not shown).

Table 7. Consumers' raw meat, seafood, and poultry storage behavior. Would you place something under the raw meat in the refrigerator (Yes/No)?

Demographic segmentation, %	India ($n = 115$)		Korea ($n = 101$)		Thailand ($n = 100$)	
	Yes	No	Yes	No	Yes	No
Total	30.4	69.6	31.7	68.3	28.0	72.0
Gender						
Male			3.0	11.9	5.0	22.0
Female	30.4	69.6	28.7	56.4	23.0	50.0
Age						
<35	**7.0**	**21.7**	**7.9**	**26.7**	**4.0**	**9.0**
>35	**23.5**	**47.8**	**23.8**	**41.6**	**24.0**	**63.0**
Education						
Less than college	**6.1**	**15.7**	**18.8**	**19.8**	**10.0**	**13.0**
Some college courses or more	**24.3**	**53.9**	**12.9**	**48.5**	**18.0**	**59.0**
Income						
Low	**13.0**	**33.0**	7.9	15.8	**1.0**	**2.0**
Medium	**9.6**	**17.4**	10.9	21.8	**13.0**	**23.0**
High	**7.8**	**19.1**	12.9	30.7	**14.0**	**47.0**

Note: values shown in bold are statistically significantly different among those sociodemographic portions according to chi-square test for ($p < 0.05$).

3.2. Preparation Practices

The main contaminants while preparing food at home are hands, knives, and cutting boards [3]. This was supported by findings in our survey (Table 8). A total of 31% of participating Korean, 24% of Indian, and 30% of Thai consumers used the same cutting board for different foods, such as meats and vegetables and either did nothing or only wiped the cutting board between different foods. Indian consumers with higher than college education were significantly more likely ($p = 0.01$) to wash the cutting board in between cutting different foods or use a different cutting surface for different types of foods. Only a few consumers did not use a cutting surface when cutting foods (8% in India and 1% in Korea). Observational studies that would look at the washing process of the cutting board and the rate of success in removing raw meat particles from the cutting board or the order in which consumers cut up food ingredients using the same cutting board during cooking might be of interest.

Table 8. Consumers' practices related to the use of cutting boards. When you are cutting various types of food such as meat, vegetables, eggs, bread, do you usually use: (Select one). Available options included: The same cutting surface (counter, plate, cutting board) and wipe or wash it at the end (1); The same cutting surface (counter, plate, cutting board) and wipe it between uses (2); The same cutting surface (counter, plate, cutting board) and wash it between uses (3); A different cutting surface for each type of food (4); Do not use a cutting surface (5).

Demographic segmentation, %	India (n = 115)					Korea (n = 97)					Thailand (n = 100) *			
	1	2	3	4	5	1	2	3	4	5	1	2	3	4
Total	12.2	12.2	44.3	22.6	8.7	19.6	12.3	36.1	31.0	1.0	12.0	18.0	43.0	27.0
Gender														
Male						4.1	1.0	5.2	5.2		7.0	4.0	11.0	5.0
Female	12.2	12.2	44.3	22.6	8.7	15.5	11.3	30.9	25.8	1.0	5.0	14.0	32.0	22.0
Age														
<35	1.7	3.5	16.5	4.3	2.6	6.2	3.1	12.4	14.4		2.0	4.0	6.0	1.0
>35	10.4	8.7	18.3	18.3	6.1	13.4	9.3	23.7	16.5	1.0	10.0	14.0	37.0	26.0
Education														
Less than college	**4.3**	**1.7**	**5.2**	**5.2**	**5.2**	9.3	6.2	12.4	9.3		3.0	7.0	8.0	5.0
Some college courses or more	**7.8**	**10.4**	**39.1**	**17.4**	**3.5**	10.3	6.2	23.7	21.6	1.0	9.0	11.0	35.0	22.0
Income														
Low	**7.8**	**6.1**	**13.9**	**11.3**	**7.0**	4.1	6.2	9.3	2.1			1.0		2.0
Medium	**2.6**	**5.2**	**12.2**	**5.2**	**1.7**	7.2	2.1	11.3	13.4		4.0	4.0	20.0	8.0
High	**1.7**	**0.9**	**18.3**	**6.1**		8.2	4.1	15.5	15.5	1.0	8.0	13.0	23.0	17.0

Notes: values shown in bold are statistically significantly different among those sociodemographic portions according to chi-square test for (p < 0.05). * None of the consumers in Thailand selected the option "Do not use a cutting surface".

Washing hands with soap and water after handling raw meat or eggs seemed to be the favored approach for more than half of surveyed consumers in India and Thailand, and a third of consumers in Korea (Table 9). However Sudershan et al. [36] found that most Indian women would wash their hands with water, but not with soap. Premakumari et al. [37] found 81% of consumers in India would wash their hands before eating. However, access to clean water may be restricted. A total of 22% of consumers in India, 38% in Korea, and 32% in Thailand would wash their hands with water, but not use soap (Table 9). According to DeDonder et al. [38] this would be insufficient to avoid cross-contamination. In Thailand, female respondents were more likely to wash their hands using soap than were male respondents. Furthermore, what consumers reported in our study may not reflect their actual practices. In a survey of hand-washing, and a follow-up observational study in Korea, about 60% of female college students claimed to use soap while washing hands, but only 0.9% used soap during the observational phase [39].

Table 9. Consumer practice of handling raw meat, poultry, seafood, or eggs. What was the first thing you did immediately after you handled these raw foods? Available options included: I cut up some other foods (1); I got other foods ready for cooking, but did not cut them up (2); I picked up a pot or pan to cook food (3); I wiped my hands off with a paper towel, dish cloth, or on my apron or clothing (4); Continued cooking without wiping, rinsing, or washing hands (5); Rinsed off my hands, but did not use soap (6); Washed hands with soap and water (7); Do not prepare raw meat, poultry, seafood, or eggs (8).

Demographic segmentation, %	India (n = 115)					Korea (n = 101)							Thailand (n = 100)						
	1	2	4	6	7	1	2	3	4	5	6	7	1	2	3	4	6	7	8
Total	0.9	2.6	7.8	22.6	66.1	4.0	3.0	8.9	5.0	3.0	38.7	37.6	1.0	3.0	2.0	7.0	32.0	52.0	3.0
Gender																			
Male						**1.0**		2.0			4.0	7.9		**3.0**		**5.0**	10.0	9.0	
Female	0.9	2.6	7.8	22.6	66.1	3.0	3.0	6.9	5.0	3.0	34.7	29.7	**1.0**		**2.0**	**2.0**	**22.0**	**43.0**	**3.0**
Age																			
<35	0.9		3.5	4.3	20.0			4.0	1.0		14.9	14.9		1.0			5.0	7.0	
>35		2.6	4.3	18.3	46.1	1.0		5.0	4.0	3.0	23.8	22.8	1.0	2.0	2.0	7.0	27.0	45.0	3.0
Education																			
Less than college		1.7	2.6	3.5	13.9	2.0	2.0	4.0	3.0	3.0	13.9	10.9		1.0			7.0	13.0	1.0
Some college courses or more	0.9	0.9	5.2	19.1	52.2	2.0	1.0	5.0	2.0		24.8	26.7	1.0	2.0	2.0	6.0	25.0	39.0	2.0
Income																			
Low	0.9	2.6	5.2	10.4	27.0	2.0	1.0	2.0	2.0	2.0	7.9	6.9						3.0	
Medium			0.9	7.8	18.3	1.0	1.0	2.0	2.0	1.0	9.9	15.8		1.0		3.0	9.0	21.0	1.0
High			1.7	4.3	20.9	1.0		5.0	1.0		20.8	14.9		2.0		4.0	23.0	28.0	2.0

Notes: values shown in bold are statistically significantly different among those sociodemographic portions according to chi-square test for ($p < 0.05$); none of the consumers selected options 3, 5, or 8 in India, 8 in Korea, and 5 in Thailand.

A total of 75% of surveyed consumers in India, 46% in Korea, and 48% in Thailand reported washing raw poultry before cooking, while 16% of Indian and 11% of Korean consumers reported washing raw eggs (Table 10). In addition, 48% of Thai consumers reported always washing both poultry and eggs. According to the Food Safety and Inspection Service (FSIS) [40] washing raw poultry or meats and eggs is not recommended as it is likely to cross-contaminate kitchen surfaces and utensils, which may lead to food-borne illnesses. It is reasonable, however, for the consumer to expect keeping food safe by washing poultry and eggs, especially if food has been acquired from a market or directly from farmer and visibly is not clean. In addition, readily available information sources, such as cookbooks or cooking shows often recommend washing chicken and other meat.

Table 10. Raw poultry and eggs preparation practice. Do you wash raw poultry/eggs before cooking them? Available options included: Yes, I always wash raw poultry (1); Yes, I always wash eggs (2); Yes, I always wash poultry and eggs (3); Sometimes—if they seem dirty (4); No (5).

Demographic segmentation, %	India (n = 133) *				Korea (n = 101) **				Thailand (n = 100)				
	1	2	4	5	1	2	4	5	1	2	3	4	5
Total	75.2	16.5	5.3	3.0	46.6	11.9	24.8	16.9	48.0	3.0	45.0	3.0	1.0
Gender													
Male					5.0	3.0	4.0	3.0	15.0		9.0	2.0	1.0
Female	75.2	16.5	5.3	3.0	41.6	8.9	20.8	13.9	33.0	3.0	36.0	1.0	
Age													
<35	22.6	1.5	1.5	0.8	15.8	1.0	12.9	5.0	7.0		5.0		1.0
>35	52.6	15.0	3.8	2.3	30.7	10.9	11.9	11.9	41.0	3.0	40.0	3.0	
Education													
Less than college	**22.5**	**3.0**		**1.5**	14.9	5.0	9.9	8.9	10.0	1.0	12.0		
Some college courses or more	**57.9**	**13.5**	**5.3**	**1.5**	31.7	6.9	14.9	7.9	38.0	2.0	33.0	3.0	1.0
Income													
Low	33.8	3.0	2.3	2.3	8.9	3.0	6.9	5.0	3.0				
Medium	20.3	6.8	0.8	0.8	14.9	4.0	9.9	4.0	11.0	1.0	24.0		
High	21.1	6.8	2.3		22.8	5.0	7.9	7.9	34.0	2.0	21.0	3.0	1.0

Notes: values shown in bold are statistically significantly different among those sociodemographic portions according to chi-square test for (p < 0.05). * Option 3 in India was "I do not cook raw poultry or eggs" and none of the consumers in India selected option 3. ** None of the consumers in Korea selected the option "Yes, I always wash raw poultry and eggs".

Temperature plays an important part in storing home-made mayonnaise and home-made salads with mayonnaise [41]. Those authors found that *Salmonella enteritidis* was able to grow rapidly in salads and mayonnaise that were stored at 25 °C, but not those stored at 10 °C. *Listeria monocytogenes*, however is able to grow in contaminated pasta and egg salads, even in cold storage [42]. FSIS [43] recommends cold perishable foods be refrigerated within 2 h from preparation. According to this study the majority of surveyed consumers would practice this behavior in India, Korea, and Thailand (Table 11). However, a large portion (38%) of Indian consumers reported leaving leftovers from a freshly prepared salad that contains mayonnaise or eggs unrefrigerated for more than 4 h. This behavior likely is associated with local food culture and habits. This is an area where local food authorities need to further educate consumers.

Table 11. Leftover handling practice. The last time you had leftovers from a freshly prepared salad that contained eggs or mayonnaise, how long did you let the leftovers sit at room temperature before you put them in the refrigerator or ate them later without refrigeration? Available options included: 1 hour or less (1); More than 1 hour, but less than 2 hours (2); More than 2 hours, but less than 3 hours (3); More than 3 hours, but less than 4 hours (4); 4 hours or more (5).

Demographic segmentation, %	India (n = 115)					Korea (n = 100) *				Thailand (n = 100)				
	1	2	3	4	5	1	2	3	5	1	2	3	4	5
Total	34.8	13.9	7.0	6.1	38.3	71.0	16.0	8.0	6.0	74.0	12.0	7.0	4.0	3.0
Gender														
Male						11.0	4.0			**16.0**	**8.0**	**2.0**	**1.0**	
Female	34.8	13.9	7.0	6.1	38.3	60.0	12.0	8.0	6.0	**58.0**	**4.0**	**5.0**	**3.0**	**3.0**
Age														
<35	11.3	4.3	0.9	4.3	7.8	24.0	4.0	2.0	5.0	**5.0**	**6.0**	**2.0**		
>35	23.5	9.6	6.1	1.7	30.4	47.0	12.0	6.0	1.0	**69.0**	**6.0**	**5.0**	**4.0**	**3.0**
Education														
Less than college	3.5	2.6	2.6	2.6	10.4	**22.0**	**8.0**	**7.0**	**2.0**	17.0	4.0	2.0		
Some college courses or more	31.3	11.3	4.3	3.5	27.8	**49.0**	**8.0**	**1.0**	**4.0**	57.0	8.0	5.0	4.0	3.0
Income														
Low	18.3	7.0	4.3	5.2	11.3	**11.0**	**6.0**	**5.0**	**2.0**	3.0				
Medium	8.7	4.3	0.9		13.0	**27.0**	**2.0**	**1.0**	**3.0**	26.0	6.0	2.0	1.0	1.0
High	7.8	2.6	1.7	0.9	13.9	**33.0**	**8.0**	**2.0**	**1.0**	45.0	6.0	5.0	3.0	2.0

Notes: values shown in bold are statistically significantly different among those sociodemographic portions according to chi-square test for ($p < 0.05$). * No consumers in Korea selected option 4.

Sudershan *et al.* [44] pointed out the limited number of studies that address consumer food safety related behavior characteristics in India. Furthermore, studies have found low levels of food safety knowledge among consumers. Sudershan *et al.* [36] looked at women with children younger than five years old in India. These authors found that food safety related practices are mostly taught from mother to daughter, and are considered important. However, there were behaviors identified that may cause foodborne illnesses. For example, in India, all eggs are purchased at room temperature because small and large grocery stores keep eggs at room temperature due to a lack of sufficient cold storage facilities. It has been reported, in India, that over 99% of food and grocery is sold by traditional retailers (Kirana stores, street hawkers, and wet market stall operators) and only 5% of all poultry output is marketed in processed form [45]. In the Indian and Thai context, refrigeration is all the more necessitated by tropical climatic conditions, which are conducive to faster microbial growth. A study on food safety practices in India revealed that over 80% of the consumers stored cooked foods at room temperature because only 19% owned refrigerators. Usage of refrigerators for storing leftover cooked non-vegetarian food was important among literate respondents and those with high standards of living [46].

According to Henley *et al.* [9] consumers with different cultural backgrounds may have specific approaches to food handling. These authors found that even though there are some overall hazardous behaviors, such as not using a thermometer and thawing foods at room temperature, there were culture-specific characteristics as well. An example of this behavior would include cutting poultry into smaller, bite-sized pieces in some cultures that could lead to cross-contamination issues. Our results confirmed the potentially contaminating behavior, such as washing raw eggs and poultry, and using the same cutting surface for multiple foods without washing in between.

The limitations of this study include a relatively small geographic representation of surveyed consumers in each country. The approximately 100 consumers in each country came from only one or two cities, which is common in many surveys. Some previous studies, in these and other regions have included a smaller number of respondents or participants [38,46]. Nevertheless, these results should not be considered as representative of the whole countries of India, Korea, or Thailand. One potential perceived limitation is the low number of men surveyed. However, this was caused by the cultural background of Asian countries, where male and female roles are somewhat rigidly defined, and men may not purchase, prepare, or store foods. Low, or nonexistent, male participant rates have been observed in other studies conducted with minorities [9] or in Asian countries [17,46].

Although limitations are present, the data serves as a first look at what likely are problem areas for food safety handling of poultry and eggs—lack of refrigerated storage for eggs, cross contamination issues during preparation, and longer than recommended times for leftovers held at room temperature in some populations. The study also suggests that food safety education may be needed, especially with younger populations that have less exposure to preparation and storage and are still developing life-long food safety habits. In addition, the collection of data in a uniform manner across three Asian countries enables immediate comparison of results across countries.

4. Conclusions

This study compared consumers' purchase, storage, handling, and preparation of poultry products and eggs in three Asian countries: India, Korea, and Thailand. The results indicated similar patterns in purchase and storage behaviors among consumers in all three countries. For example most consumers would store raw eggs in the refrigerator, but some stored cooked eggs at room temperature. In addition most consumers froze raw meat, poultry, or seafood. However, some consumers who stored raw meat in the refrigerator would do this on the top or middle shelf and this could potentially lead to meat juice contaminating other foods. Most consumers claimed to wash hands with soap and water; however there were portions of consumers who did not. Consumers reported practices that support cross-contamination during cooking, such as washing raw poultry and eggs, inadequately washing cutting surfaces between foods, and not refrigerating dishes that could become a food hazard. Health education in food safety must be designed to suit prevailing socioeconomic conditions and specific cultural groups in different countries. The findings from this study indicate a need for raising consumer awareness about food safety issues. The similarity of these findings across three Asian countries suggests food safety educational efforts may benefit from collaborative efforts among universities, health educators, and food safety authorities.

Acknowledgments

The authors wish to thank Sandria Godwin, Tennessee State University, for her help in providing an initial questionnaire on which this survey was based.

Conflicts of Interest

The authors declare no conflict of interest.

References

1. Forsythe, R.H. Food safety: A global perspective. *Poult. Sci.* **1996**, *75*, 1448–1454.
2. World Health Organization. Five Keys to Safer Food Manual. Available online: http://www.who.int/foodsafety/publications/consumer/manual_keys.pdf (accessed on 30 October 2013).
3. Kennedy, J.; Nolan, A.; Gibney, S.; O'Brien, S.; McMahon, M.A.S.; McKenzie, K.; Healy, B.; McDowell, D.; Fanning, S.; Wall, P.G. Determinants of cross-contamination during home food preparation. *Br. Food. J.* **2011**, *113*, 280–297.
4. Evans, T. At the International Egg Commission's Annual Marketing and Production Conference in Guadalajara, Mexico, Morten Ernst of Sanovo International A/S, Highlighted Some of the Factors that Will Increase the Usage of Egg Products in Asia. The Future Prospects for the Industry Look Promising. Available online: www.wattagnet.com/PrintPage.aspx?id=4048 (accessed on 30 October 2013).
5. Global Poultry Trends 2013—Asia Produces One Third of World'S Broilers. Available online: http://www.thepoultrysite.com/articles/2928/global-poultry-trends-2013-asia-produces-onethird-of-worlds-broilers (accessed on 18 November 2013).
6. Global Poultry Trends 2013—Asian Chicken Meat Consumption Trends. Availabe online: http://www.thepoultrysite.com/articles/4/breeding-and-reproduction/1751/global-poultry-trends-rising-human-population-and-per-capita-consumption-in-asia-boost-total-chicken-demand (accessed on 18 November 2013).
7. Sampers, I.; Berkvens, D.; Jacxsens, L.; Ciocci, M.-C.; Dumoulin, A.; Uyttendaele, M. Survey of Belgian consumption patterns and consumer behavior of poultry meat to provide insight in risk factors for campylobacteriosis. *Food Control* **2012**, *26*, 293–299.
8. Al-Sakkaf, A. Evaluation of food handling practice among New Zealanders and other developed countries as a main risk factor for campylobacteriosis rate. *Food Control* **2012**, *27*, 330–337.
9. Henley, S.C.; Stein, S.E.; Quinlan, J.J. Identification of unique food handling practices that could represent food safety risks for minority consumers. *J. Food Prot.* **2012**, *75*, 2050–2054.
10. Booth, R.; Hernandez, M.; Baker, E.L.; Grajales, T.; Pribis, P. Food safety attitudes in college students: A structural equation modeling analysis of a conceptual model. *Nutrients* **2013**, *5*, 328–339.
11. Liu, R.; Pieniak, A.; Verbeke, W. Consumers' attitudes and behavior towards safe food in China: A review. *Food Control* **2013**, *33*, 93–104.
12. Lazzarin Uggioni, P.; Salay, E. Consumer knowledge concerning safe handling practices to prevent microbiological contamination in commercial restaurants and socio-demographic characteristics, Campinas/SP/Brazil. *Food Control* **2012**, *26*, 331–336.
13. Ergönül, B. Consumer awareness and perception to food safety: A consumer analysis. *Food Control* **2013**, *32*, 461–471.

14. Lazou, T.; Georgiadis, M.; Pentieva, K.; McKevitt, A.; Iossifidou, E. Food safety knowledge and food-handling practices of Greek university students: A questionnaire-based survey. *Food Control* **2012**, *28*, 400–411.

15. Sudershan, R.V.; Kumar, R.N.; Kashinath, L.; Bhaskar, V.; Polasa, K. Microbiological hazard identification and exposure assessment of poultry products sold in various localities of Hyderabad, India. *ScientificWorldJournal* **2012**, *2012*, doi:10.1100/2012/736040.

16. Rathod, P.; Landge, S.; Nikam, T.R.; Hatey, A.A. Preferential poultry meat consumption and cooking patterns in Bijapur district of Karnataka. *J. Vet. Public Health* **2011**, *9*, 39–42.

17. Subba Rao, G.M.; Sudershan, R.V.; Rao, P.; Vishnu Vardhana Rao, M.; Polasa, K. Food safety knowledge, attitudes and practices of mothers: Findings from focus group studies in South India. *Appetite* **2007**, *49*, 441–449.

18. Yoon, E.; Seo, S.H. Differences on perceptions and attitudes towards food safety based on behavioral intention to prevent foodborne illness among middle school students in Seoul. *Korean J. Food Cook. Sci.* **2012**, *28*, 149–158.

19. Integrating Sensory Evaluation to Product Development. An Asian Perspective. Available online: http://www.utdallas.edu/~herve/abdi-ProceedingSpise2012.pdf (accessed on 6 October 2013).

20. Choi, J.H.; Lee, E.S.; Lee, Y.J.; Lee, H.S.; Chang, H.J.; Lee, K.E.; Yi, N.Y.; Ahn, Y.; Kwak, T.K. Development of food safety and nutrition education contents for the elderly: By focus group interview and Delphi technique. *Korean J. Community Nutr.* **2012**, *17*, 167–181.

21. Kim, J.G. Studies on the food hygiene & safety knowledge, attitudes, and practices of kitchen employees in school food-service programs—Part 1. *Korean J. Environ. Health* **2004**, *30*, 173–183.

22. Yi, N.Y.; Lee, K.Y.; Park, J.Y. Evaluation of food service workers' food safety knowledge and practices at senior welfare centers. *Korean J. Food Cook. Sci.* **2009**, *25*, 677–689.

23. Jung, H.K. Consumer Survey and Hazard Analysis for the Improvement of Food Hygiene and Safety in Purchase. Master Thesis, Korea University, Seoul, Korea, 2011.

24. Cha, M.; Park, J.R. Identification of college students' food safety awareness and perceived barriers to proper food handling practices. *J. Food Sci. Nutr.* **2005**, *10*, 74–80.

25. Cha, M.; Park, J.R. Knowledge and attitudes of food safety among hospitality and culinary students. *J. Food Sci. Nutr.* **2005**, *10*, 68–73.

26. Vindigni, S.M.; Srijan, A.; Wongstitwilairoong, B.; Marcus, R.; Meek, J.; Riley, P.L.; Mason, C. Prevalence of foodborne microorganisms in retail foods in Thailand. *Foodborne Pathog. Dis.* **2007**, *4*, 208–215.

27. Langiano, E.; Ferrara, M.; Lanni, L.; Viscardi, V.; Abbatecola, A.M.; de Vito, E. Food safety at home: Knowledge and practices of consumers. *Z. Gesundh. Wiss.* **2012**, *20*, 47–57.

28. Song, Y.J.; Yu, H.J. The covariance structural analysis of perceived risk on food safety consciousness and food safety pursuit between Seoul and Shanghai consumers: Focused on food consumption. *J. Consum. Stud.* **2008**, *19*, 215–244.

29. Godwin, S.; Coppings, R.J. Analysis of consumer food-handling practices from grocer to home including transport and storage of selected foods. *J. Food Distrib. Res.* **2005**, *36*, 55–62.

30. Song, B.R.; Kim, Y.J.; Yoon, H.; Lim, J.S.; Seo, K.H.; Heo, E.J.; Park, H.J.; Wee, S.H.; Moon, O.K.; Oh, S.M.; Moon, J.S. Evaluation on freshness and microbiological quality for eggs collected from grocery stores in Korea. *J. Prev. Vet. Med.* **2013**, *37*, 59–65.

31. The Ministry of Food and Drug Safety (MFDS). *Standards for Processing and Ingredients Specifications of Livestock Products*; Ministry of Food and Drug Safety: Chungcheongbuk-do, Korea, 2013.

32. Crump, J.A.; Luby, S.P.; Mintz, E.D. The global burden of typhoid fever. *Bull. World Health Org.* **2004**, *82*, 346–353.

33. Lublin, A.; Sela, S. The impact of temperature during the storage of table eggs on the viability of *Salmonella enterica* serovars Enteritidis and Virchow in the eggs. *Poult. Sci.* **2008**, *87*, 2208–2214.

34. World Health Organization; Food and Agriculture Organization of the United Nations. Microbiological Risk Assessment Series 2. Risk assessments of *Salmonella* in Eggs and Broiler Chickens—2. Available online: http://www.fao.org/docrep/005/y4392e/y4392e00.htm (accessed on 30 October 2013).

35. Godwin, S.L.; Chen, F.C.; Chambers, E., IV; Coppings, R.; Chambers, D. A comprehensive evaluation of temperatures within home refrigerators. *Food Prot. Trends* **2007**, *27*, 168–173.

36. Sudershan, R.V.; Subba Rao, G.M.; Polasa, K. Women and food safety—Some perspectives from India. *Reg. Health Forum* **2009**, *13*, 11–13.

37. Premakumari, S.; Kowsalya, S.; Kalai Selvi, E. Knowledge, attitude and practices of consumers towards food safety. *Indian J. Nutr. Diet.* **2012**, *49*, 409–416.

38. DeDonder, S.; Jacob, C.J.; Surgeoner, B.V.; Chapman, B.; Phebus, R.; Powell, D.A. Self-reported and observed behavior of primary meal preparers and adolescents during preparation of frozen, uncooked, breaded chicken products. *Br. Food J.* **2009**, *111*, 915–929.

39. Kim, J.G.; Kim, J.S. A study on the hand-washing awareness and practices of female university students. *J. Food Hyg. Saf.* **2009**, *24*, 128–135.

40. Washing Food: Does It Promote Food Safety? Available online: http://www.fsis.usda.gov/wps/portal/fsis/topics/food-safety-education/get-answers/food-safety-fact-sheets/safe-food-handling/washing-food-does-it-promote-food-safety/washing-food (accessed on 30 October 2013).

41. Kurihara, K.; Mizutani, H.; Nomura, H.; Takeda, N.; Imai, C. Behavior of *Salmonella enteritidis* in home-made mayonnaise and salads. *Jpn. J. Food Microbiol.* **1994**, *11*, 35–41.

42. Hwang, C.A.; Marmer, B.S. Growth of *Listeria monocytogenes* in egg salad and pasta salad formulated with mayonnaise of various pH and stored at refrigerated and abuse temperatures. *Food Microbiol.* **2007**, *24*, 211–218.

43. Leftovers and Food Safety. Available online: http://www.fsis.usda.gov/wps/portal/fsis/topics/food-safety-education/get-answers/food-safety-fact-sheets/safe-food-handling/leftovers-and-food-safety/ct_index (accessed on 30 October 2013).

44. Sudershan, R.V.; Rao, P.; Polasa, K. Food safety research in India: A review. *Asian J. Food Agro-Ind.* **2009**, *2*, 412–433.

45. Reardon, T.; Gulati, A. *The Rise of Supermarkets and Their Development Implications. International Experience Relevant to India*; International Food Policy Research Institute: New Delhi, India, 2008.

46. Sudershan, R.V.; Subba Rao, G.M.; Rao, P.; Vishnu Vardhana Rao, M.; Polasa, K. Food safety related perceptions and practices of mothers—A case study in Hyderabad, India. *Food Control* **2008**, *19*, 506–513.

5

Past, Present and Future of Sensors in Food Production

Catherine C. Adley

Microbiology Laboratory, Department of Chemical and Environmental Sciences, University of Limerick, Limerick, Ireland; E-Mail: Catherine.adley@ul.ie

Abstract: Microbial contamination management is a crucial task in the food industry. Undesirable microbial spoilage in a modern food processing plant poses a risk to consumers' health, causing severe economic losses to the manufacturers and retailers, contributing to wastage of food and a concern to the world's food supply. The main goal of the quality management is to reduce the time interval between the filling and the detection of a microorganism before release, from several days, to minutes or, at most, hours. This would allow the food company to stop the production, limiting the damage to just a part of the entire batch, with considerable savings in terms of product value, thereby avoiding the utilization of raw materials, packaging and strongly reducing food waste. Sensor systems offer major advantages over current systems as they are versatile and affordable but need to be integrated in the existing processing systems as a process analytical control (PAT) tool. The desire for good selectivity, low cost, portable and usable at working sites, sufficiently rapid to be used at-line or on-line, and no sample preparation devices are required. The application of biosensors in the food industry still has to compete with the standard analytical techniques in terms of cost, performance and reliability.

Keywords: food; pathogens; biosensors

1. Introduction

The consumer is dependent on quality food manufacturing processes. Contaminating microorganisms may enter and reach the end-product through raw materials, air in the processing plant area, process surfaces, or factory personnel. Spoilage bacteria may also build up in high numbers in processing equipment and develop into biofilm. The sources of spoilage bacteria are numerous, however personnel and the environment being the most prevalent.

Microbial management during the food processing operations is strategic for preventing contamination and for improving the product safety, quality and production hygiene. Built in mechanisms for in-process sampling points and frequency is necessary. Risk assessment tools like Hazard Analysis Critical Control Points (HACCP) can be used to detect areas of a process that are at risk of contamination [1], in addition, approaches such as Failure Modes and Effects Analysis (FMEA), can be implemented as outlined in a salmon processing company [2]. European Union (EU) member state companies must adhere to the rules laid out on Food Hygiene Legislation [3]. This legislation lays out rules on food hygiene through both general requirements and more specific rules, including the layout of premises, temperature control, HACCP, equipment, transport of food, waste, personal hygiene and training of food handling personnel. Specific hygiene rules for food of animal origin are also in the legislation [4]. These regulations are updated and changed on an ongoing basis.

Contamination screening during food processing operations would allow the food company to preventively stop the production, thus limiting the damage to just a part of the entire batch with considerable savings in terms of product value. Many contamination events are from biofilm formation and result from ineffective cleanings and disinfection processes [5]. The downstream processing of food cannot always prevent microorganism from entering the systems and many types of equipment cannot be sterilized, hence process management is vital [6]. The testing of food quality has in the past mainly dealt with the characterization of chemical contamination of the food product and testing has included physicochemical, biological and serological test techniques (*i.e.*, chromatography, spectrophotometry, electrophoresis, titration and others). Chemicals are generally analyzed using gas chromatography (GC) or high pressure liquid chromatography (HPLC). These methods are carried out to separate the components of a complex sample and identify them through specific types of detectors. Common detectors used include, flame ionization (FID) and thermal conductivity (TCD) for GC; ultraviolet light (UV), fluorescence (FL) or mass spectrometry (MS) for HPLC.

Microbiological testing has been based on traditional "growth" based methods. These methods relied on nutrient media and have provided the basis for quantitative microbial assay for microbial safety and quality product release. The time required to get results using these techniques is long and forward processing decisions and confirming manufacturing processes are static, results that may take days are now deemed to be inadequate.

Analytical methods to detect food borne pathogens are still evolving. There has been a surge in rapid microbial methods in the literature but in general they break down into three main categories: Qualitative methods (ATP bioluminescence, electrochemical measurements, micro-calorimetry); Quantitative methods (flow cytometery, direct epifluorescence technology) and identification methods (fatty acid analysis, ribotyping, polymerase chain reaction (PCR)). Newer emerging technologies include Raman spectroscopy, direct laser based detection, quantitative Real Time PCR and sensors and lab on chip (LOC) methods. Newer mass spectroscopy (MS) innovative methods such as matrix assisted laser desorption ionization time of flight (MALDI-TOF), surface enhanced laser desorption ionization time of flight (SELDI-TOF) and Fourier transfer infrared (FT-IR) mass spectroscopy (MS) methods have emerged. However, these MS methods rely on using isolated colonies as starting materials. Nucleic acid amplification methodologies such as PCR, ribotyping and gene sequencing burst on the commercial scene and have proved to have some sustainability. As patents for commercial system (electrochemical mostly) expire, new players are entering the market.

Applications for monitoring technologies range across the food industry. These monitoring technologies encompass, process control (the moisture content of the food; viscosity and texture); pH and conductivity (acidity and salt content); sugar content (glucose and sucrose are the main sugars monitored); food freshness including the detection of microbes (*Escherichia coli*, *Salmonella*, *etc.*) and the detection of microbial toxins (liquid and gas); ingredient freshness (milk, meat, *etc.*); frying oil (viscosity and chemical make-up) and food quality including taste (electronic nose).

2. Food Borne Pathogens

The European Food Safety Authority (EFSA) and the European Centre for Disease Prevention and Control (ECDC) analyzed the information submitted by 27 European Union Member States on the occurrence of zoonoses and food-borne outbreaks in 2011 [7]. The term zoonoses cover infections and diseases that are naturally transmissible either directly or indirectly, for example via contaminated foodstuffs, between animals and humans.

Campylobacteriosis, with 220,209 human cases confirmed in 2011, was the most reported zoonosis in the EU with broiler meat being the most documented source of infection [7]. Salmonellosis cases have shown a decrease with a total of 95,548 confirmed cases in 2011, down from 101,037 confirmed cases in 2010 [7]. This reduction is attributed to successful *Salmonella* control programmes in poultry populations. The bulk of *Salmonella* that has been detected has come from meat and products thereof. Recent updated Directive EU 218/2014 [8] enhances the process hygiene criterion for *Salmonella* in pig carcases. Numbers of confirmed human case of listeriosis have decreased to 1476 [7]. *Listeria* was rarely detected above the legal safety limit for ready-to-eat foods. Nine thousand, four hundred and eighty-five confirmed cases of verotoxigenic *Escherichia coli* (VTEC) infection were described in 2011, representing an increase of 159.4% when compared with 2010 [7]. This was as a result of the large outbreak that happened, primarily in Germany, of Shiga toxin-producing *E. coli*/verotoxigenic *E. coli* (STEC/VTEC) that caused 54 deaths.

A total of 5648 food-borne outbreaks were reported in the European Union in 2011. These outbreaks resulted in 69,553 confirmed human cases, 7125 hospitalisations and 93 deaths [7]. The majority of the reported outbreaks were found to be caused by *Salmonella*, bacterial toxins, *Campylobacter* and viruses; however, the outbreak with most human cases was caused by STEC/VTEC and associated with sprouted seeds in Germany and France. The food sources most associated with these outbreaks were eggs and egg products, followed by mixed foods and fish and its products [7]. The full surveillance report for 2011 [7] including data in table format for each country can be obtained from the EFSA webpage. It must be remembered that the report relies on full compliance for reporting by the EU member states and some states are more diligent and established than others to date.

3. Biosensors

An analysis of the word "sensors" in the ISI Web of Science showed 433,020 hits for sensors from 1945–2014, however if one screens for food borne pathogens within this cohort only 47 articles are listed [9]. The first sensor developed, detected glucose using the enzyme glucose oxidase immobilized on a platinum electrode [10]. The first commercial glucose sensor was from the Yellow Springs Instrument (Model 23 YSI) and it reached the market in 1974. The instrument directly measured whole blood glucose levels from a 25 μL with a ±2% accuracy. The US Food and Drug Administration (FDA) have identified the YSI Model 23A and subsequent designs as the reference standard for measuring glucose [11]. Later antibodies in conjunction with optical transducers were developed for real time bioaffinity monitors. Blood glucose measurement still comprises about 85% of the world market for biosensors.

The biosensor market is highly competitive and is driven mainly by the medical and pharmaceutical sector. Market analysis in 2010, estimate that global revenues for biosensors will demonstrate robust growth and exceed $14 billion mark in 2016, with 47 different end user applications [12]. The bulk of the market in 2009 was for glucose sensors and toxicity testing, food borne pathogens including *E coli*, *Salmonella, Listeria*, is a small percentage of this market [12,13]. The growth in the market will be from security and biodefense, environmental monitoring, home diagnostics and process industry market sectors. Further developments of sensors are likely in the following areas: inherent accuracy, capability, intelligence, reliability, smaller sizes, power consumption, packaging, lower costs, and the elimination of lead. Despite the vast number of publications and reports, the field of biosensors comprises two broad categories (1) sophisticated, high throughput laboratory machines capable of rapid accurate measurement of complex biological interactions and components and (2) easy to use portable devices for use by non-specialists for in situ or home monitoring. Further developments are expected to be in the areas of Micro-Electro-Mechanical Systems (MEMS) and nanotechnologies. Sensors developed for industries such as the motor industry are been translated to human heart and motion monitoring.

A key feature of the biosensor market is the large number of industrial alliances and licencing agreements. New approaches including molecular imprinting polymers (MIP) [14] as generic alternatives to antibodies, which allow selected functional monomers to self-assemble around a target analyte, is expanding sensor applications. The resulting MIP structures contains cavities which reflect both the shape and chemical functionality of the target species [15] with advances in reusable (up to 30 times) molecular templates developing [16,17]. During the last few years, mass-sensitive acoustic transducers, in particular the quartz crystal microbalance (QCM), have become very popular in combination with imprinted polymers [18]. There have also been recent serious discussions about harnessing the capabilities of smart phones as sensing tools [19].

Market challenges include, regulatory compliance, extended product lifecycles, reduced product development time, and product safety [12,13]. A significant number of reviews on sensors are available [20–22]; some are specific to food borne pathogens [23,24] and in specific application such as endotoxins [25], mycotoxins [26]; species specific reviews on *Campylobacter* spp. [27], *E. coli* non 0157 [28], recent trends in antibody sensors [29] and other reviews which deal with pesticides [30], milk [31], food processing [32]; nanomaterials [33,34]; conducting polymers [35] and molecular imprinted polymers [18].

4. Biosensor Component

A biosensor is an analytical device that converts a biological response into a detectable measurable signal. A number of stages must be realised in developing a biosensor (Figure 1). *Transduction, signal generation* (increase of signal or reduction of noise); *fluidic design* (sample injection and drainage, concentration of sample, reduction of sample consumption, increase of analyte transport, reduction in detection time); *surface immobilization chemistry* (analyte capture efficiency, elimination of nonspecific binding); *detection format* (direct binding, sandwich type binding, competitive binding) and *data analysis* (extraction of information regarding analyte concentration, binding kinetics) [36]. Taking all of these considerations together a biosensor is made up of three components: the sensor material base has traditionally being made of metal, glass, polymer or even paper, onto which a bioreceptor is coupled. The bioreceptor (antibodies, enzymes, nucleic acid aptamers or single stranded DNA, cellular structures/cells, biomimetic and bacteriophage (phage) [24], is coupled in the sensor through a number of immobilizing techniques which can be physical or chemical. Chemical groups that are reactive can include functional groups such as carboxyl, –COOH; amine; –NH_2; and hydroxyl, –OH. As environmental factors can affect biological materials making them very sensitive, they can easily lose their activity when forced to interact with the solid surface. The methodology for surface attachment of the probe is the most important step in fabrication of biosensors and requires a high level of control over the surface chemistry present.

Figure 1. Components of a biosensor.

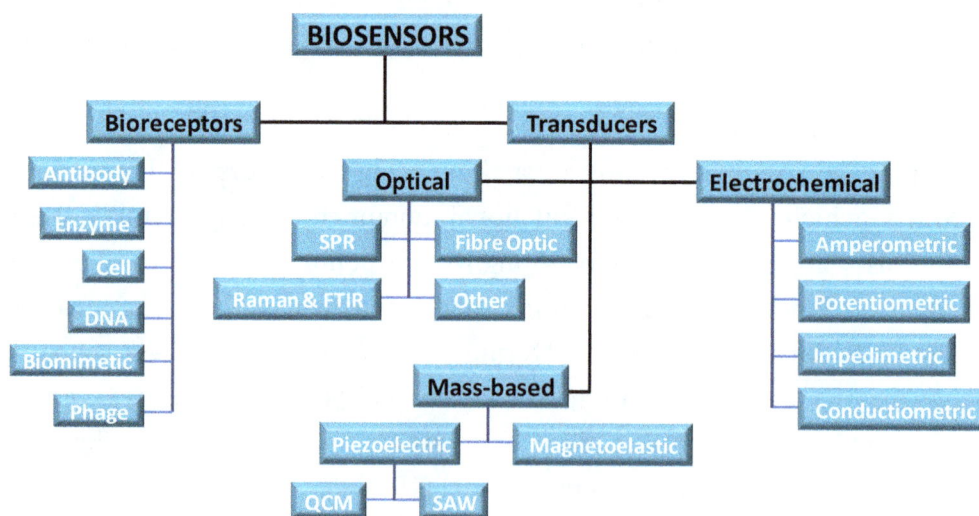

The trend in biosensors to date include, enzyme, antibody or antigen based biosensors; gene based sensors and whole cell sensor. Enzyme-based biosensors dominate the market and are mostly based on electrochemical transduction systems with glucose oxidase sensors dominating the market, the other focus are on chemical determinants (e.g., toxins, pesticides). However, many conjugated polymer based biosensors rely on indirect detection of the target analyte, usually a fluorescently labelled compound and this is especially true for biomolecular macromolecules such as proteins. Fluorescent sensors using boronic acid as a ligand, in a non-enzymatic approach for the detection of saccharides have found applications in microbial detection, as polysaccharides are a component of the bacterial cell membrane [37].

The third component which is vital is the transducing element. The bioreceptor should bring about a physio-chemical change that is measurable in close proximity to the transducer when it engages the target analyte. This change must produce a measurable signal that is proportionate to the concentration of the bioreceptor/target interaction. The signal can be measured by different techniques such as electrochemical, optical techniques, *etc.* (Figure 1). The sensor surface should be in an inactive or passive state when a measurement is not being conducted. For reusable sensors, after the measurement is completed, the target species is expelled by an external stimulus and the surface returns to its inactive form.

5. Sensor Materials

The sensor surface can be made of metal, polymer, glass or paper. Conducting polymers are polymer materials with metallic and semiconductor characteristics, a combination of properties not exhibited by any other known material. A key property of a conductive polymer is the presence of conjugated double bonds along the backbone of the polymer. In conjugation, the bonds between the carbon atoms are alternately single and double. The most common types of conjugated polymers are poly(acetylene)s, poly(pyrrole)s, poly(thiophene)s, poly(terthiophene)s, poly(aniline)s, poly(fluorine)s, poly(3-alkylthiophene)s, polytetrathiafulvalenes, polynapthalenes, poly(*p*-phenylene sulfide), poly(*p*-phenylenevinylene)s, poly(3,4 ethylenedioxythiophene), polyparaphenylene, polyazulene, polyparaphenylene sulfide, polycarbazole and polydiaminonaphthalene. They have found extensive use in the creation of electrochemical sensors such as potentiometric, amperometric and conductometric sensors [38]. Polyaniline followed by polypyrrole and polythiopene are the most used [39]. The structure of polypyrrole is shown in Figure 2.

Figure 2. Structure of polypyrrole.

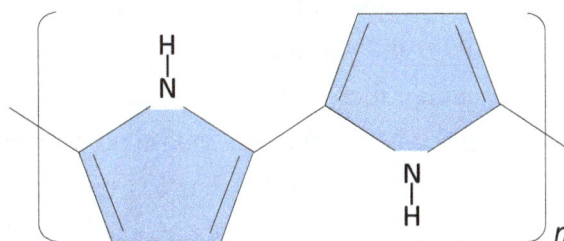

In work carried by our group we developed a polymer nanocomposite sensors using polyethylene adipate (PEA) for a gas sensor for quantification of bacterial cultures [40]. In addition we used polypyrrole in the detection of *Bacillus cereus* [41], developing unique DNA primers which could differentiate between the *B. cereus* group spp., in spiked milk [42].

Metal-organic framework (MOF) materials have recently been explored as chemical sensors. MOF's are extended crystalline structures where the metal cations or clusters of cations ("nodes") are connected by multitopic organic "struts" or "linker" ions or molecules [43]. Their importance is in their tunability and structural diversity. Improving detection by coupling MOF's with vibrational spectroscopy such as surface enhanced Raman scattering (SERS) has shown additional promise. MOF's have been recently applied in the development of glucose sensors in a non-enzymatic approach [44] and for the detection of dipicolinic acid (pyridine-2,6-dicarboxylic acid) a unique compound in bacterial spores [45]. Paper based sensors emerged as an alternative surface for sensors. Paper is thin lightweight and flexible. The main constituents of paper are cellulose fibres. The paper absorbs and transports liquids by capillary force without additional mechanical assistance; however they still suffer from limitations, including accuracy and sensitivity [46]. Among the patterning methods employed for deposition of functional materials on paper substrates, the inkjet printing method was advanced due to its ability to deposit precise amount of materials rapidly and ability to perform computer-controlled printing on specific locations [47]. Using paper and species specific enzymes with a colorimetric reporting system has been outlined for select food borne pathogens (*L. monocytogenes*, *E. coli* 0157:H7 and *S.* Typhimurium) with a reduced enrichment time and a LOD of 10^4 CFU/mL [48].

Nanomaterials show similar dimensions to biomolecules like proteins and DNA. The integration of nanomaterials with biomaterials has developed into a study called nanobiomaterials. Nanostructured biomaterials have been projected to be the next stage in development of many devices, including in sensor technology with unique capabilities for data collection, processing and recognition with minimal false positive counts. Carbon nanotubes (CNT's) are conducting, act as electrodes, and generate electrochemiluminescence (ECL) in aqueous solutions. They can be derivatized with functional groups (carboxylic, carbonyl and hydroxyl) that allow immobilisation of biomolecules either through covalent or non-covalent bonding [49]. The variety and range of sensor materials can be seen from using double layer gold nanoparticles and chitosan to detect *Bacillus cereus* [50] in an electrochemical immunosensor approach with a detection limit of 10.0 CFU/mL, in pure culture. Colloidal gold is one of the most studied nanomaterial available for biosensors, albeit it is expensive for large scale applications. Multiplexing using a carbon screen printed array to detect *E. coli* 0157:H7 and *E. sakazakii* (*Chronobacter*) and multiwalled carbon nanotubes with horse radish peroxidase (HRP) gave a LOD of 3.27×10^3 CFU/mL and 4.5×10^3 CFU/mL respectively [51]. Quantum dot nanoparticles and anti-*Salmonella* polyclonal antibodies immobilised by streptavidin biotin binding achieved a detection limits of 4×10^3 CFU/mL in food extracts, using a custom built fluorometer to detect the fluorescent light [34]. Oligonucleotides immobilised on nanopillar arrays of silicon was fabricated to target ssDNA and measuring the refractive index with an ellipsometer, as a new approach in a label free optical sensor [52]. A selection of immunosensors has been reported for food borne pathogens including *E. coli* 0157:H7 using modified graphene paper and gold nanoparticles with antibody and biotin streptavidin system with a detection limit of 1.5×10^2 CFU/mL [53]. A screen printed carbon electrode/carbon nanotube was developed to detect *E. sakazakii* in the range of 10^3–10^9 CFU/mL and a detection limit of 7.7×10^{-1} CFU/mL with long term storage capabilities [54]. However limitation due to *E. sakazakii* growth in milk powder after addition of water and delayed use was highlighted as a limitation. *Cronobacter* is now the officially recognised bacterial genus name for *Enterobacter*. A stable label-free electrochemical impedance immunosensor for the detection of *Salmonella* Typhimurium in milk was developed by immobilising anti-*Salmonella* antibodies onto gold nanoparticles and poly(amidoamine)-multiwalled carbon nanotubes-chitosan nanocomposite film modified glassy carbon electrode. A detection limit of 5.0×10^2 CFU/mL was reported [55]. The application of a quartz crystal microbalance (QCM) instrument with a microfluidic system for the rapid and real time detection of *Salmonella* Typhimurim using immobilised anti-*Salmonella* antibody and gold-nanoparticles gave a sensitivity with a limit of detection (LOD) 10–20 CFU/mL compared to direct and sandwich assay (1.83×10^2 CFU/mL and 1.01×10^2 CFU/mL, respectively) [56]. Reviews on nanomaterials and biosensors as diagnostic tools and in food applications are available [33,57,58].

Nobel metals (e.g., gold, silver platinum, *etc.*) nanoparticles have been a focus. Numerous techniques to synthesis these nanoparticles and to control their properties (their size, shape and homogeneity) have been demonstrated. These techniques include both chemical methods such as chemical reduction, photochemical reduction, co-precipitation and hydrolysis, and physical methods such as laser ablation, grinding and vapor deposition [21]. Examples of food borne pathogen Nano Metal Particle (NMP) based sensors including electrical/electrochemical with gold NP to detect *E. coli* 0157:H7 in food samples at a LOD of 5.3×10^2 CFU/mL [59] and *Salmonella* in pork samples with a detection limit of 1.0×10^2 CFU/mL [60].

6. Sensor Designs

The technique used for the physical or chemical fixation of bioreceptor which can be cells, organelles, enzymes, or other proteins (e.g., monoclonal antibodies) onto a solid support, or into a solid matrix or retained by a membrane, is used in order to increase their stability. Methods used can be physical retention or chemical binding.

Adsorption is a physical method of immobilization. Many substances can adsorb enzymes and other biological materials on their surfaces for example alumina, charcoal, clay, cellulose, kaolin, silica gel, glass, collagen, carbon pellets and advanced material such as carbon nanotubes (CNTs). A simple procedure is when microbial cells are immobilized by simple absorption by placing the cells on a porous cellulose membrane. Generating pastes such as when enzymes or tissue are mixed with graphite powder and liquid paraffin.

Entrapment, physical method of immobilization: Entrapment means physical enclosure of biomolecule in a small space. Inert membranes have been used to provide close contact between the biomaterial and transducer. Types of membranes used include cellulose acetate (dialysis membrane); polycarbonate (Nucleopore), synthetic non-permselective material; Collagen, a natural protein; PTFE: polytetrafluoroethylene (trade name Teflon) and is a synthetic polymer selectively permeable to gases. Nafion, (a Dupont material), which is biocompatible and shown to be stable in cell culture and the human body. Polymeric gels can be used and prepared in a solution containing the biomaterial. Chemical polymers such as calcium alginate, carrageenan, polyacrylamide, and sol-gel (Sol-gel, is a glassy silica produced by polymerization of silicate monomers).

Bonding and cross linking: a number of bonding mechanisms have been used including covalent bonding. A covalent bond exists between two atoms if they share electrons between them. The Biotin-Avidin bond is one of the strongest known non-covalent bonds. Avidin is a terameric protein that forms a highly specific binding site for Biotin. Sulphur compounds are known for their reactivity to metals and this absorb readily to the noble metals. Thiolised DNA can be attached to gold via different methods.

Transducing element: the transducing element must produce a measurable signal that is proportionate to the concentration of the analyte/bioreceptor. Transducers can be divided into optical, electrochemical and mass based (Figure 1).

Optical transducers can be subdivided into light absorption, fluorescence/phosphorescence, reflectance, refractive index, bio/chemiluminiscence.

In reflectance three widely used methods are Surface Plasmon resonance (SPR), total internal reflection fluorescence (TIFR) and attenuated total reflectance (ATR). SPR has found some commercial instruments being developed by Biacore [61] for vitamin and antibiotic analysis of food. Using a polyclonal antibody against *L. monocytogenes* and a subtractive inhibition assay carried out with a BIAcore 3000 biosensor with a sensitivity of 1×10^5 cells/mL comparable to ELISA tests has been reported [62]. Biosensing Instruments Ltd. [63] has developed an endotoxin detector also using SPR. Using a custom built SPR sensor based on ATR method and glass chips coated in gold and with streptavidin for biotinylated antibody binding for selected species (*E. coli* 0157:H7; *S. choleraesusi* serotype Typhimurium, *L. monocytogenes* and *C. jejuni*) provided limits of detection ranging from 3.4×10^3 to 1.2×10^5 CFU/mL. Both single and mixtures of the four species gave comparable results [64].

Fiber optic biosensors [65] and their application in food quality and safety [66] have been reviewed. Significant results in food matrixes to detect *Salmonella*, *E. coli* and *Listeria* was obtained, using streptavidin coated optical waveguides immobilized with biotinylated polyclonal antibodies in a multiplex reaction. The limit of detection for the sensor was ~10^3 CFU/mL after 2 h for all pathogens [67]. However enrichment for 18 h was an initial step.

Electrochemical transduction methods can be subdivided based on the measured parameter: amperometric (current), potentiometric (potential), impedimetric (impedance) and conductometric. The amperometric sensors have a superior sensitivity and better linear range than potentiometric devices and the most successful commercially. Most work has been done on amperometric and potentiometric biosensors with little work being devoted to conductometric biosensors [68]. Modern electrochemical techniques have low detection limits (10^{-7}–10^{-9} M or 30 ppb) for gaseous compounds [69]. A range of detector components (antibody, DNA) have been used in the detection of *Campylobacter* spp. using both ampermometric and impedimetric transducers [27]. Electrochemical enzyme-based biosensors have dominated the market in the food sector including newer amperometric nanoparticles glucose sensors, based on hydrogel heterostructures with a response time of 3 s and sensitivity as high as 96.1 $\mu A \cdot mM^{-1} \cdot cm^{-2}$ [70].

Mass based transducers: mass sensitive biosensors are suitable for very sensitive detection, in which the transduction is based on detecting a small changes in mass. The two main types of mass based sensors are (1) bulk wave (BW) or quartz crystal microbalance (QCM) and (2) surface acoustic wave (SAW).

However, the detection of foodborne pathogens based on piezoelectric sensors are not versatile. A quartz crystal microbalance (QCM) immunosensor in the direct detection of *S.* Typhimurium in a chicken meat sample was demonstrated [71] which showed that the resonant frequency and motional resistance were proportional to the cell concentration in the range of 10^5–10^8 and 10^6–10^8 cells/mL, respectively. The detection limit was lowered to 10^2 cells/mL by using anti-*Salmonella*-magnetic beads. A QCM is a real mass sensor belonging to a wider class of inertial mass sensors [72].

Acoustic wave sensors (AWS) monitor the change in oscillation frequency when the device responds to the input stimulus. The global AWS device market is expected to reach €1.8 billion by 2016 [73]. AWS can be subdivided into: (1) bulk acoustic wave resonators (BAW); (2) Flexural-plate-wav-resonators (FPW); (3) Surface acoustic wave resonators (SAW); and (4) shear-horizontal acoustic plate mode resonators (SAW). A review of SAW for the detection of pathogens is available [74]. An interesting SAW application is its use in an intelligent food packaging humidity monitoring system, consisted of a ZnO surface acoustic wave sensor directly built on the protein zein (a prolamine protein found in maize (corn)), measuring humidity for food freshness/protection [75]. Bulk acoustic wave have been used to detect proteins and DNA. Some applications in food to detect *E. coli* 0157:H7, *Salmonella* and *Listeria* have been summarised [76].

The overall features of a good sensor includes: *Selectivity*: the biosensor must be highly selective for the target analyte and have little or no cross reactivity with moieties that have a chemical structure similar to that of the target analyte. *Sensitivity*: the biosensor should be able to measure in the range of interest for a given target analyte with little in the way of additional steps such as pre cleaning and pre concentration of the samples. *Linearity of response*: the linear response range of the system should cover the same concentration range over which the target analyse is to be measured. *Reproducibility of signal response*: when samples having same concentrations are analyzed several times, they should give same response. *Quick response time and recovery time*: the time it takes for the biosensor to respond to the selected analyte should be quick enough so that real time monitoring can take place in an efficient manner. The recovery time of the sensor should be as small possible for reusability of the biosensor system. *Stability and operating life*: as such most of the biological compounds are unstable in different biochemical and environmental conditions [32].

7. Microbial Sensing

In order to detect microorganism in a liquid or solid sample multiple approaches have been undertaken. For the extensive amount of research generated there is limited commercial output. Approaches taken have been diverse from whole cell to cellular components.

Microbial Whole Cell Biosensors

Sensors to detect whole cell bacteria have been slow to come to market, as microbial cells are complex and a sensor prefers a simple matrix in order to work efficiently. Microbial cells because of their low cost, long lifetime and wide range of suitable pH and temperature, have been explored. Some of the basic limitations of microbial biosensors as compared to enzyme sensors have been their long response time, low sensitivity and detection limits. Their slow response has been attributed to diffusional problems associated with the cell membranes. Systems reported using whole cell as sensors for ethanol in the food fermentation industry has commercial interest and multiple approaches have been taken and reviewed [77]. Using *Acetobacter aceti* and its respiratory membrane bound enzyme Alcohol dehydrogenase catalytic activity for ethanol measurement was an initial approach [78]. In many cases whole cell microorganisms have been used to detect chemical components such as environmental pesticides. Genetic engineered *Pseudomonas putida* JS444 was constructed to display organophosphorus hydrolase (OPH) activity on a dissolved oxygen electrode to detect synthetic organophosphate compounds (OP). In optimal condition it measured as low as 55 ppb for paraoxon, a potent acetylcholinesterase-inhibiting insecticides, without interference from other common pesticides [79].

Using genetically engineered microorganisms and enzymes is now the norm, including fusion proteins for tailoring sensors for specific purposes. In a new configuration for Biological Oxygen

Demand (BOD) used to detect pollution problems, a chronamperometric response system, employed a double mediator system coupled with ferricyanide and a lipohhilic mediatator mendaione (synthetic compound) and *Saccharomyces cerevisiae* [80]. *P. syringae* was used as the biocatalyst to also measure BOD in water samples with a response time of 3–5 min, the biocatalyse was placed between cellulose and Teflon membranes [81]. A comprehensive list of electrochemical, conductometric, potentiometric whole cell microbial biosensors targeting a range of chemicals has been reviewed [82]. Commercialisation of whole cell biosensors has proved to be slow due to problems fabricating the whole cell to the appropriate surface and the stability of the microorganism. In the food processing industry applications of microbial whole cell biosensors in pathogen detection have not been embraced.

8. Nucleic Acid Sensors

Nucleic acid sensors have been the focus of much research. Several gene sensing detection methods for food borne pathogens have been developed with optical, electrochemical, mass sensitive and microgravimetioc techniques [24,83] and with multiplex PCR approached [84,85].

In the nucleic acid sensor, a DNA or RNA target is detected through the hybridization reaction between DNA or RNA and ssDNA sensing element. Examples of early DNA-based biosensor for *E. coli*, using PCR and piezoelectric quartz crystals was demonstrated to detect 23 cells per 100 mL water samples with application in public beach water quality regulations [86]. Other reports included using embedded *E. coli* DNA-uidA gene in polypyrrole [87] and in real time using a quart crystal microbalance using the eaeA gene (104bp) of *E. coli* 0157:H7 [88]. *Salmonella* spp. were the target using DNA streptavidin modified magnetic beads and electrochemical detection [89]. *L. monocytogenes* was detected using a magneto electrochemical luminescence PCR detection platform which gave a detection limit of 500 fb/µL genome DNA in 1 h [90]. The detection of *E. coli*, *Bacillus subtilis*, *B. atrophaeus* and *L. innocua* in meat juices demonstrated a detection limit of 500 CFU/*E. coli* in one working day [91] using esterase and an amplification based DNA array sensor. To enable large scale screening procedures, new multiplex analytical formats are being developed, and these allow the detection and/or identification of more than one pathogen in a single analytical run, thus cutting assay times and costs [92].

Microfluidic strategies coupled with electrochemical transducers have produced miniaturised devices. The lab-on-chip includes electrodes, hybridisation, washing and response. Label free detection using synthesised target DNA and real DNA samples from *S. choleraesusi* in dairy food was measured in real time [93]. The ability for microfluidic and multiplexing was demonstrated in an integrated system using gold nanoparticle labels for detection of *E. coli* and *B. subtilis* [94].

Real time detections is still a goal and coupled with PCR showed early developments [95]. A microchip with integrated modules for performing cell lysis, PCR, and quantitative analysis of DNA amplicons in a single step has been described for a lab-on-chip detection of *E. coli* O157:H7 and *Bacillus subtilis* [96]. This system however, demonstrated the classic shortcoming of temperature control in the PCR reaction. The application of loop meditated isothermal amplification (LAMP), has been demonstrated for *E. coli* and *S. aureus* using target genes amplified with LAMP using ruthenium hexamine as the intercalating electrochemical indicator [97].

The development of aptasensors has shown increased promise. Aptamers are DNA or RNA molecules that are selected from random pools and engineered through repeated rounds of *in vitro* selection based on their ability to bind other molecules; they can bind nucleic acid, proteins, small organic compounds, and even entire organisms. There are two main classes of aptamers—nucleic (DNA and RNA) aptamers and peptide aptamers. DNA and RNA aptamers typically consist of between 20 and 80 nucleotides. Aptamers have many advantages compared to antibodies as they can be produced easily and inexpensively. They are simple to modify chemically, label with different reporter molecules, to integrate into different analytical methods and can be coupled to different transduction systems [98]. Applications in food safety control have been reviewed [99], and in real food situations, e.g., *E. coli* using a potentiometric aptamer based biosensor with detection of 6 CFU/mL in milk and 25 CFU/mL in apple juice [100]. *Vibrio cholera* was detected at 0.85 ng/μL genomic DNA; DNAzyme aptamers for *Salmonella paratyphi* using nanotubes and fluorescence [101]. The design of the aptamers was carried out using SELEX (Systematic Evolution of Ligands by Exponential Enrichment). The engineering of aptamers using SELEX has caused recent excitement in the field of sensors since their discovery [98], and their applications have been explored in designing biomarkers, to treat cancer and in specific pathogen detection [102]. Gene-sensing methods gave initially very high hopes for rapid on line systems. Limitations include extraction of the DNA, dead cell detection even with the use of RNA to determine viability, the complex matrix of food, all provided ample false negatives results.

9. Sensors Using Bacteriophage

Since their discovery by Twort and d'H´erelle, bacteriophages have not been universally exploited as control agents of disease. Although used in the former Soviet Union extensively, they did not translate into viable infection control options until recently. Bacteriophages are specific for certain bacteria and using this selectivity, phage typing has been extensively developed to differentiate between diverse strains of particular species of bacteria. Phage typing exploits their ability to specifically recognize molecules on the surface of the bacteria, to infect the cells and ultimately lyse their host. Phage as a detection system has come into the limelight comprehensively reviewed by Tawil *et al.* [103]; Schmelcher and Loessner [104].

10. Companies Developing and Producing Biosensors

The commercialization of biosensors lies in glaring contrast to the promise that is shown in the research literature. The global bio chip market is expected to reach US$11.4 billion by 2018 with a compound annual growth rate (CAGR) of 18.6% during 2012–2018. Biochip instruments are expected to exert the highest support to the industry with a CAGR of 20%. The microarray segment accounts for nearly 70% of the industry value [12,13].

However, rapid, lab on chip hand held systems are not forthcoming. Some systems are available including Nanosphere's VeriGene Enteric Pathogens (EP)—a single use self-contained microfluidic cassette [105]. 3M has developed a number of systems for pathogen detection including *Salmonella* which received the Association of Analytical Communities (AOAC) Official Methods of Analysis Validation and an equivalent system for *Listeria*. 3M uses isothermal amplification of nucleic acids sequences with bioluminescence to detect the amplification. In addition there is the 3M™ Microbial

Luminescence System (MLS) to detect the presence of microbial ATP in ultra-high treated (UHT) and extended shelf life (ESL) dairy end products [106].

Neogen [107] have a number of commercial food safety systems, the ANSR to detect *Salmonella* and *Listeria* uses isothermal amplification technology. The Reveal® test system is an immunoassay with chromatography but requires enrichment. The NeoSeek™, targets seven STEC/*E.coli* strains, using enrichment with next day results. A mass spectrometry-based multiplexing system is the technology used. Their GeneQuence® detection assays utilize DNA hybridisation technology in a microwell format to detect *Salmonella*, *Listeria*, or *Listeria monocytogenes* and can run up to 372 samples at a time fully automated [107]. Serosep [108] have an EntericBio human stool samples to detect food borne pathogen which can be applied to food matrices. VereFoodborne™ is a nucleic acid-based, device, combines multiplex PCR and microarray hybridization to detect, differentiate and identify 15 of the major food pathogens [109], but has not achieved diagnostic validation. PDS Biophage Pharma [110] has advertised two systems PDS® biosensor for total bacterial detection and bacTrapping which has phage on paramagnetic beads with magnetic separation. DetScan from Elice [111] is an electrochemical based sensor. Stratophase Ltd. (UK) Ranger™ Probe is an optical structure around a silicon chip that gives real-time, in-line bioprocess monitoring and fermentation control in food industries [112].

There are a number of prominent real-time PCR cycler manufactures which designed instruments for research with low capacities and others for high-throughput applications, Most employs fluorescent probes for detection with Quantitative PCR, multiplexing, HRM (high resolution melting), these include Roche, Agilient, Biacore, BioRad, Life Technologies-Applied Biosystems, A comprehensive listing of real time PCR instruments is available [113].

11. Conclusions

In determining the microbial control parameters in food, the spatial heterogeneity of the food matrix is not always taken into account. Microbial behaviour can be influenced by variables such as porosity, viscoelastic properties and the physicochemical attributes of foods, such as pH, water activity and the ability of nutrients and/or metabolites to diffuse. The microorganisms themselves can be influenced by the spatial and temporal heterogeneity of bacteria, the variability in the physiological stage in the cells, and the succession of the microbial community in time will all affect the sensors ability to detect. In addition, the stability and longevity of the sensing biomolecules under conditions in the field also need to be considered, e.g., is the sensor affected by temperature ranges, the presence of other chemicals and particulates? Simplified sample preparation procedures and separation techniques to selectively fractionate bacteria is also a limiting factor in sensor technologies.

Food inspecting agencies worldwide have a zero tolerance policy for the serious food borne pathogen organisms (*Salmonella*, *E. coli* 0157, *etc.*) presence in food. This zero tolerance must be the target for any new biosensor in its design and development to incorporate an inclusivity and exclusivity of detection in the systems. Sensor technology development has favored home diagnostics: point of care testing in healthcare; research laboratories; security and biodefense. The food industry has not embraced rapid method applications in food production and processing. Cost, performance and reliability have still to be addressed.

Acknowledgements

Thanks to Michael Ryan for critical review of the manuscript.

Conflicts of Interest

The authors declare no conflict of interest.

References

1. Food Quality and Safety Systems—A Training Manual on Food Hygiene and the Hazard Analysis and Critical Control Point (HACCP) System. Food and Agriculture Organisation of the United Nations, Rome, 1998. Available online: http://www.fao.org/docrep/W8088E/W8088E00.htm (accessed on 19 January 2014).

2. Arvanitoyannis, I.S.; Varzakas, T.H. Application of ISO 22000 and failure mode and effect analysis (FEMA) for industrial processing of salmon: A case study. *Crit. Rev. Food Sci. Nutr.* **2008**, *48*, 411–429.

3. Regulation (EC) No 852/2004 of the European Parliament and of the Council of 29 April 2004 on the Hygiene of Foodstuffs. *Off. J. Eur. Union* **2004**, *L139*, 1.

4. Regulation (EC) No 853/2004 of the European Parliament and of the Council of 29 April 2004 Laying Down Specific Hygiene Rules for Food of Animal Origin. *Off. J. Eur. Union* **2004**, *L139*, 55.

5. Chmielewski, R.A.N.; Frank, J.F. Biofilm formation and control in food processing facilities. *Compr. Rev. Food Sci. Food Saf.* **2006**, *2*, 22–32.

6. Cappitelli, F.; Polo, A.; Villa, F. Biofilm formation in food processing environments is still poorly understood and controlled. *Food Eng. Rev.* **2014**, *6*, 1–2, 29.

7. The European Union Summary Report on Trends and Sources of Zoonoses, Zoonotic Agents and Food-Borne Outbreaks in 2011. *EFSA J.* **2013**, *11*, 3129, doi:10.2903/j.efsa.2013.3129.

8. Commission Regulation (EU) No 218/2014 of 7 March 2014 Amending Annexes to Regulations (EC) No 853/2004 and (EC) No 854/2004 of the European Parliament and of the Council and Commission Regulation (EC) No 2074/2005. *Off. J. Eur. Union* **2014**, *L69*, 95.

9. Web of Science™. Available online: www.webofknowledge.com (accessed on 9 January 2014).

10. Clark, L.C., Jr.; Lyons, C. Electrode systems for continuous monitoring in cardiovascular surgery. *Ann. N. Y. Acad. Sci.* **1962**, *102*, 29–45.

11. Chemical Heritage Foundation. Available online: http://www.chemheritage.org/discover/collections/collection-items/scientific-instruments/ysi-blood-glucose-analyzer-model-23a.aspx (accessed on 23 March 2014).

12. Thusu, R. Strong Growth Predicted for Biosensor Market. Available online: http://www.sensorsmag.com/specialty-markets/medical/strong-growth-predicted-biosensors-market-7640 (accessed on 23 March 2014).

13. Thusu, R. Sensors Facilitating Health Monitoring. Available online: http://www.sensorsmag.com/specialty-markets/medical/sensors-facilitate-health-monitoring-8365 (accessed on 23 March 2014).

14. Turiel, E.; Martin-Esteban, A. Molecularly imprinted polymers for sample preparation: A review. *Anal. Chim. Acta* **2010**, *668*, 87–99.

15. Haupt, K.; Mosbach, K. Molecularly imprinted polymers and their use in biomimetic sensors. *Chem. Rev.* **2000**, *100*, 2495–2504.

16. Poma, A.; Turner, A.P.; Piletsky, S.A. Advances in the manufacture of MIP nanoparticles. *Trends Biotechnol.* **2010**, *28*, 629–637.

17. Poma, A.; Guerreiro, A.; Whitcombe, M.J.; Piletska, E.V.; Turner, A.P.F.; Piletsky, S.A. Solid-phase synthesis of molecularly imprinted polymer nanoparticles with a reusable template—"Plastic antibodies". *Adv. Funct. Mater.* **2013**, *23*, 2821–2827.

18. Vasapollo, G.; del Sole, R.; Mergola, L.; Lazzoi, M.R.; Scardino, A.; Scorrano, S.; Mele, G. Molecularly imprinted polymers: Present and future prospective. *Int. J. Mol. Sci.* **2011**, *12*, 5908–5945.

19. Ozcan, A. Mobile phones democratize and cultivate next-generation imaging, diagnostics and measurement tools. *Lab Chip* **2014**, *14*, 3187–3194.

20. Turner, A.P. Biosensors: Sense and sensibility. *Chem. Soc. Rev.* **2013**, *42*, 3184–3196.

21. Doria, G.; Conde, J.; Veigas, B.; Giestas, L.; Almeida, C.; Assuncao, M.; Rosa, J.; Baptista, P.V. Noble metal nanoparticles for biosensing applications. *Sensors* **2012**, *12*, 1657–1687.

22. Gehring, A.G.; Tu, S.I. High-throughput biosensors for multiplexed food-borne pathogen detection. *Annu. Rev. Anal. Chem. (Palo Alto Calif.)* **2011**, *4*, 151–172.

23. Arora, P.; Sindhu, A.; Dilbaghi, N.; Chaudhury, A. Biosensors as innovative tools for the detection of food borne pathogens. *Biosens. Bioelectron.* **2011**, *28*, 1–12.

24. Velusamy, V.; Arshak, K.; Korostynska, O.; Oliwa, K.; Adley, C. An overview of foodborne pathogen detection: In the perspective of biosensors. *Biotechnol. Adv.* **2010**, *28*, 232–254.

25. Das, A.P.; Kumar, P.S.; Swain, S. Recent advances in biosensor based endotoxin detection. *Biosens. Bioelectron.* **2014**, *51*, 62–75.

26. Vidal, J.C.; Bonel, L.; Ezquerra, A.; Hernandez, S.; Bertolin, J.R.; Cubel, C.; Castillo, J.R. Electrochemical affinity biosensors for detection of mycotoxins: A review. *Biosens. Bioelectron.* **2013**, *49*, 146–158.

27. Yang, X.; Kirsch, J.; Simonian, A. *Campylobacter* spp. detection in the 21st century: A review of the recent achievements in biosensor development. *J. Microbiol. Methods* **2013**, *95*, 48–56.

28. Wang, F.; Yang, Q.; Kase, J.A.; Meng, J.; Clotilde, L.M.; Lin, A.; Ge, B. Current trends in detecting non-O157 Shiga toxin-producing *Escherichia coli* in food. *Foodborne Pathog. Dis.* **2013**, *10*, 665–677.

29. Holford, T.R.; Davis, F.; Higson, S.P. Recent trends in antibody based sensors. *Biosens. Bioelectron.* **2012**, *34*, 12–24.

30. Liu, S.; Zheng, Z.; Li, X. Advances in pesticide biosensors: Current status, challenges, and future perspectives. *Anal. Bioanal. Chem.* **2013**, *405*, 63–90.

31. Mortari, A.; Lorenzelli, L. Recent sensing technologies for pathogen detection in milk: A review. *Biosens. Bioelectron.* **2014**, *60*, 8–21.

32. Thakur, M.S.; Ragavan, K.V. Biosensors in food processing. *J. Food Sci. Technol.* **2013**, *50*, 625–641.

33. Ezzati Nazhad Dolatabadi, J.; de la Guardia, M. Nanomaterial-based electrochemical immunosensors as advanced diagnostic tools. *Anal. Methods* **2014**, *6*, 3891–3900.

34. Kim, G.; Park, S.B.; Moon, J.-H.; Lee, S. Detection of pathogenic *Salmonella* with nanobiosensors. *Anal. Methods* **2013**, *5*, 5717–5723.

35. Arshak, K.; Velusamy, V.; Korostynska, O.; Oliwa-Stasiak, K.; Adley, C. Conducting polymers and their applications to biosensors: Emphasizing on foodborne pathogen detection. *Sensors J. IEEE* **2009**, *9*, 1942–1951.

36. Fan, X.; White, I.M.; Shopova, S.I.; Zhu, H.; Suter, J.D.; Sun, Y. Sensitive optical biosensors for unlabeled targets: A review. *Anal. Chim. Acta* **2008**, *620*, 8–26.

37. Amin, R.; Elfeky, S.A. Fluorescent sensor for bacterial recognition. *Spectrochim. Acta Mol. Biomol. Spectros.* **2013**, *108*, 338–341.

38. Faridbod, F.; Norouzi, P.; Dinarvand, R.; Ganjali, M.R. Developments in the field of conducting and non-conducting polymer based potentiometric membrane sensors for ions over the past decade. *Sensors* **2008**, *8*, 2331–2412.

39. Inzelt, G. Rise and rise of conducting polymers. *J. Solid State Electrochem.* **2011**, *15*, 1711–1718.

40. Arshak, K.; Adley, C.; Moore, E.; Cunniffe, C.; Campion, M.; Harris, J. Characterisation of polymer nanocomposite sensors for quantification of bacterial cultures. *Sens. Actuators B Chem.* **2007**, *126*, 226–231.

41. Velusamy, V.; Arshak, K.; Korostynska, O.; Oliwa, K.; Adley, C. Conducting polymer based DNA biosensor for the detection of the *Bacillus cereu*s group species. In *SPIE7315, Sensing for Agriculture and Food Quality and Safety*; Kim, M.S., Tu, S.-I., Chao, K., Eds.; International Society for Optics and Photonics: Orlando, FL, USA, 2009.

42. Oliwa-Stasiak, K.; Kolaj-Robin, O.; Adley, C.C. Development of Real-Time PCR assays for detection and quantification of *Bacillus cereus* group species: Differentiation of *B. weihenstephanensis* and rhizoid *B. pseudomycoides* isolates from milk. *Appl. Environ. Microbiol.* **2011**, *77*, 80–88.

43. Kreno, L.E.; Leong, K.; Farha, O.K.; Allendorf, M.; van Duyne, R.P.; Hupp, J.T. Metal-organic framework materials as chemical sensors. *Chem. Rev.* **2012**, *112*, 1105–1125.

44. Wei, C.T.; Li, X.; Xu, F.G.; Tan, H.L.; Li, Z.; Sun, L.L.; Song, Y.H. Metal organic framework-derived anthill-like Cu@carbon nanocomposites for nonenzymatic glucose sensor. *Anal. Methods* **2014**, *6*, 1550–1557.

45. Xu, H.; Rao, X.; Gao, J.; Yu, J.; Wang, Z.; Dou, Z.; Cui, Y.; Yang, Y.; Chen, B.; Qian, G. A luminescent nanoscale metal-organic framework with controllable morphologies for spore detection. *Chem. Commun.* **2012**, *48*, 7377–7379.

46. Rozand, C. Paper-based analytical devices for point-of-care infectious disease testing. *Eur. J. Clin. Microbiol. Infect. Dis.* **2014**, *33*, 147–156.

47. Wu, A.; Gu, Y.; Beck, C.; Iqbal, Z.; Federici, J.F. Reversible chromatic sensor fabricated by inkjet printing TCDA-ZnO on a paper substrate. *Sens. Actuators B Chem.* **2014**, *193*, 10–18.

48. Jokerst, J.C.; Adkins, J.A.; Bisha, B.; Mentele, M.M.; Goodridge , L.D.; Henry, C.S. A paper based analytical device for the colorimetric detection of food borne pathogens. In Proceedings of the 15th International Conference on Miniaturized Systems for Chemistry and Life Sciences, Seattle, DC, USA, 2–6 October 2011.

49. Vaseashta, A.; Dimova-Malinovska, D. Nanostructured and nanoscale devices, sensors and detectors. *Sci. Tech. Adv. Mater.* **2005**, *6*, 312–318.

50. Kang, X.; Pang, G.; Chen, Q.; Liang, X. Fabrication of *Bacillus cereus* electrochemical immunosensor based on double-layer gold nanoparticles and chitosan. *Sens. Actuators. B Chem.* **2013**, *177*, 1010–1016.

51. Dou, W.; Tang, W.; Zhao, G. A disposable electrochemical immunosensor arrays using 4-channel screen-printed carbon electrode for simultaneous detection of *Escherichia coli* O157:H7 and *Enterobacter sakazakii. Electrochim. Acta* **2013**, *97*, 79–85.

52. Chen, J.-K.; Zhou, G.-Y.; Chang, C.-J.; Cheng, C.-C. Label-free detection of DNA hybridization using nanopillar arrays based optical biosensor. *Sens. Actuators. B Chem.* **2014**, *194*, 10–18.

53. Wang, Y.; Ping, J.; Ye, Z.; Wu, J.; Ying, Y. Impedimetric immunosensor based on gold nanoparticles modified graphene paper for label-free detection of *Escherichia coli* O157:H7. *Biosens. Bioelectron.* **2013**, *49*, 492–498.

54. Zhang, X.; Dou, W.C.; Zhan, X.J.; Zhao, G.Y. A novel immunosensor for *Enterobacter sakazakii* based on multiwalled carbon nanotube/ionic liquid/thionine modified electrode. *Electrochim. Acta* **2012**, *61*, 73–77.

55. Dong, J.; Zhao, H.; Xu, M.; Ma, Q.; Ai, S. A label-free electrochemical impedance immunosensor based on AuNPs/PAMAM-MWCNT-Chi nanocomposite modified glassy carbon electrode for detection of *Salmonella* Typhimurium in milk. *Food Chem.* **2013**, *141*, 1980–1986.

56. Salam, F.; Uludag, Y.; Tothill, I.E. Real-time and sensitive detection of *Salmonella* Typhimurium using an automated quartz crystal microbalance (QCM) instrument with nanoparticles amplification. *Talanta* **2013**, *115*, 761–767.

57. Pérez-López, B.; Merkoçi, A. Nanomaterials based biosensors for food analysis applications. *Trends Food Sci. Technol.* **2011**, *22*, 625–639.

58. Gilmartin, N.; O'Kennedy, R. Nanobiotechnologies for the detection and reduction of pathogens. *Enzyme Microb. Technol.* **2012**, *50*, 87–95.

59. Chen, S.-H.; Wu, V.C.H.; Chuang, Y.-C.; Lin, C.-S. Using oligonucleotide-functionalized Au nanoparticles to rapidly detect foodborne pathogens on a piezoelectric biosensor. *J. Microbiol. Methods* **2008**, *73*, 7–17.

60. Yang, G.J.; Huang, J.L.; Meng, W.J.; Shen, M.; Jiao, X.A. A reusable capacitive immunosensor for detection of *Salmonella* spp. Based on grafted ethylene diamine and self-assembled gold nanoparticle monolayers. *Anal. Chim. Acta* **2009**, *647*, 159–166.

61. Biacore. Available online: https://www.biacore.com/lifesciences/index.html (accessed on 14 January 2014).

62. Leonard, P.; Hearty, S.; Quinn, J.; O'Kennedy, R. A generic approach for the detection of whole *Listeria monocytogenes* cells in contaminated samples using surface plasmon resonance. *Biosens. Bioelectron.* **2004**, *19*, 1331–1335.

63. Biosensing Instruments Ltd. Available online: www.biosensingusa.com (accessed on 16 April 2014).

64. Taylor, A.D.; Ladd, J.; Yu, Q.; Chen, S.; Homola, J.; Jiang, S. Quantitative and simultaneous detection of four foodborne bacterial pathogens with a multi-channel SPR sensor. *Biosens. Bioelectron.* **2006**, *22*, 752–758.

65. Leung, A.; Shankar, P.M.; Mutharasan, R. A review of fiber-optic biosensors. *Sens. Actuators. B Chem.* **2007**, *125*, 688–703.

66. Narsaiah, K.; Jha, S.N.; Bhardwaj, R.; Sharma, R.; Kumar, R. Optical biosensors for food quality and safety assurance—A review. *J. Food Sci. Technol.* **2012**, *49*, 383–406.

67. Ohk, S.H.; Bhunia, A.K. Multiplex fiber optic biosensor for detection of *Listeria monocytogenes*, *Escherichia coli* O157:H7 and *Salmonella* enterica from ready-to-eat meat samples. *Food Microbiol.* **2013**, *33*, 166–171.

68. Adley, C.; Ryan, M.P. Conductometric biosensor for high throughput screening of pathogens in food. In *High Throughput Screening for Food Safety Assessment: Biosensor Technologies, Hyperspectral Imaging and Practical Applications*; Bhunia, A.K., Kim, M.S., Taitt, C.R., Eds.; Elsevier: Cambridge, UK, 2014; pp. 315–326.

69. Yang, X.; Zitova, A.; Kirsch, J.; Fergus, J.W.; Overfelt, R.A.; Simonian, A.L. Portable and remote electrochemical sensing system for detection of tricresyl phosphate in gas phase. *Sens. Actuators. B Chem.* **2012**, *161*, 564–569.

70. Zhai, D.; Liu, B.; Shi, Y.; Pan, L.; Wang, Y.; Li, W.; Zhang, R.; Yu, G. Highly sensitive glucose sensor based on Pt nanoparticle/polyaniline hydrogel heterostructures. *ACS Nano* **2013**, *7*, 3540–3546.

71. Su, X.-L.; Li, Y. A QCM immunosensor for *Salmonella* detection with simultaneous measurements of resonant frequency and motional resistance. *Biosens. Bioelectron.* **2005**, *21*, 840–848.

72. Mecea, V.M. Is quartz crystal microbalance really a mass sensor? *Sens. Actuators A Phys.* **2006**, *128*, 270–277.

73. BCC Research. Surface Acoustic Wave (SAW) Devices: Technologies and Global Markets. Available online: http://www.bccresearch.com/report/download/report/ias039a (accessed on 11 May 2014).

74. Rocha-Gaso, M.I.; March-Iborra, C.; Montoya-Baides, A.; Arnau-Vives, A. Surface generated acoustic wave biosensors for the detection of pathogens: A review. *Sensors* **2009**, *9*, 5740–5769.

75. Reyes, P.I.; Li, J.; Duan, Z.; Yang, X.; Cai, Y.; Huang, Q.; Lu, Y. ZnO surface acoustic wave sensors built on zein-coated flexible food packages. *Sens. Lett.* **2013**, *11*, 539–544.

76. Sharma, H.; Mutharasan, R. Review of biosensors for foodborne pathogens and toxins. *Sens. Actuators. B Chem.* **2013**, *183*, 535–549.

77. Barthelmebs, L.; Calas-Blanchard, C.; Istamboulie, G.; Marty, J.-L.; Noguer, T. Biosenosrs as analytical tools in food fermentation industry. In *Bio-Farms for Nutraceuticals; Functional Food and Safety Control by Biosensors*; Giardia, M.T., Rea, G., Berra, B., Eds.; Landes Bioscience: Austin, TX, USA, 2010; pp. 293–307.

78. Ikeda, T.; Kato, K.; Maeda, M.; Tatsumi, H.; Kano, K.; Matsushita, K. Electrocatalytic properties of *Acetobacter aceti* cells immobilized on electrodes for the quinone-mediated oxidation of ethanol. *J. Electroanal. Chem.* **1997**, *430*, 197–204.

79. Lei, Y.; Mulchandani, P.; Chen, W.; Mulchandani, A. Direct determination of *p*-nitrophenyl substituent organophosphorus nerve agents using a recombinant *Pseudomonas putida* JS444-modified clark oxygen electrode. *J. Agric. Food Chem.* **2004**, *53*, 524–527.

80. Nakamura, H.; Suzuki, K.; Ishikuro, H.; Kinoshita, S.; Koizumi, R.; Okuma, S.; Gotoh, M.; Karube, I. A new BOD estimation method employing a double-mediator system by ferricyanide and menadione using the eukaryote *Saccharomyces cerevisiae*. *Talanta* **2007**, *72*, 210–216.

81. Kara, S.; Keskinler, B.; Erhan, E. A novel microbial BOD biosensor developed by the immobilization of *P. syringae* in micro-cellular polymers. *J. Chem. Technol. Biotechnol.* **2009**, *84*, 511–518.

82. Su, L.; Jia, W.; Hou, C.; Lei, Y. Microbial biosensors: A review. *Biosens. Bioelectron.* **2011**, *26*, 1788–1799.

83. Paniel, N.; Baudart, J.; Hayat, A.; Barthelmebs, L. Aptasensor and genosensor methods for detection of microbes in real world samples. *Methods* **2013**, *64*, 229–240.

84. Patterson, A.S.; Hsieh, K.; Soh, H.T.; Plaxco, K.W. Electrochemical real-time nucleic acid amplification: Towards point-of-care quantification of pathogens. *Trends Biotechnol.* **2013**, *31*, 704–712.

85. Pedrero, M.; Campuzano, S.; Pingarron, J.M. Electroanalytical sensors and devices for multiplexed detection of foodborne pathogen microorganisms. *Sensors* **2009**, *9*, 5503–5520.

86. Sun, H.; Zhang, Y.; Fung, Y. Flow analysis coupled with PQC/DNA biosensor for assay of *E. coli* based on detecting DNA products from PCR amplification. *Biosens. Bioelectron.* **2006**, *22*, 506–512.

87. Rodriguez, M.I.; Alocilja, E.C. Embedded DNA-polypyrrole biosensor for rapid detection of *Escherichia coli*. *IEEE Sens. J.* **2005**, *5*, 733–736.

88. Wu, V.C.; Chen, S.H.; Lin, C.S. Real-time detection of *Escherichia coli* O157:H7 sequences using a circulating-flow system of quartz crystal microbalance. *Biosens. Bioelectron.* **2007**, *22*, 2967–2975.

89. Lermo, A.; Campoy, S.; Barbe, J.; Hernandez, S.; Alegret, S.; Pividori, M.I. *In situ* DNA amplification with magnetic primers for the electrochemical detection of food pathogens. *Biosens. Bioelectron.* **2007**, *22*, 2010–2017.

90. Zhu, X.; Zhou, X.; Xing, D. Nano-magnetic primer based electrochemiluminescence-polymerase chain reaction (NMPE-PCR) assay. *Biosens. Bioelectron.* **2012**, *31*, 463–468.

91. Pöhlmann, C.; Wang, Y.; Humenik, M.; Heidenreich, B.; Gareis, M.; Sprinzl, M. Rapid, specific and sensitive electrochemical detection of foodborne bacteria. *Biosens. Bioelectron.* **2009**, *24*, 2766–2771.

92. Mairhofer, J.; Roppert, K.; Ertl, P. Microfluidic systems for pathogen sensing: A review. *Sensors* **2009**, *9*, 4804–4823.

93. Berdat, D.; Martin Rodriguez, A.C.; Herrera, F.; Gijs, M.A. Label-free detection of DNA with interdigitated micro-electrodes in a fluidic cell. *Lab Chip* **2008**, *8*, 302–308.

94. Yeung, S.W.; Lee, T.M.; Cai, H.; Hsing, I.M. A DNA biochip for on-the-spot multiplexed pathogen identification. *Nucleic Acids Res.* **2006**, *34*, e118.

95. Fang, T.H.; Ramalingam, N.; Xian-Dui, D.; Ngin, T.S.; Xianting, Z.; Lai Kuan, A.T.; Peng Huat, E.Y.; Hai-Qing, G. Real-time PCR microfluidic devices with concurrent electrochemical detection. *Biosens. Bioelectron.* **2009**, *24*, 2131–2136.

96. Jha, S.K.; Chand, R.; Han, D.; Jang, Y.C.; Ra, G.S.; Kim, J.S.; Nahm, B.H.; Kim, Y.S. An integrated PCR microfluidic chip incorporating aseptic electrochemical cell lysis and capillary electrophoresis amperometric DNA detection for rapid and quantitative genetic analysis. *Lab Chip* **2012**, *12*, 4455–4464.

97. Ahmed, M.U.; Nahar, S.; Safavieh, M.; Zourob, M. Real-time electrochemical detection of pathogen DNA using electrostatic interaction of a redox probe. *Analyst* **2013**, *138*, 907–915.

98. Ellington, A.D.; Szostak, J.W. *In vitro* selection of RNA molecules that bind specific ligands. *Nature* **1990**, *346*, 818–822.

99. Amaya-Gonzalez, S.; de-los-Santos-Alvarez, N.; Miranda-Ordieres, A.J.; Lobo-Castanon, M.J. Aptamer-based analysis: A promising alternative for food safety control. *Sensors* **2013**, *13*, 16292–16311.

100. Zelada-Guillen, G.A.; Bhosale, S.V.; Riu, J.; Rius, F.X. Real-time potentiometric detection of bacteria in complex samples. *Anal. Chem.* **2010**, *82*, 9254–9260.

101. Ning, Y.; Li, W.; Duan, Y.; Yang, M.; Deng, L. High specific DNAzyme-aptamer sensor for *Salmonella paratyphi* a using single-walled nanotubes-based dual fluorescence-spectrophotometric methods. *J. Biomol. Screen.* **2014**, *19*, 1099–1106.

102. Jyoti, A.; Vajpayee, P.; Singh, G.; Patel, C.B.; Gupta, K.C.; Shanker, R. Identification of environmental reservoirs of nontyphoidal salmonellosis: Aptamer-assisted bioconcentration and subsequent detection of *Salmonella* Typhimurium by quantitative polymerase chain reaction. *Environ. Sci. Technol.* **2011**, *45*, 8996–9002.

103. Schmelcher, M.; Loessner, M.J. Application of bacteriophages for detection of foodborne pathogens. *Bacteriophage* **2014**, *4*, e28137.

104. Tawil, N.; Sacher, E.; Mandeville, R.; Meunier, M. Bacteriophages: Biosensing tools for multi-drug resistant pathogens. *Analyst* **2014**, *139*, 1224–1236.

105. Nanosphere. Available online: www.nanosphere.us (accessed on 10 May 2014).

106. 3M. Available online: http://www.3m.com/ (accessed on 12 May 2014).

107. Neogen. Available online: www.neogen.com (accessed on 12 May 2014).

108. Serosep. Available online: www.serosep.com (accessed on 12 May 2014).

109. Veredus Laboratories. Available online: www.vereduslabs.com (accessed on 1 May 2014).

110. BIO phage PHARMA Inc. Available online: http://www.biophagepharma.net/index.php/en/ (accessed on 10 May 2014).

111. Easy Life Science. Available online: http://www.elice.fr/ (accessed on 1 May 2014).

112. STRATOPHASE™. Available online: http://www.stratophase.com/ (accessed on 11 April 2014).

113. Available Real-Time PCR Cyclers. Available online: http://cyclers.gene-quantification.info/ (accessed on 12 May 2015).

6

The Composition and Biological Activity of Honey: A Focus on Manuka Honey

José M. Alvarez-Suarez [1,2,*], Massimiliano Gasparrini [1], Tamara Y. Forbes-Hernández [1,2], Luca Mazzoni [1] and Francesca Giampieri [3,*]

[1] Department of Odontostomatologic and Specialized Clinical Sciences, Faculty of Medicine and Surgery, Polytechnic University of Marche, Avenue Ranieri 65, Ancona 60100, Italy; E-Mails: m.gasparrini@univpm.it (M.G.); tamara.forbe@gmail.com (T.Y.F.-H.); l.mazzoni@univpm.it (L.M.)

[2] Department of Nutrition and Health, International Iberoamerican University (UNINI), Avenue Adolfo Ruiz Cortines 112, Torres de Cristal L 101 A-3, Campeche 24040, Mexico

[3] Department of Agricultural, Food and Environmental Sciences, Polytechnic University of Marche, Via Ranieri 65, Ancona 60100, Italy

[*] Authors to whom correspondence should be addressed; E-Mails: j.m.alvarez@univpm.it (J.M.A.-S.); f.giampieri@univpm.it (F.G.)

Abstract: Honey has been used as a food and medical product since the earliest times. It has been used in many cultures for its medicinal properties, as a remedy for burns, cataracts, ulcers and wound healing, because it exerts a soothing effect when initially applied to open wounds. Depending on its origin, honey can be classified in different categories among which, monofloral honey seems to be the most promising and interesting as a natural remedy. Manuka honey, a monofloral honey derived from the manuka tree (*Leptospermum scoparium*), has greatly attracted the attention of researchers for its biological properties, especially its antimicrobial and antioxidant capacities. Our manuscript reviews the chemical composition and the variety of beneficial nutritional and health effects of manuka honey. Firstly, the chemical composition of manuka honey is described, with special attention given to its polyphenolic composition and other bioactive compounds, such as glyoxal and methylglyoxal. Then, the effect of manuka honey in wound treatment is described, as well as its antioxidant activity and other important biological effects.

Keywords: manuka honey; polyphenolic composition; wound treatments; antimicrobial activity

---

1. Introduction

Honey is a sweet and flavorful natural product, which is consumed for its high nutritive value and for its effects on human health, with antioxidant, bacteriostatic, anti-inflammatory and antimicrobial properties, as well as wound and sunburn healing effects [1]. Honey is produced by bees from plant nectars, plant secretions and excretions of plant-sucking insects. Concerning its nutrient profile, it represents an interesting source of natural macro- and micro-nutrients, consisting of a saturated solution of sugars, of which fructose and glucose are the main contributors, but also of a wide range of minor constituents, especially phenolic compounds [2,3]. The composition of honey is rather variable and depends primarily on its floral source; seasonal and environmental factors can also influence its composition and its biological effects. Several studies have shown that the antioxidant potential of honey is strongly correlated not only with the concentration of total phenolics present, but also with the color, with dark colored honeys being reported to have higher total phenolic contents and, consequently, higher antioxidant capacities [3–6].

According to the origin, honey can be classified in different categories as follows: (1) blossom honey, obtained predominantly from the nectar of flowers (as opposed to honeydew honey); (2) honeydew honey, produced by bees after they collect "honeydew" (secretions of insects belonging to the genus, *Rhynchota*), which pierce plant cells, ingest plant sap and then secrete it again; (3) monofloral honey, in which the bees forage predominantly on one type of plant and which is named according to the plant; and (4) multifloral honey (also known as polyfloral) that has several botanical sources, none of which is predominant, e.g., meadow blossom honey and forest honey.

It is has been suggested that many of the medicinal properties of plants can be transmitted through honey, so that honey could be used as a vehicle for transporting plant medicinal properties [3]. Within monofloral honey, manuka honey, a dark honey, has greatly attracted the attention of the international scientific community for its biological properties, especially for its antimicrobial and antioxidant capacities. This honey is derived from the manuka tree, *Leptospermum scoparium*, of the Myrtaceae family, which grows as a shrub or a small tree throughout New Zealand and eastern Australia [7]. In traditional medicine, different extracts of the manuka tree are used as sedatives and wound-healing remedies. Moreover, manuka honey itself has long been employed for clearing up infections, including abscesses, surgical wounds, traumatic wounds, burns and ulcers of different etiology [8]. Currently, the main bioactive compounds in manuka honey and the mechanisms responsible for their biological activities are being studied. These studies would support the increased use of manuka honey in skin medicine, and they can also be the basis for the isolation and purification of compounds for the development of bio-pharmaceutical products with antimicrobial properties and wound healing properties; these new findings could represent an added economic value that can favor also the beekeepers in their productions.

This review focuses on the phytochemical composition of manuka honey and on its biological effects. An overview of the most abundant phytochemicals is presented, with particular attention to recent evidence on its antimicrobial activity and its impact on wound treatments, as well as on its antioxidant capacity.

2. Chemical Composition

Polyphenolic characterization has proven to be suitable for the differentiation of the floral origin of honeys [9], and therefore, flavonoids could represent a valid botanical marker for honey [10], being closely related with their antioxidant capacity. The qualitative and quantitative difference in flavonoid contents of manuka honey determined in diverse studies may represent the result of the different extraction and detection methods applied, and this limit makes the data available in the literature difficult to compare. The major compounds identified are represented in Table 1. Several studies have determined that the major flavonoids in manuka honey are: pinobanksin, pinocembrin and chrysin, while luteolin, quercetin, 8-methoxykaempferol, isorhamnetin, kaempferol and galangin have been also identified in minor concentration [11–13].

Regarding phenolic acids and volatile norisoprenoids constituents, Oelschlaegel *et al.* [13] detected different profiles in manuka honey attributed to three chemotypes of *L. scoparium* in New Zealand. The first group was characterized by high levels of 4-hydroxybenzoic acid, dehydrovomifoliol and benzoic acid yields, the second one by high concentrations of kojic acid and 2-methoxybenzoic acid and the third group by high contents of syringic acid, 4-methoxyphenyllactic acid and methyl syringate. According to the determined average amounts, phenylacetic acid, phenyllactic acid, 4-methoxyphenyllactic acid, leptosin and methyl syringate were the dominating compounds [14,15].

Methyl syringate (MSYR) and leptosin (the novel glycoside of MSYR, methyl syringate 4-*O*-β-D-gentiobiose) (Figure 1) are the active compounds from manuka honey to which its myeloperoxidase (MPO)-activity inhibition is ascribed. Although the biological activities and biosynthetic pathway/origin of the glycoside are still unknown, it may be a good chemical marker for the purity of manuka honey [7].

Other constituents of interest found in manuka honey are: different 1,2-dicarbonyl compounds, such as glyoxal (GO), 3-deoxyglucosulose (3-DG) and methylglyoxal (MGO). These compounds are typically formed during the Maillard reaction or caramelization reactions as degradation products from reducing carbohydrates, and they have been identified as important contributors to the non-peroxide antibacterial activity [13,16,17].

From the nutritional point of view, the physiological significance resulting from the uptake of MGO and other 1,2-dicarbonyl compounds must be a topic of further investigations. MGO and glycation compounds resulting from the reaction of MGO with amino acid side chains of lysine or arginine, respectively, have been identified *in vivo* and are associated with complications of diabetes and some neurodegenerative diseases, although the role of these compounds in the pathogenesis of different diseases have not yet been fully understood [16].

Table 1. Most common compounds identified in manuka honey.

Phenolic Acid and Flavonoids	Ref.	Other Compounds	Ref.
Caffeic acid	[12,13]	Phenyllactic acid	[13]
Isoferulic acid	[12]	4-Methoxyphenolactic acid	[13]
p-Coumaric acid	[12]	Kojic acid	[13]
Gallic acid	[13,17]	5-Hydroxymethylfurfural	[13]
4-Hydrobenzoic acid	[13]	2-Methoxybenzoic acid	[13]
Syringin acid	[13]	Phenylacetic acid	[13]
Quercetin	[12,17]	Methyl syringate	[13]
Luteolin	[12,13]	Dehydrovomifoliol	[13]
8-Methoxykaempferol	[12]	Leptosin	[13]
Pinocembrin	[12]	Glyoxal	[13,16]
Isorhamnetin	[12,17]	Methylglyoxal	[13,16]
Kaempferol	[12]	3-Deoxyglucosulose	[13,16]
Chrysin	[12]	-	-
Galangin	[12]	-	-
Pinobanksin	[12]	-	-

Figure 1. Chemical structures of methyl syringate and leptosin.

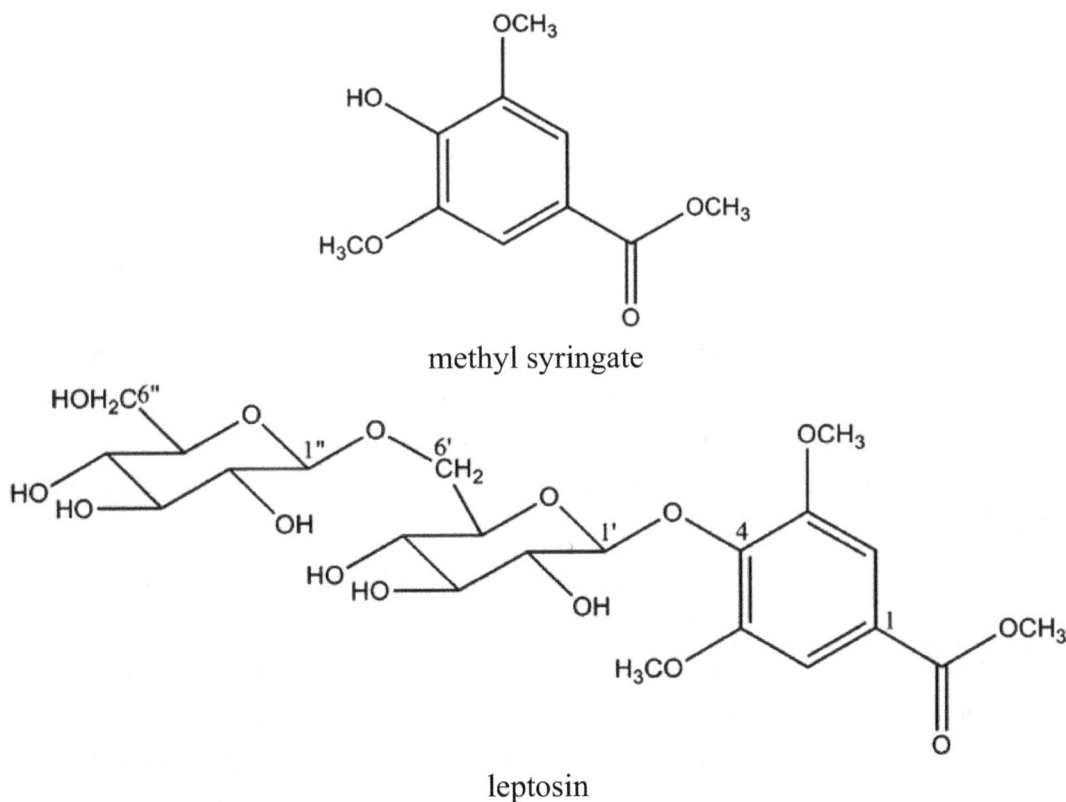

methyl syringate

leptosin

3. Use of Manuka Honey in Wound Treatments

The importance of honey in the field of wound treatments has been well known since ancient times. This healing property is related to the antioxidant and antibacterial activity that honey offers, maintaining a moist wound condition, and to the high viscosity that provides a protective barrier on the wound, preventing microbial infection. Its immunological activity is relevant also for wound repair, exerting the same time pro- and anti-inflammatory effects [18–23]. Normal wound healing is a complex process composed of a series of overlapping events (coagulation, inflammation, cell proliferation, tissue remodeling) in which the damaged tissue is gradually removed and replaced by restorative tissues [24]. While normal inflammation resolves within 1–2 days as the neutrophil number decreases, the accumulation of these cells in the wound site contributes to a disordered network of regulatory cytokines, leaving the wound in a chronic state of inflammation [25]. In these chronic wounds, bacterial cells predominantly exist as biofilms, where cells are embedded within a matrix of polysaccharides and other components that limit the availability of antibiotics for wound healing. Furthermore, the emergence of bacterial resistance to multiple antibiotics has worsened the problem of chronic wound biofilm treatment [26].

Current therapeutic products widely used in wound care (silver sulfadiazine (SSD), hydrogel, hydrocolloid and alginate dressings impregnated with silver) are considered useful for limiting bacterial infections, even if excessive use of ionic silver has generated some concern regarding the development of bacterial resistance [27,28]; this situation, in recent years, has stimulated modern medicine to focus attention on natural products with antimicrobial activity and their use in clinical practice. The low cost and absence of the antimicrobial resistance risk of natural products, such as honey, aloe vera or curcumin, are the major arguments for implementing natural products in wound treatment [29]. Although it is an ancient topical treatment for wounds, honey has been currently established in conventional medicine as a licensed medical device, either combined into sterile dressings or sterilized in tubes [30].

The healing time decrease after honey treatment can be explained through a dual effect on the inflammatory response. Firstly, honey prevents a prolonged inflammatory response suppressing the production and propagation of inflammatory cells at the wound site; secondly, it stimulates the production of proinflammatory cytokine, allowing normal healing to occur [31] and stimulating the proliferation of fibroblasts and epithelial cells [32,33]. The effect of honey and its components on the production of inflammatory cytokines has been evaluated in primary human monocytes cells [34]. In these studies, it was shown that manuka honey stimulated the production of inflammatory cytokines TNF-α, IL-1β or IL-6 via a TLR4-dependent mechanism. For the first time, a 5.8-kDa component responsible for cytokine induction in human monocytes via TLR4 was isolated from manuka honey [35].

Microorganisms that colonize a burn wound originate from the patient's gastrointestinal and respiratory flora, from endogenous skin or from contaminated external sources (soil, water, air) [36]. The topical application of honey rapidly clears wound infection, promoting the healing process of deep surgical infected wounds [37–39], also when they do not respond to conventional antibiotic and antiseptic therapy [37]. Furthermore, in burn wounds, honey application decreases the wound area, exerts an antibacterial effect and promotes better re-epithelialization compared to hydrofiber silver or SSD treatment. Moreover, the anti-inflammatory action of honey decreases damage caused by free radicals that result from inflammation, preventing further necrosis [40].

The antibacterial nature of honey depends on different factors acting singularly or synergistically, the most salient of which are phenolic compounds, wound pH, H_2O_2, pH of honey and osmotic pressure exerted by the honey itself [3]. It has been documented that the pronounced antibacterial activity of manuka honey directly originates from the MGO it contains (Figure 2A) [16].

This non-peroxide antibacterial activity due to the presence of MGO is called the unique manuka factor (UMF) [16].

Figure 2. (**A**) Chemical structures of the methylglyoxal (MGO). (**B**) Homology model of defensin-1 from *Apis mellifera*. The model of the mature protein (residues 44–94) was obtained using the experimentally-resolved structure of lucifensin from *Lucilia sericata* (PDB ID: 2LLD) as a template. Alignment and modeling was performed using the Swiss Model server [41]; the figure has been obtained through PyMOL Molecular Graphics System, Version 1.5.0.4, Schrödinger, LLC (Portland, OR, USA).

A   B

Other antimicrobial compounds in honeys include bee defensin-1 (Figure 2B), various phenolic compounds and complex carbohydrates [1,2]. The combination of these diverse assaults may account for the inability of bacteria to develop resistance to honey, in contrast to the rapid induction of resistance observed with conventional single-component antibiotics [42,43]. A few studies have examined the antimicrobial effect of manuka honey, showing that it is active against a range of bacteria, including Group A *Streptococcus pyogenes*, *Streptococcus mutans*, *Proteus mirabilis*, *Pseudomonas aeruginosa*, *Enterobacter cloacae* and *Staphylococcus aureus* [44–47]. A list of microorganisms that have been found to be sensitive to manuka honeys is shown in Table 2. Furthermore, no resistant bacteria (*Escherichia coli*, MRSA, *Pseudomonas aeruginosa* and *Staphylococcus epidermidis*) have been isolated after exposure of wound isolates to sub-inhibitory concentrations of manuka honey [42,43]. This seems to be very likely due, at least in part, to differences in the levels of the principle antibacterial components in the honey, MGO and hydrogen peroxide, which varies with the floral and geographic source of nectar, honey storage time and conditions and any other possible treatment that could affect it. Anti-biofilm activity was highest in the honey blend that contained the highest level of manuka-derived honey; the effectiveness of the different manuka-type honeys tested increased with MGO content, although the same level of MGO, with or without sugar, could not eradicate biofilms. This suggests that additional factors in these manuka-type honeys are responsible for their potent anti-biofilm activity [48].

Table 2. List of microorganisms that have been found to be sensitive to manuka honeys [49].

Gram Positive Strains	Gram Negative Strains
Streptococcus pyogenes	*Stenotrophomonas maltophilia*
Coagulase negative staphylococci	*Acinetobacter baumannii*
Methicillin-resistant *Staphylococcus aureus (MRSA)*	*Salmonella enterica* serovar *typhi*
Streptococcus agalactiae	*Pseudomonas aeruginosa*
Staphylococcus aureus	*Proteus mirabilis*
Coagulase-negative *Staphylococcus aureus (CONS)*	*Shigella flexneri*
Hemolytic streptococci	*Escherichia coli*
Enterococcus	*Enterobacter cloacae*
Streptococcus mutans	*Shigella sonnei*
Streptococcus sobrinus	*Salmonella typhi*
Actinomyces viscosus	*Klebsiella pneumonia*
-	*Stenotrophomonas maltophilia*
-	*Burkholderia cepacia*
-	*Helicobacter pylori*
-	*Campylobacter* spp.
-	*Porphyromonas gingivalis*

Manuka honey has been shown to eradicate methicillin-resistant *Staphylococcus aureus* (MRSA) from colonized wounds and to inhibit MRSA *in vitro* by interrupting cell division. Furthermore, the presence of manuka honey restores MRSA susceptibility to oxacillin; molecular analysis indicated that it also affects the regulation of the mecR1 gene, possibly accounting for the restored susceptibility [30]. In another study, a synergistic effect between rifampicin and commercially available FDA-approved manuka honey (Medihoney, Medihoney Ltd, Slough, United Kingdom) was demonstrated on clinical *S. aureus* isolates, including MRSA strains. Unlike with rifampicin alone, in which resistance was observed after overnight incubation on plates, the combination of Medihoney and rifampicin maintained the susceptibility of *S. aureus* to rifampicin [50]. Manuka honey, therefore, seems to offer real potential in providing novel synergistic combinations with antibiotics for treating wound infections of multidrug-resistant (MDR) bacteria. It is interesting to note that the antibiotics that have shown synergy with manuka honey are from different antibiotic classes, which inhibit distinct targets, such as the 30 S ribosome, RNA polymerase, membranes and penicillin binding proteins. This finding supports the idea that honey is a complex substance, perhaps with multiple active components that affect more than one cellular target site [30].

Manuka honey is also known to have a relatively low pH (3.5–4.5), which, besides inhibiting microbial growth, stimulates the bactericidal actions of macrophages and, in chronic wounds, reduces protease activity and increases fibroblast activity and oxygenation [51–53]. Growth factors, such as TGF-β, are known to become physiologically active when subjected to an acid treatment, and the use of Medihoney demonstrates a further increase in cellular activity. This impact has been reported in the hDF-based studies and in an *in vitro* wound healing assay study, where Medihoney supplements resulted in statistically significant increases in cell proliferation and migration [25].

Finally, manuka honey has been shown to specifically decrease the inflammatory response associated with ulcerative colitis, an inflammatory intestine disease characterized by an overexpression of inflammatory cells, in embryonic kidney cell lines. The anti-inflammatory effect by the manuka honey was strongest in the presence of the Pam3CSK4 ligand, indicating that the honeys act through the TLR1/TLR2 signaling pathway. The anti-inflammatory activity of manuka honeys is therefore pathway specific [31].

4. Antioxidant Activity

In addition to antibacterial activity, honeys are known to possess strong antioxidant capacity, which acts in modulating free radical production, thus protecting cell components from their harmful action [54,55].

Manuka honey contains a high amount of phenolic compounds [14,15], as well as other phenolic compounds that have been identified with a potent capacity to reduce free radicals, providing a relevant antioxidant capacity [56,57]. For its relevant bioactive properties, it has often been used in different studies as the "gold standard" [8] to test and evaluate the antioxidant capacity of different kinds of honey from different botanical and geographical origins. Manuka honey, in fact, exhibits the highest value in terms of phenolic content and antioxidant capacity, for example compared to acacia, wild carrot and Portobello honeys [58,59], obtained, respectively, from Germany, Algeria, Saudi Arabia and Scotland. Similar results are obtained with Malaysian monofloral honeys [56] and Tualang honey, a Malaysian multifloral jungle honey [60]. The scavenger role of manuka honey against superoxide anion radicals has also been investigated through electronic paramagnetic resonance [54,61]; the results proved that the quenching properties of manuka honey could be attributed to methyl syringate [62]. Finally, manuka honey seems to exert a protective role against oxidative damage also in an *in vivo* model [57], reducing DNA damage, the malondialdehyde level and glutathione peroxidase activity in the liver of both young and middle-aged groups of rats. These effects could be mediated through the modulation of antioxidant enzyme activities (such as catalase) and through the high antioxidant capacity of its relevant total phenolic content. The results obtained suggest a possible use of manuka honey as an alternative natural supplement to improve the physiological oxidative status.

5. Other Effects

In addition to its antimicrobial and antioxidant activities, recent studies demonstrated that honey can exert anti-proliferative effects against cancer cells [62–64]. These anticancer properties can involve different processes: (1) the apoptosis of cancer cells through the depolarization of the mitochondrial membrane, (2) the inhibition of cyclooxygenase-2 by various constituents (like flavonoids), (3) the release of cytotoxic H_2O_2 and (4) the scavenging of ROS and have been correlated with the phytochemical compounds [65]. Manuka honey has been shown to possess a potent anti-proliferative effect on murine melanoma (B16.F1), colorectal carcinoma (CT26) and human breast cancer (MCF-7) cell lines in a time- and dose-dependent manner [8]. The main mechanism by which it exerts such an anti-proliferative effect is through the activation of mitochondrial apoptotic pathways, involving the stimulation of the initiator, caspase-9, which determines the activation of the executioner, caspase-3 [65]. Moreover, it induces apoptosis via the activation of PARP, the induction of DNA fragmentation and the loss of Bcl-2 expression. *In vivo*, manuka honey is also effective in: (1) decreasing the tumor volume and increasing the apoptosis of tumor cells in a mouse melanoma model; and (2) reducing colonic inflammation in inflammatory bowel disease in rats, restoring lipid peroxidation and improving antioxidant parameters [65].

Finally, in healthy individuals, manuka honey UMF 20+ has been evaluated for its safety: its consumption showed: (1) no significant effect on the allergic status of the subjects; (2) no detrimental effect in relation to advanced glycation end products, which are implicated in a number of serious diseases, including renal disease, diabetes, neurodegenerative disease and heart disease; and (3) no change in gut microbiota homeostasis, confirming its safety for healthy individuals [66].

6. Conclusions

Besides its main components, manuka honey contains a large number of other constituents in small and trace amounts, able to exert numerous nutritional and biological effects, like antimicrobial and antioxidant activities. The above information shows that in microbiological and clinical tests, manuka honey offers advantages in controlling bacterial growth and in the treatment of several health problems. The easiness of administration in wound treatment and the absence of antibiotic resistance, which instead is found with conventional antibiotics, are important characteristics for the use of this honey in the treatment of clinical wounds.

Acknowledgments

The authors wish to thank Monica Glebocki for extensively editing the manuscript.

Conflicts of Interest

The authors declare no conflict of interest.

References

1. Alvarez-Suarez, J.M.; Giampieri, F.; Battino, M. Honey as a source of dietary antioxidants: Structures, bioavailability and evidence of protective effects against human chronic diseases. *Curr. Med. Chem.* **2013**, *20*, 621–638.

2. Bogdanov, S.; Jurendic, T.; Sieber, R.; Gallmann, P. Honey for nutrition and health: A review. *Am. J. Coll. Nutr.* **2008**, *27*, 677–689.

3. Alvarez-Suarez, J.M.; Tulipani, S.; Romandini, S.; Bertoli, E.; Battino, M. Contribution of honey in nutrition and human health: A review. *Mediterr. J. Nutr. Metab.* **2010**, *3*, 15–23.

4. Alvarez-Suarez, J.M.; Tulipani, S.; Díaz, D.; Estevez, Y.; Romandini, S.; Giampieri, F.; Damiani, E.; Astolfi, P.; Bompadre, S.; Battino, M. Antioxidant and antimicrobial capacity of several monofloral Cuban honeys and their correlation with color, polyphenol content and other chemical compounds. *Food Chem. Toxicol.* **2010**, *48*, 2490–2499.

5. Alvarez-Suarez, J.M.; González-Paramás, A.M.; Santos-Buelga, C.; Battino, M. Antioxidant characterization of native monofloral Cuban honeys. *J. Agric. Food Chem.* **2010**, *58*, 9817–9824.

6. Alvarez-Suarez, J.M.; Giampieri, F.; González-Paramás, A.M.; Damiani, E.; Astolfi, P.; Martinez-Sanchez, G.; Bompadre, S.; Quiles, J.L.; Santos-Buelga, C.; Battino, M. Phenolics from monofloral honeys protect human erythrocyte membranes against oxidative damage. *Food Chem. Toxicol.* **2012**, *50*, 1508–1516.

7. Kato, Y.; Umeda, N.; Maeda, A.; Matsumoto, D.; Kitamoto, N.; Kikuzaki, H. Identification of a novel glycoside, leptosin, as a chemical marker of manuka honey. *J. Agric. Food Chem.* **2012**, *60*, 3418–3423.

8. Patel, S.; Cichello, S. Manuka honey: An emerging natural food with medicinal use. *Nat. Prod. Bioprospect.* **2013**, *3*, 121–128.

9. Anklam, E. A review of the analytical methods to determine the geographical and botanical origin of honey. *Food Chem.* **1998**, *63*, 549–562.

10. Tomas-Barberan, F.A.; Ferreres, F.; Garcia-Viguera, C.; Tomas-Lorente, F. Flavonoids in honey of different geographical origin. *Z. Lebensm. Unters. Forsch.* **1993**, *196*, 38–44.

11. Yaoa, L.; Jiang, Y.; Singanusong, R.; Datta, N.; Raymont, K. Phenolic acids in Australian Melaleuca, Guioa, Laphostemon, Banksia and Helianthus honeys and their potential for floral authentication. *Food Res. Internat.* **2005**, *38*, 651–658.

12. Chan, C.W.; Deadman, B.J.; Manley-Harris, M.; Wilkins, A.L.; Alber, D.G.; Harry, E. Analysis of the flavonoid component of bioactive New Zealand manuka (*Leptospermum scoparium*) honey and the isolation, characterisation and synthesis of an unusual pyrrole. *Food Chem.* **2013**, *141*, 1772–1781.

13. Oelschlaegel, S.; Gruner, M.; Wang, P.; Boettcher, A.; Koelling-Speer, I.; Speer, K. Classification and characterization of Manuka honeys based on phenolic compounds and methylglyoxal. *J. Agric. Food Chem.* **2012**, *60*, 7229–7237.

14. Tuberoso, C.I.; Bifulco, E.; Jerkovic, I.; Caboni, P.; Cabras, P.; Floris, I. Methyl syringate: A chemical marker of asphodel (*Asphodelus microcarpus* Salzm. et Viv.) monofloral honey. *J. Agric. Food Chem.* **2009**, *57*, 3895–3900.

15. Stephens, J.M.; Schlothauer, R.C.; Morris, B.D.; Yang, D.; Fearnley, L.; Greenwood, D.R.; Loomes, K.M. Phenolic composition and methylglyoxal in some New Zealand manuka and kanuka honeys. *Food Chem.* **2010**, *120*, 78–86.

16. Mavric, E.; Wittmann, S.; Barth, G.; Henle, T. Identification and quantification of methylglyoxal as the dominant antibacterial constituent of manuka (*Leptospermum scoparium*) honeys from New Zealand. *Mol. Nutr. Food Res.* **2008**, *52*, 483–489.

17. Adams, C.J.; Manley-Harris, M.; Molan, P.C. The origin of methylglyoxal in New Zealand manuka (*Leptospermum scoparium*) honey. *Carbohydr. Res.* **2009**, *344*, 1050–1053.

18. Mandal, M.D.; Mandal, S. Honey: Its medicinal property and antibacterial activity. *Asian Pac. J. Trop. Biomed.* **2011**, *1*, 154–160.

19. Tonks, J.; Cooper, R.A.; Jones, K.P.; Blair, S.; Parton, J.; Tonks, A. Honey stimulates inflammatory cytokine production from monocytes. *Cytokine* **2003**, *21*, 242–247.

20. Majtan, J.; Kovacova, E.; Bílikova, K.; Simuth, J. The immunostimulatory effect of the recombinant apalbumin 1-major honeybee royal jelly protein-on TNFα release. *Int. Immunopharmacol.* **2006**, *6*, 269–278.

21. Van den Berg, A.J.; van den Worm, E.; van Ufford, H.C.; Halkes, S.B.; Hoekstra, M.J.; Beukelman, C.J. An *in vitro* examination of the antioxidant and anti-inflammatory properties of buckwheat honey. *J. Wound Care* **2008**, *17*, 172–174.

22. Ahmad, A.; Khan, R.A.; Mesaik, M.A. Anti-inflammatory effect of natural honey on bovine thrombin-induced oxidative burst in phagocytes. *Phytother. Res.* **2009**, *23*, 801–808.

23. Majtan, J.; Kumar, P.; Majtan, T.; Walls, A.F.; Klaudiny, J. Effect of honey and its major royal jelly protein 1 on cytokine and MMP-9 mRNA transcripts in human keratinocytes. *Exp. Dermatol.* **2010**, *19*, 73–79.

24. Falanga, V. Wound healing and its impairment in the diabetic foot. *Lancet* **2005**, *366*, 1736–1743.

25. Sell, S.A.; Wolfe, P.S.; Spence, A.J.; Rodriguez, I.A.; McCool, J.M.; Petrella, R.L.; Garg, K.; Ericksen, J.J.; Bowlin, G.L. A Preliminary study on the potential of manuka honey and platelet-rich plasma in wound healing. *Int. J. Biomater.* **2012**, *2012*, 313781; doi:10.1155/2012/313781.

26. Engemann, J.J.; Carmeli, Y.; Cosgrove, S.E.; Fowler, V.G.; Bronstein, M.Z.; Trivette, S.L.; Briggs, J.P.; Sexton, D.J.; Kaye, K.S. Adverse clinical and economic outcomes attributable to methicillin resistance among patients with *Staphylococcus aureus* surgical site infection. *Clin. Infect. Dis.* **2003**, *36*, 592–598.

27. Percival, S.L.; Woods, E.; Nutekpor, M.; Bowler, P.; Radford, A.; Cochrane, C. Feature: Prevalence of silver resistance in bacteria isolated from diabetic foot ulcers and efficacy of silver-containing wound dressings. *Ostomy Wound Manag.* **2008**, *54*, 30–40.

28. Loh, J.V.; Percival, S.L.; Woods, E.J.; Williams, N.J.; Cochrane, C. Silver resistance in MRSA isolated from wound and nasal sources in humans and animals. *Int. Wound J.* **2009**, *6*, 32–38.

29. Davis, S.C.; Perez, R. Cosmeceuticals and natural products: Wound healing. *Clin. Dermatol.* **2009**, *27*, 502–506.

30. Jenkins, R; Cooper, R. Improving antibiotic activity against wound pathogens with manuka honey *in vitro*. *PLoS One* **2012**, *7*, e45600.

31. Tomblin, V.; Ferguson, L.R.; Han, D.Y.; Murray, P.; Schlothauer, R. Potential pathway of anti-inflammatory effect by New Zealand honeys. *Int. J. Gen. Med.* **2014**, *7*, 149–158.

32. Visavadia, B.G.; Honeysett, J.; Danford, M.H. Manuka honey dressing: An effective treatment for chronic wound infections. *Br. J. Oral Maxillofac. Surg.* **2008**, *46*, 55–56.

33. Tonks, A.; Cooper, R.A.; Price, A.J.; Molan, P.C.; Jones, K.P. Stimulation of TNF-alpha release in monocytes by honey. *Cytokine* **2001**, *14*, 240–242.

34. Riches, D.W. Macrophage involvement in wound repair, remodeling and fibrosis. In *The Molecular and Cellular Biology of Wound Repair*; Clarke, R., Ed.; Plenum Press: New York, NY, USA, 1996; pp. 95–141.

35. Tonks, J.; Dudley, E.; Porter, N.G.; Parton, J.; Brazier, J.; Smith, E.L.; Tonks, A. A 5.8-kDa component of manuka honey stimulates immune cells via TLR4. *J. Leukoc. Biol.* **2007**, *82*, 1147–1155.

36. Hern, T.T.; Rosliza, A.R.; Siew, H.G.; Ahmad, S.H.; Siti, A.H.; Siti, A.S.; Kirnpal-Kaur, B.S. The antibacterial properties of Malaysian tualang honey against wound and enteric microorganisms in comparison to manuka honey. *BMC Complement. Altern. Med.* **2009**, *9*, 1–8.

37. Ahmed, A.K.; Hoekstra, M.J.; Hage, J.; Karim, R.B. Honey-medicated dressing: Transformation of an ancient remedy into modern therapy. *Ann. Plast. Surg.* **2003**, *50*, 143–148.

38. Shupp, J.W.; Nasabzadeh, T.J.; Rosenthal, D.S.; Jordan, M.H.; Fidler, P.; Jeng, J.C. A review of the local pathophysiologic bases of burn wound progression. *J. Burn Care Res.* **2010**, *31*, 849–873.

39. Nisbet, H.O.; Nisbet, C.; Yarim, M.; Guler, A.; Ozak, A. Effects of three types of honey on cutaneous wound healing. *Wounds* **2010**, *22*, 275–283.

40. Molan, P.C. Potential of honey in the treatment of wounds and burns. *Am. J. Clin. Dermatol.* **2001**, *2*, 13–19.

41. Swiss Model Server. Available online: http://swissmodel.expasy.org (accessed on 8 May 2014).

42. Blair, S.E.; Cokcetin, N.N.; Harry, E.J.; Carter, D.A. The unusual antibacterial activity of medical-grade Leptospermum honey: Antibacterial spectrum, resistance and transcriptome analysis. *Eur. J. Clin. Microbiol. Infect. Dis.* **2009**, *28*, 1199–1208.

43. Cooper, R.A.; Jenkins, L.; Henriques, A.F.; Duggan, R.S.; Burton, N.F. Absence of bacterial resistance to medical-grade manuka honey. *Eur. J. Clin. Microbiol. Infect. Dis.* **2010**, *29*, 1237–1241.

44. Alandejani, T.; Marsan, J.G.; Ferris, W.; Slinger, R.; Chan, F. Effectiveness of honey on *Staphylococcus aureus* and *Pseudomonas aeruginosa* biofilms. *Otolaryngol. Head Neck Surg.* **2008**, *139*, 114–118.

45. Maddocks, S.E.; Jenkins, R.E.; Rowlands, R.S.; Purdy, K.J.; Cooper, R.A. Manuka honey inhibits adhesion and invasion of medically important wound bacteria *in vitro*. *Future Microbiol.* **2013**, *8*, 1523–1536.

46. Maddocks, S.E.; Lopez, M.S.; Rowlands, R.S.; Cooper, R.A. Manuka honey inhibits the development of *Streptococcus pyogenes* biofilms and causes reduced expression of two fibronectin binding proteins. *Microbiology* **2012**, *158*, 781–790.

47. Majtan, J.; Bohova, J.; Horniackova, M.; Klaudiny, J.; Majtan, V. Anti-biofilm effects of honey against wound pathogens proteus mirabilis and enterobacter cloacae. *Phytother. Res.* **2013**, *28*, 69–75.

48. Lu, J.; Turnbull, L.; Burke, C.M.; Liu, M.; Carter, D.A.; Schlothauer, R.C.; Whitchurch, C.B.; Harry, E.J. Manuka-type honeys can eradicate biofilms produced by *Staphylococcus aureus* strains with different biofilm-forming abilities. *Peer J.* **2014**, *25*, e326, doi:10.7717/peerj.326.

49. Ahmed, S.; Othman, H.N. Review of the medicinal effects of Tualang honey and a comparison with manuka honey. *Malays. J. Med. Sci.* **2013**, *20*, 6–13.

50. Muller, P.; Alber, D.G.; Turnbull, L.; Schlothauer, R.C.; Carter, D.A.; Whitchurch, C.B.; Harry, E.J. Synergism between medihoney and rifampicin against methicillin-resistant *Staphylococcus aureus* (MRSA). *PLoS One* **2013**, *8*, e57679.

51. Lusby, P.E.; Coombes, A.; Wilkinson, J.M. Honey: A potent agent for wound healing? *J. Wound Ostomy Cont. Nurs.* **2002**, *29*, 295–300.

52. Al-Waili, N.S.; Salom, K.; Al-Ghamdi, A.A. Honey for wound healing, ulcers, and burns; data supporting its use in clinical practice. *ScientificWorldJournal* **2011**, *11*, 766–787.

53. Gethin, G.T.; Cowman, S.; Conroy, R.M. The impact of Manuka honey dressings on the surface pH of chronic wounds. *Int. Wound J.* **2008**, *5*, 185–194.

54. Henriques, A.; Jackson, S.; Cooper, R.; Burton, N. Free radical production and quenching in honeys with wound healing potential. *J. Antimicrob. Chemother.* **2006**, *58*, 773–777.

55. Alzahrani, H.A.; Boukraa, L.; Bellik, Y.; Abdellah, F.; Bakhotmah, B.A.; Kolayli, S.; Sahin, H. Evaluation of the antioxidant activity of three varieties of honey from different botanical and geographical origins. *Glob. J. Health Sci.* **2012**, *4*, 191–196.

56. Moniruzzaman, M.; Sulaiman, S.A.; Khalil, M.I.; Gan, S.H. Evaluation of physicochemical and antioxidant properties of sourwood and other Malaysian honeys: A comparison with manuka honey. *Chem. Cent. J.* **2013**, *7*, 138, doi:10.1186/1752-153X-7-138.

57. Jubri, Z.; Rahim, N.B.; Aan, G.J. Manuka honey protects middle-aged rats from oxidative damage. *Clinics* **2013**, *68*, 1446–1454.

58. Alzahrani, H.A.; Alsabehi, R.; Boukraâ, L.; Abdellah, F.; Bellik, Y.; Bakhotmah, B.A. Antibacterial and antioxidant potency of floral honeys from different botanical and geographical origins. *Molecules* **2012**, *17*, 10540–10549.

59. Schneider, M.; Coyle, S.; Warnock, M.; Gow, I.; Fyfe, L. Anti-microbial activity and composition of manuka and portobello honey. *Phytother. Res.* **2013**, *27*, 1162–1168.

60. Khalil, M.I.; Alam, N.; Moniruzzaman, M.; Sulaiman, S.A.; Gan, S.H. Phenolic acid composition and antioxidant properties of Malaysian honeys. *J. Food Sci.* **2011**, *76*, 921–928.

61. Inoue, K.; Murayama, S.; Seshimo, F.; Takeba, K.; Yoshimura, Y.; Nakazawa, H. Identification of phenolic compound in manuka honey as specific superoxide anion radical scavenger using electron spin resonance (ESR) and liquid chromatography with coulometric array detection. *J. Sci. Food Agric.* **2005**, *85*, 872–878.

62. Fukuda, M.; Kobayashi, K.; Hirono, Y.; Miyagawa, M.; Ishida, T.; Ejiogu, E.C.; Sawai, M.; Pinkerton, K.E.; Takeuchi, M. Jungle honey enhances immune function and antitumor activity. *Evid. Based Complement. Alternat. Med.* **2011**, *2011*, 908743, doi:10.1093/ecam/nen086.

63. Ghashm, A.A.; Othman, N.H.; Khattak, M.N.; Ismail, N.M.; Saini, R. Antiproliferative effect of Tualang honey on oral squamous cell carcinoma and osteosarcoma cell lines. *BMC Complement. Altern. Med.* **2010**, *10*, 49, doi:10.1186/1472-6882-10-49.

64. Swellam, T.; Miyanaga, N.; Onozawa, M.; Hattori, K.; Kawai, K.; Shimazui, T.; Akaza, H. Antineoplastic activity of honey in an experimental bladder cancer implantation model: *In vivo* and *in vitro* studies. *Int. J. Urol.* **2003**, *10*, 213–219.

65. Forbes-Hernández, T.Y.; Giampieri, F.; Gasparrini, M.; Mazzoni, L.; Quiles, J.L.; Alvarez-Suarez, J.M.; Battino, M. The effects of bioactive compounds from plant foods on mitochondrial function: A focus on apoptotic mechanisms. *Food Chem. Toxicol.* **2014**, *68*, 154–182.

66. Wallace, A.; Eady, S.; Miles, M.; Martin, H.; McLachlan, A.; Rodier, M.; Willis, J.; Scott, R.; Sutherland, J. Demonstrating the safety of manuka honey UMF 20 in a human clinical trial with healthy individuals. *Br. J. Nutr.* **2010**, *103*, 1023–1028.

7

Effect of Particle Orientation during Thermal Processing of Canned Peach Halves: A CFD Simulation

Adreas Dimou [†]**, Nikolaos G. Stoforos** [†,]*** and Stavros Yanniotis** [†]

Department of Food Science and Human Nutrition, Agricultural University of Athens, Athens 11855, Greece; E-Mails: dimouadreas@gmail.com (A.D.); yanniotis@aua.gr (S.Y.)

[†] These authors contributed equally to this work.

* Author to whom correspondence should be addressed; E-Mail: stoforos@aua.gr

Abstract: The objective of this work was to apply Computational Fluid Dynamics (CFD) to study the effect of particle orientation on fluid flow, temperature evolution, as well as microbial destruction, during thermal processing of still cans filled with peach halves in sugar syrup. A still metal can with four peach halves in 20% sugar syrup was heated at 100 °C for 20 min and thereafter cooled at 20 °C. Infinite heat transfer coefficient between heating medium and external can wall was considered. Peach halves were orderly placed inside the can with the empty space originally occupied by the kernel facing, in all peaches, either towards the top or the bottom of the can. In a third situation, the can was placed horizontally. Simulations revealed differences on particle temperature profiles, as well as process F values and critical point location, based on their orientation. At their critical points, peach halves with the kernel space facing towards the top of the can heated considerably slower and cooled faster than the peaches having their kernel space facing towards the bottom of the can. The horizontal can case exhibited intermediate cooling but the fastest heating rates and the highest F process values among the three cases examined. The results of this study could be used in designing of thermal processes with optimal product quality.

Keywords: Computational Fluid Dynamics; natural convection; thermal processing; liquid/particulate; peaches; particle orientation; canning; modeling

1. Introduction

Thermal processing is a widely used and extensively studied method for food preservation. Undesirable quality degradation that inevitably accompanies the targeted destruction of pathogens and spoilage agents during thermal processing of foods, calls for optimum and accurate design of a thermal process. The scientific principles for designing safe thermal processes pioneered by Ball and his colleagues [1,2] at the beginning of the previous century. These principles formed also the basis for quality retention calculations during a thermal process, calculations led by Stumbo [3,4], initiated a number of new thermal process calculation methodologies [5], among which we should mention the one presented by Hayakawa [6], and served as guide for analyzing novel preservation processes as, for example, with the case of high hydrostatic pressure processing of foods [7].

Modeling and simulation of food processes is the core of process optimization. Computational Fluid Dynamics (CFD) has been efficiently used for simulation of food processes [8,9]. A number of CFD studies refer to thermal processing of liquid/particulate systems. The majority of these investigations were focused on the analysis of fluid motion and temperature evolution. Examples include thermal processing studies on pineapple slices in juice [10], solid particles in water [11], peas in water [12] and asparagus in brine [13]. The use of CFD for microbial destruction calculations during thermal processing (for table olives in brine) has been also reported [14].

A number of processing schedules, depending on can size, processing method (stationary vs. rotary systems) method of vacuum formation (thermal vs. mechanical) etc., have been reported in the literature for peach canning [15]. Thus, for example, processing for 20 to 25 min is recommended for syrup packed peach halves in 401 × 411 cans still processed in boiling water, when exhausted to a temperature of 71.1 °C (160 °F). Cans using mechanical means to obtain vacuum require 5 to 10 min longer heating time due to their lower initial temperature. Alternatively, achieving a minimum temperature of 87.8 °C (190 °F) at the end of the heating cycle, at the critical point of the product, before air cooling, or of 90.6 °C (195 °F) before water cooling, is proposed [15]. Knowledge of heat transfer characteristics, as well as of the location of the critical point, that is, the point that receives the least effect, as far as microbial destruction is concerned, of the heat treatment, is essential in designing thermal processes for such products.

Particle orientation can significantly influence liquid motion and heat transfer rates in liquid/particulate systems during natural convection heating. The objective of this work was to apply Computational Fluid Dynamics in studying the flow field and the temperature profile, as well as microbial inactivation, in thermally processed still cans filled with peach halves in sugar syrup.

2. Experimental Section

A tin can with dimensions of 7.5 cm in diameter and 10.5 cm in height filled with syrup (20% sugar) and four peach halves was used in all simulations. Peach halves were orderly placed inside the can. In two cases, the can was placed vertically having the peach halves with the empty space originally occupied by the kernel facing, in all peaches, either towards the top (termed hereafter "upward") or the bottom of the can (termed hereafter "downward"). In a third case, the vertical can having the peach halves with the empty space originally occupied by the kernel facing towards the bottom of the container was placed horizontally (turned 90 degrees clockwise, termed hereafter "sideward"). The system was designed with mid-plane symmetry. The three dimensional arrangement and orientation of cans and peach halves in the cans are shown in Figure 1. Note that the x–z plane at the presentation of the horizontal can on Figure 1 is reversed, in order for the geometry of the empty space originally occupied by the kernel to be visible.

Figure 1. Three dimensional arrangement and orientation of cans and peach halves in the cans.

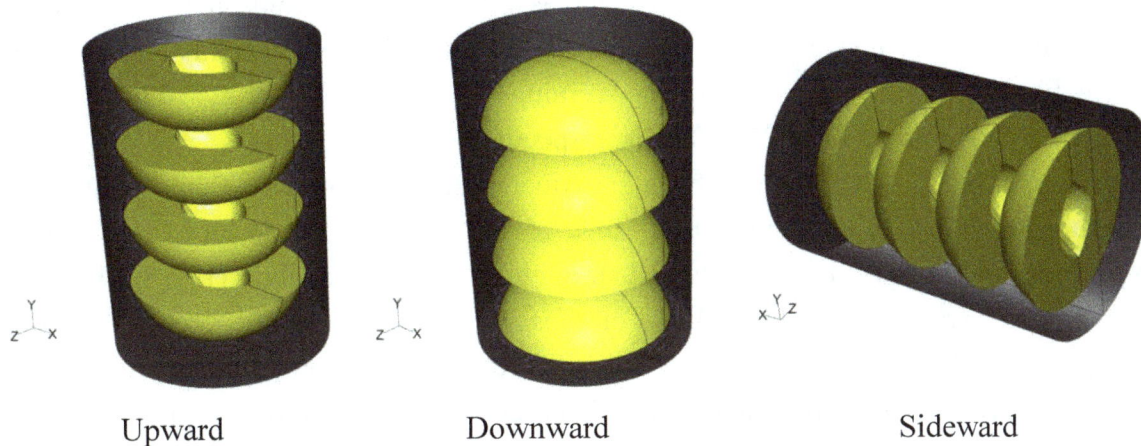

Upward   Downward   Sideward

The can with its contents were initially at rest and at uniform temperature (20 °C), heated in a medium at constant temperature (100 °C) for 20 min and cooled at constant medium temperature of 20 °C. The internal heat transfer coefficient between the filling liquid (syrup) and internal can wall (based on natural convection heating of the syrup) is expected to create the main resistance to heat transfer. Thus, infinite heat transfer coefficient between heating or cooling medium and external can wall was considered. Negligible resistance to heat transfer of the metal wall, and no slip at the container's wall, was further assumed. Only heat transfer from the can to the syrup and from the syrup to peaches was considered.

For the simulation of the fluid field and the temperature evolution in the product (peaches and syrup), the partial differential equations of mass, momentum and energy conservation [16] were solved using the software FLUENT 6.3.26 (Fluent Ansys Inc., Canonsburg, PA, USA, 2006) with 3D, double precision, pressure-based, laminar flow (due to the Rayleigh number being less than 10^8 during the heating and cooling cycle). Mid-plane symmetry (as depicted on Figure 1) was assumed. Axi-symmetric simulation could not be applied for sideward case. The time step used was equal to 0.5 s, the Courant number was equal to 0.5, the algorithm of pressure-velocity coupling was "Coupled" and the model used for the discretization of pressure was "PRESTO!". For the discretization of momentum and energy equations, the model "First Order Upwind" was selected. Preliminary trials showed no differences in temperature predictions between first- and second- order upwind models. In FLUENT nomenclature, the internal surface of the can, as well as the external surface of the peach halves, was defined as wall. The volume of the peach halves was considered as solid. The volume between the peach halves and the can, occupied by the syrup, was considered as fluid. The volumes of the syrup and the peach halves were designed in Gambit® 2.3.16.

Peach shapes were designed to resemble, as close as possible, the real product geometry. Thus peach halves were designed as semi oblate spheroids with 6.0 cm in large diameter and 4.0 cm in small diameter. The kernels were designed as semi ellipsoids with 2.0 cm in the middle axis (corresponding to kernel's depth) and 3.0 cm in the largest axis and 1.5 cm in the smallest axis, resulting in uneven thickness of the peaches (1.5 cm in the x axis, 2.25 cm in the z axis, and 1 cm in the y axis, Figure 1).

The shape of the grid was "Tet/Hybrid" for both peach halves and syrup volumes with the option of 18% "Shortest edge" of the software. The latter means that a value equal to the 18% of the size of the smallest edge existing in the entire geometry is used to calculate the total number of intervals at all edges. Gambit® calculates the number of intervals on any edge by dividing the size of the particular edge by this value. For the half can volume (due to the assumed symmetry only half of the can was simulated) and for the case of "upward" peach halves, 74,770 cells (for both peach halves and syrup) were used for the discretization and solution of the governing equations, while for the "downward" and the "sideward" peach halves 73,364 and 73,590 cells, respectively were employed.

The thermo-physical properties of 20% (w/v) sugar aqueous solution obtained from literature were used for the syrup properties. In the temperature range of 20 °C to 130 °C, values for density (ρ), viscosity (μ) and specific heat (C_p) were obtained through on-line calculations [17], while thermal conductivity (k) values were read from [18]. Based on those values, each property was expressed as a function of temperature (T, in °C) through a second or third order polynomial equation (derived in Microsoft® Office Excel) as shown on Table 1. Peach properties from literature [19] were assumed constant throughout the process and equal to the mean values presented on Table 1.

Table 1. Physical and thermal properties of liquid and solids used in simulations.

	Syrup (20% w/v sugar)	Peach
ρ (kg/m³)	$1.0868 \times 10^3 - 2.4377 \times 10^{-1} \times T - 2.4201 \times 10^3 \times T^2$ ($R^2 = 0.9985$)	1022 ± 3
μ (Pa·s)	$3.3817 \times 10^{-3} - 8.2745 \times 10^{-5} \times T + 8.3164 \times 10^{-7} \times T^2 - 2.8823 \times 10^{-9} \times T^3$ ($R^2 = 0.9955$)	-
C_p (J/kg·K)	$3.6839 \times 10^3 + 1.4510 \times 10^0 \times T + 3.4965 \times 10^{-4} \times T^2$ ($R^2 = 0.9980$)	3992 ± 7
k (W/m·K)	$5.0486 \times 10^{-1} + 1.5929 \times 10^{-3} \times T - 5.0000 \times 10^{-6} \times T^2$ ($R^2 = 0.9999$)	0.58 ± 0.09

Based on the classical thermobacteriological approach [20], the F value of the process was calculated through Equation (1):

$$F_{T_{ref}}^{z} = \int_0^t 10^{\frac{T(t)-T_{ref}}{z}} dt \tag{1}$$

where the $F_{T_{ref}}^{z}$ value is defined as the equivalent processing time of a hypothetical thermal process at a constant reference temperature, T_{ref}, that produces the same effect (in terms of microbial, or of any other heat sensitive substance destruction) as the actual thermal process, the z value is the temperature difference required to achieve a tenfold change of the decimal reduction time of the heat sensitive substance, and T being the temperature (at a particular point in the product) and t the processing time.

Calculations were based on a z value of 11.5 °C (indicative to *Clostridium butiricum*, [21]) and a reference temperature of 90 °C, appropriate for thermal treatment of acid foods. A User Defined

Function (UDF) was written and imported to FLUENT in order to calculate F values at every point inside the container where temperature values were calculated. For the completion of the calculation procedure described so far, computational time was about 3 days using a personal computer with Intel® Core TM 2, CPU 660 @ 2.40 GHz and 2048 MB RAM.

3. Results and Discussion

The model used in the present investigation was validated for various cases, such as asparagus (two experimental set-ups in triplicate), table olives (duplicate experiments) and peach halves (single run) in liquid, processed in still cans, in earlier works [13,14,22]. In brief, and for the case of peach halves, the temperature inside (at predescribed locations) the bottom and the central peach halves, in a can containing three peach halves, closely packed, in 20% sugar syrup, heated in boiling water, was measured with type K thermocouples and compared (successfully) to model predictions (Figure 2). Validation of model's predictions with experimental data for the setup with four peach halves analyzed in the present study was not performed and this could be considered as limitation of the present work.

Figure 2. Experimental and simulated temperature data at the interior of the central (**A**) and the bottom peach half (**B**), in reference to the inserted picture, in a can containing three peach halves in 20% sugar syrup.

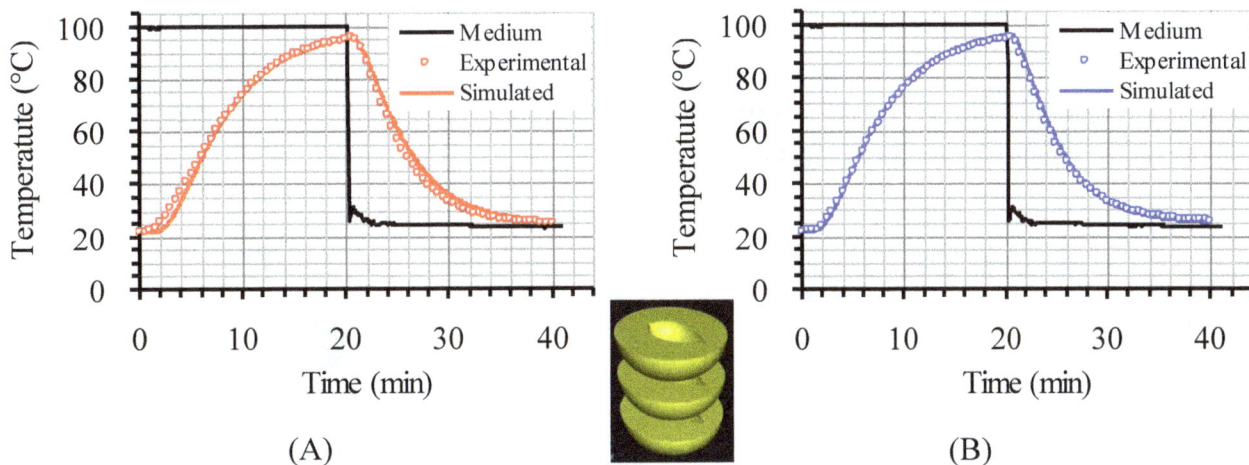

(A)　　(B)

3.1. Velocity Profile

Velocity pathways, indicative to fluid motion due to buoyancy, are presented on Figures 3–5, 30 s after the beginning of heating and 30 s after the beginning of cooling, for the cases studied. In all cases, during heating, hot syrup close to can vertical walls moves towards the top of the container forcing the syrup on the top of the container to move downwards at the interior of the container between the solid peaches. Thirty seconds after the onset of heating, indicative velocities (maximum calculated values) were 2.568 cm/s, 2.760 cm/s and 2.147 cm/s for the "upward", "downward" and "sideward" cases, respectively. Hot syrup heats the cold peach halves when in contact. The whole process is repeated until no temperature differences within the syrup exist.

During cooling the direction of fluid motion is reversed, for the syrup being moving downwards near the wall because of the fluid's cooling and upwards in the interior. Thirty seconds after the onset of cooling, maximum calculated velocities were 2.838 cm/s, 2.709 cm/s and 2.301 cm/s for the "upward", "downward" and "sideward" cases, respectively.

Figure 3. Velocity pathlines after 30 s of heating (left) and after 30 s of cooling (right) for the "upward" case (velocity in cm/s).

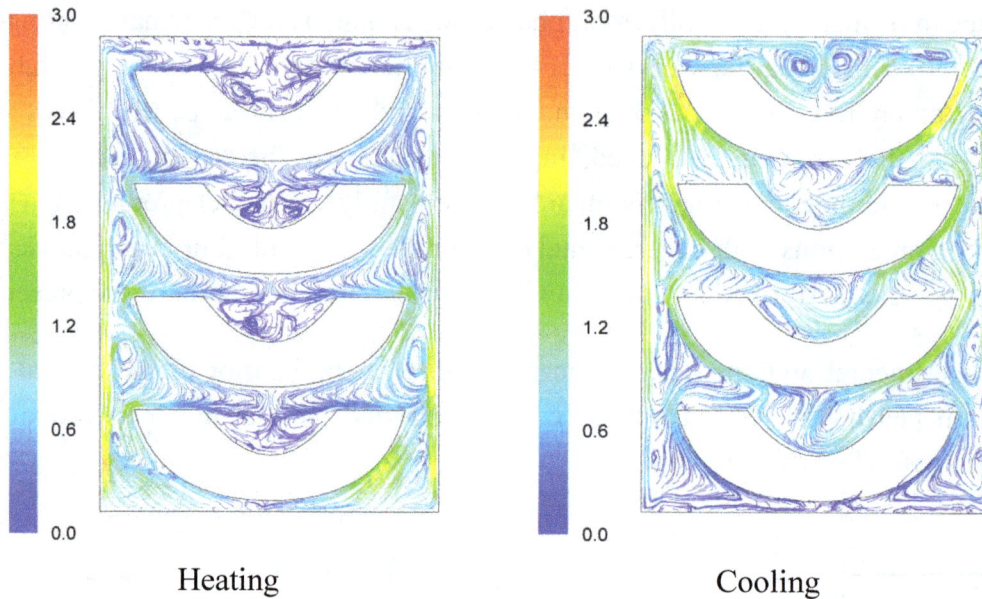

Heating   Cooling

Figure 4. Velocity pathlines after 30 s of heating (left) and after 30 s of cooling (right) for the "downward" case (velocity in cm/s).

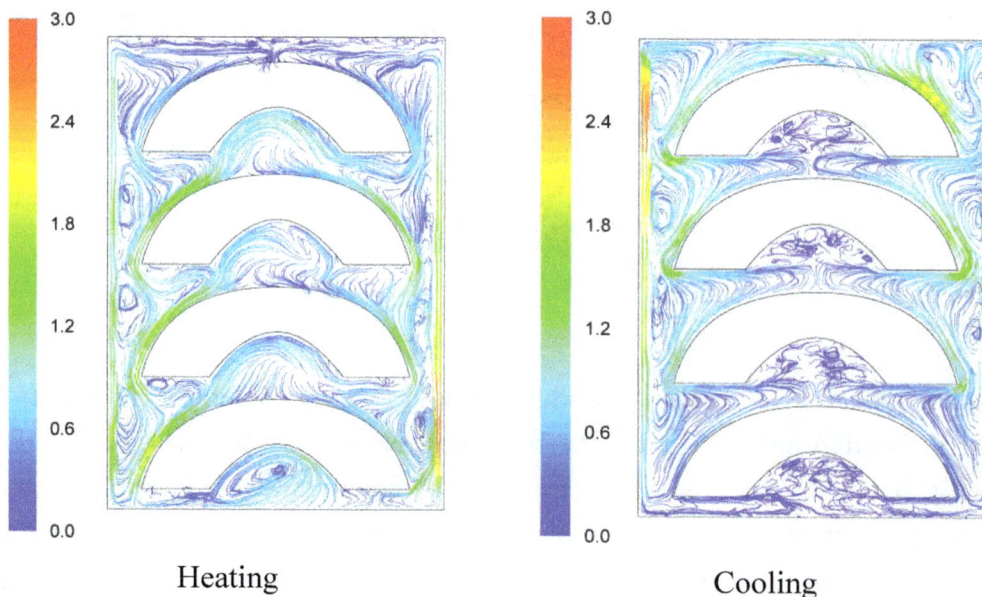

Heating   Cooling

Figure 5. Velocity pathlines after 30 s of heating (left) and after 30 s of cooling (right) for the "sideward" case (velocity in cm/s).

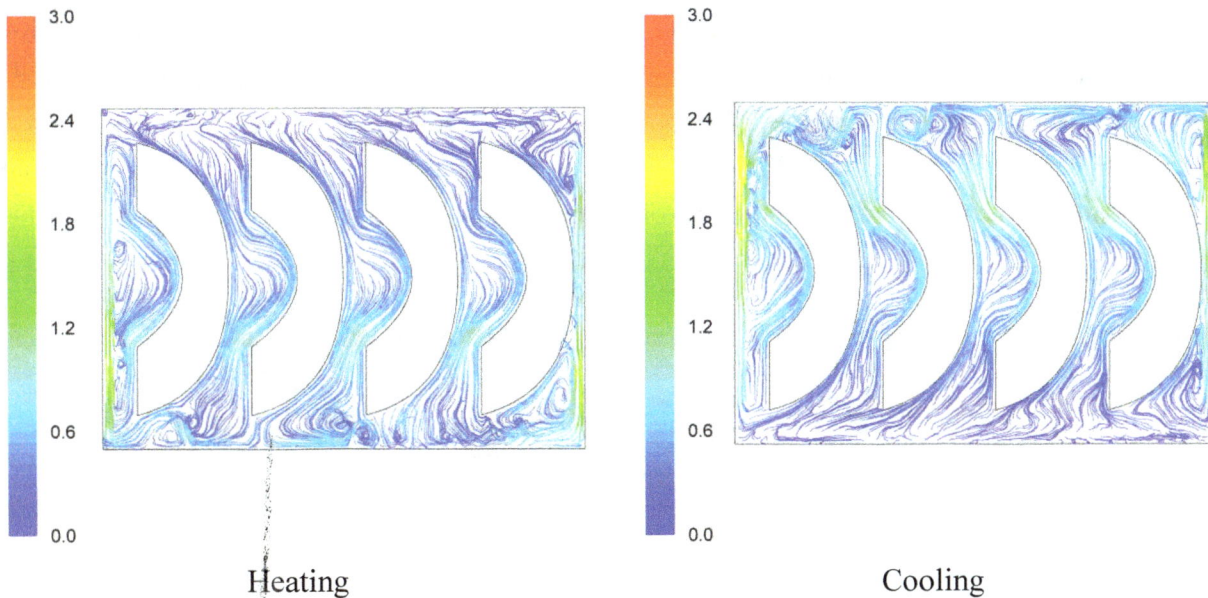

Heating   Cooling

3.2. Temperature Profile

The temperature evolution during heating and cooling of the product is indicatively shown on Figures 6–8 at selected processing times, for the three cases studied. Heating and cooling rates are a direct consequence of the syrup motion which, by its turn, is greatly affected by the orientation of the can (vertical *vs.* horizontal) and the arrangement of the peach halves within the can (having the empty space originally occupied by the kernel facing either towards the top or the bottom of the can). The contents of the horizontal can ("sideward" case) were heated at comparable rates or slightly faster compared to the "downward" case of the vertical container (as can be seen on Figures 7 and 8 after 5 min of heating) with the latter (the "downward" case) heating faster compared to the "upward" case (as can be seen on Figures 6 and 7 after 5 min of heating). During cooling, slower rates were observed for the "downward" case, followed by the "sideward" and "upward" cases which cool in comparable rates (see Figures 6–8 after 5 min of cooling).

Differences between the "upward" and the "downward" cases in heating and cooling rates are related to the corresponding differences in fluid flow. During heating, fluid velocities at the cavity of the peach halves for the "upward" case are low (Figure 3) resulting in a rather stagnant, slow heated pool of liquid. For the "downward" case, natural convection currents at the peach cavities (Figure 4) result to higher heating rates compared to the "upward" case. During cooling, descending natural convection currents at the peach cavities for the "upward" case (Figure 3) result to higher cooling rates compared to the "downward" case, where hot fluid is trapped into the peach cavity (Figure 4). Thus, the "upward" case heats faster and cools slower than the "downward" case, resulting to more intense thermal treatment. The unrestricted motion of the fluid on both sides of the peach halves for the "sideward" case (Figure 5) results in higher heat transfer rates from fluid to the particles during both heating and cooling compared to the other two cases.

Based on temperature calculations at each grid point, the slowest heating zone during the heating cycle and the slowest cooling zone during the cooling cycle of the thermal process were identified. Thus, for example, the exact rectangular coordinates (x, y, z) in mm, for the slowest heating point for the cases studied were: $(0, -40.0, 0)$ for the "upward", $(-7.6, -35.8, -7.9)$ for the "downward" and $(-35.6, -13.5, 0)$ for the "sideward" facing peach halves, respectively. Note that the origin, $x = 0$, $y = 0$ and $z = 0$ refer to the geometric center of the can, while in reference to Figures 6–8, positive x is measured as we move from the center to the right, positive y is measured as we move from the center to the top and positive z is measured as we move from the center outwards of the paper (see Figure 1).

Figure 6. Temperature (°C) contours for peach halves in syrup at different heating and cooling times during heating (at 100 °C) and cooling (at 20 °C) for the "upward" case.

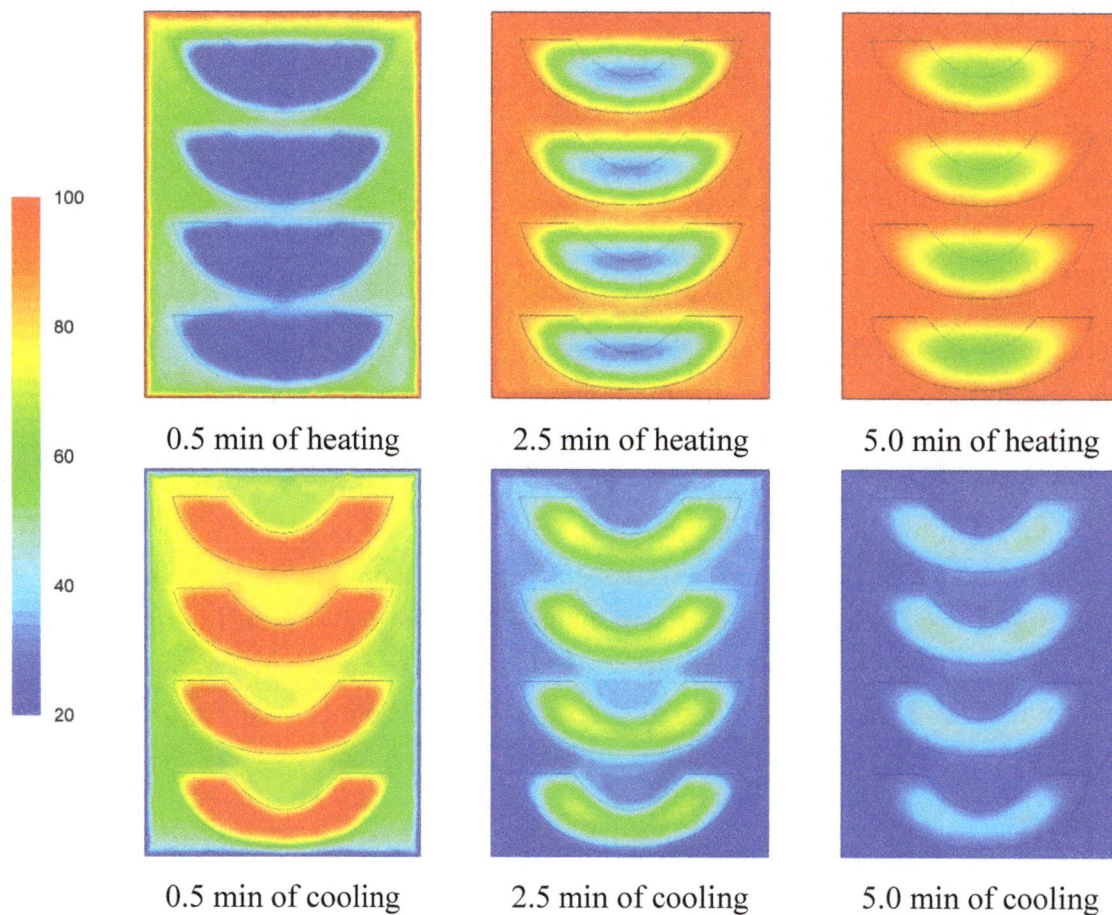

Figure 7. Temperature (°C) contours for peach halves in syrup at different heating and cooling times during heating (at 100 °C) and cooling (at 20 °C) for the "downward" case.

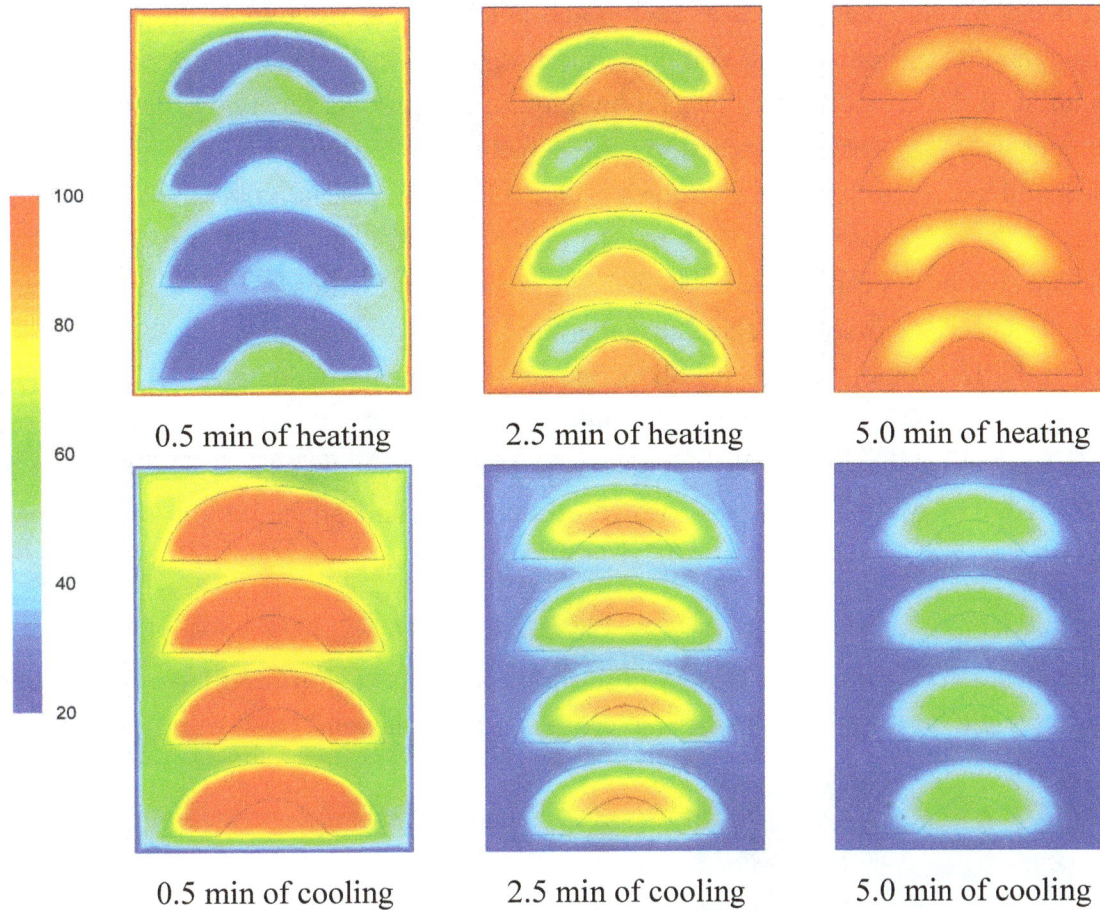

0.5 min of heating	2.5 min of heating	5.0 min of heating
0.5 min of cooling	2.5 min of cooling	5.0 min of cooling

Figure 8. Temperature (°C) contours for peach halves in syrup at different heating and cooling times during heating (at 100 °C) and cooling (at 20 °C) for the "sideward" case.

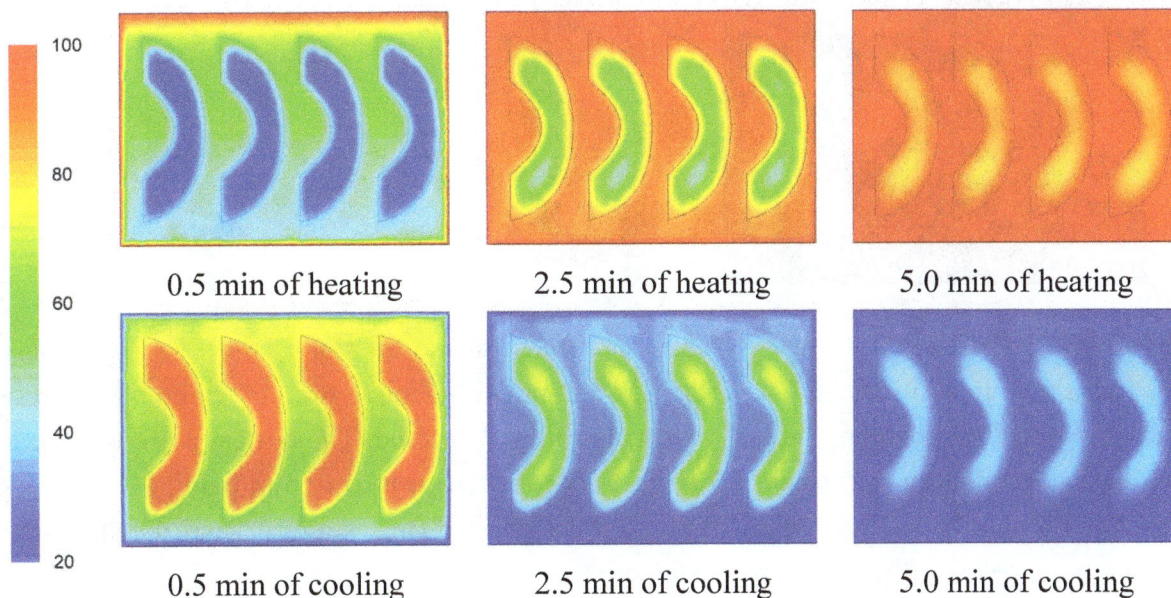

0.5 min of heating	2.5 min of heating	5.0 min of heating
0.5 min of cooling	2.5 min of cooling	5.0 min of cooling

3.3. F *Value Distribution*

Based on process F value calculations at each nodal point, the distribution of the F values within the product throughout the whole process could be assessed. The distribution of the F values as well as the precise location of the critical zone, that is the region in the product receiving the least effects of the process, in terms of microbial destruction, are depicted on Figures 9–11 for the three peach arrangements examined. The process F values ranges for the peach halves for the "upward" "downward" and "sideward" cases, were 34.9 min to 116.4 min, 66.1 min to 119.9 min, and 73.9 min to 128.1 min, respectively. As can be seen from the left pictures on Figures 9–11, the highest F values were calculated at the surface of the peaches located at the top part of the can.

If we focus at the interior of the peach halves, the lowest F values calculated for the four peach halves of the "upward" case were 34.9 min, 37.8 min, 40.9 min and 40.6 min as we move from the bottom towards the top of the can (right picture on Figure 9). Note that these F values are rather narrowly distributed. The critical point was located at the peach half at the bottom of the can, with exact coordinates (x, y, z) in mm being $(0, -40.0, 0)$. Note that for this case the critical point coincides with the slowest heating point discussed earlier.

Figure 9. F value (in min) distribution within the peach halves (left) and location of the critical point (right) at the end of the thermal process for the "upward" case.

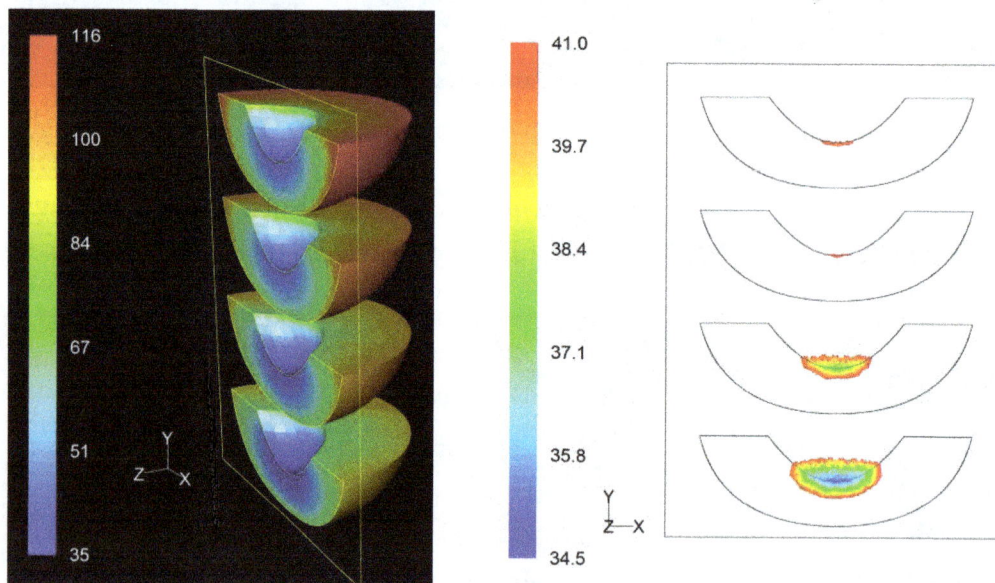

For the "downward" case, the lowest F values calculated at the interior of the four peach halves peach halves were 66.1 min, 69.2 min, 74.2 min and 81.3 min as we move from the bottom towards the top of the can (Figure 10). Note that these F values are rather dispersed. The critical point was located at the peach half at the bottom of the can, with exact coordinates (x, y, z) in mm being $(-10.2, -35.9, -9.1)$. Note that for this case the location of the critical point differs slightly from the location of slowest heating point $(-7.6, -35.8, -7.9)$ presented earlier.

Figure 10. F value (in min) distribution within the peach halves (left) and location of the critical point (right) at the end of the thermal process for the "downward" case.

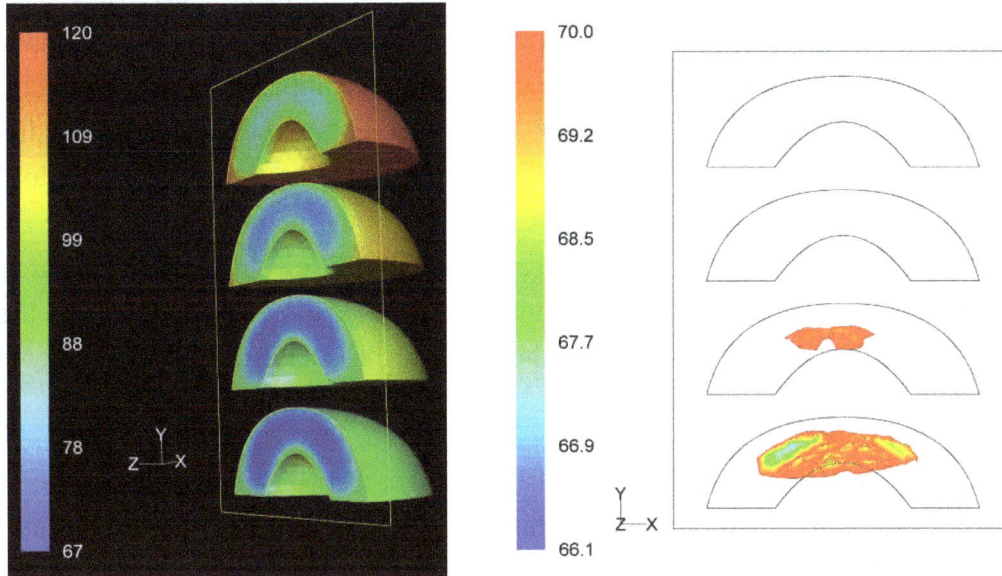

Finally, for the "sideward" case, the lowest F values calculated at the interior of the four peach halves were 73.9 min, 74.6 min, 74.8 min and 75.2 min as we move from left to right on the right picture on Figure 11. Note that these F values are practically identical. The critical point was located at the first peach half from the left, with exact coordinates (x, y, z) in mm being $(-35.6, -13.5, 0)$. As for the "upward" case, for this case also the critical point coincides with the slowest heating point.

Figure 11. F value (in min) distribution within the peach halves (left) and location of the critical point (right) at the end of the thermal process for the "sideward" case.

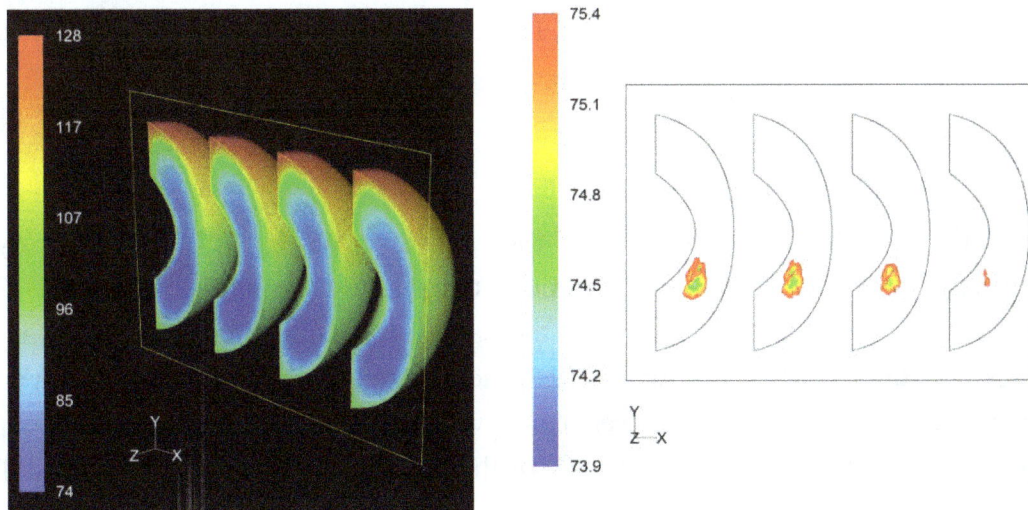

For the simulations presented, heating for 20 min at 100 °C (followed by cooling at 20 °C) was used. This process schedule resulted to high F process values for all cases. A realistic target F value can be set by requiring a six-log reduction of *Clostridium butyricum* population, characterized by a decimal reduction time at 90 °C equal to 1.1 min [23]. This leads to a target $F_{90°C}^{11.5°C}$ value of 6.6 min. Given the existence of a distribution of F values, if one had to calculate heating times for a particular microbial destruction, then appropriate calculations should be performed at the critical point for each case in order to avoid underprocessing.

At the critical point, for the three cases studied, the "upward" case resulted to lower heat transfer rates during heating as illustrated on Figure 12. Furthermore, during cooling, the temperature at the critical point for the "upward" case dropped noticeably faster compared to the other two cases (Figure 12). This led to the lowest F values, at the critical point, within the three cases studied. Thus, for a given target F value, the "upward" case will require longer heating times compared to the other two cases studied. Comparing the temperature profiles, at the critical points, for the other two cases, one can note the slightly faster heating and cooling for the "sideward" case compared to the "downward" case. Due to the fact that at the beginning of the cooling the temperatures at the critical points for the two cases were comparable, the faster heating rates associated with the "sideward" case caused the accumulation of higher F values for this case.

Figure 12. Comparison of temperature profiles at the critical point during thermal processing of peach halves of different orientations in syrup (20% sugar) in stationary metal cans.

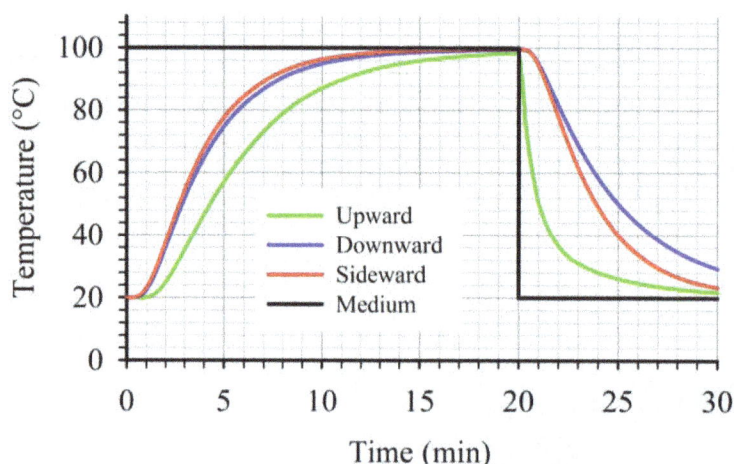

4. Conclusions

The CFD simulations of thermal processing of four peach halves in 20% w/v sugar syrup in stationary cans gave insights as far as fluid motion, temperature evolution, microbial F value distribution and location of the critical point. For the same processing time, the "sideward" case recorded the highest and the most uniform, within the peach halves, F process values. Comparing the two cases of the vertical can with the peach halves placed with the empty space originally occupied by the kernel facing towards the top (the "upward" case) or the bottom of the can (the "downward" case) slower heating but faster cooling rates were calculated for the "upward" case, fact that led to lower F process values for the "upward" case. The F value at the critical points within each peach half for the "upward" case was more uniform compared to the "downward" case.

For the "upward" case, the critical point was located at the upper part (where the kernel was located) of the peach half at the bottom of the can. For the "downward" case, the critical point was located at the interior of the peach half at the bottom of the can. Finally, for the "sideward" case the critical point was at the lower portion (at about 1/3 of the peach diameter) of the first peach half from the left (that is, the peach half having the empty space originally occupied by the kernel facing towards from the can bottom). A number of parameters, such as, heating time and temperature, target F value and microbial z value, size and number of peach halves, particle arrangement—including random placement, can be further investigated based on the presented principles and approach.

Author Contributions

Adreas Dimou has done the experimental and computational work. Nikolaos G. Stoforos has contributed to the design and analysis of the experimental and theoretical work and has written the paper. Stavros Yanniotis is the supervisor of Adreas Dimou. He has initiated the work and has contributed to the design, analysis and execution of the experimental and theoretical work.

Conflicts of Interest

The authors declare no conflict of interest.

References

1. Bigelow, W.D.; Bohart, G.S.; Richardson, A.C.; Ball, C.O. *Heat Penetration in Processing Canned Foods*; Bulletin No. 16-L; Research Laboratory, National Canners Association: Washington, DC, USA, 1920.
2. Ball, C.O. *Thermal Process Time for Canned Food*; Bulletin No. 37; National Research Council: Washington, DC, USA, 1923; Volume 7, Part 1.
3. Stumbo, C.R. New procedures for evaluating thermal processes for foods in cylindrical containers. *Food Technol.* **1953**, *7*, 309–315.
4. Stumbo, C.R. *Thermobacteriology in Food Processing*, 2nd ed.; Academic Press, Inc.: New York, NY, USA, 1973.
5. Stoforos, N.G.; Noronha, J.; Hendrickx, M.; Tobback, P. A critical analysis of mathematical procedures for the evaluation and design of in-container thermal processes for foods. *Crit. Rev. Food Sci. Nutr.* **1997**, *37*, 411–441.
6. Hayakawa, K. Experimental formulas for accurate estimation of transient temperature of food and their application to thermal process evaluation. *Food Technol.* **1970**, *24*, 1407–1418.
7. U.S. Food and Drug Administration (FDA). Kinetics of Microbial Inactivation for Alternative Food Processing Technologies—Overarching Principles: Kinetics and Pathogens of Concern for All Technologies. Available online: http://www.fda.gov/Food/FoodScienceResearch/SafePracticesforFoodProcesses/ucm100198.htm (accessed on 27 February 2014).
8. Norton, T.; Sun D.W. An overview of CFD applications in the food industry. In *Computational Fluid Dynamics in Food Processing*; Sun, D.W., Ed.; CRC Press: Boca Raton, FL, USA, 2007; pp. 1–41.
9. Yanniotis, S.; Stoforos, N.G. Modelling food processing operations with CFD: A review. *Scientia Agric. Bohem.* **2014**, *45*, 1–10.

10. Abdul Ghani, A.G.; Farid, M.M. Using the computational fluid dynamics to analyze the thermal sterilization of solid-liquid food mixture in cans. *Innov. Food Sci. Emerg. Technol.* **2006**, *7*, 55–61.

11. Rabiey, L.; Flick, D.; Duquenoy, A. 3D simulations of heat transfer and liquid flow during sterilization of large particles in a cylindrical vertical can. *J. Food Eng.* **2007**, *82*, 409–417.

12. Kiziltaş, S.; Erdoğdu, F.; Palazoğlu, K. Simulation of heat transfer for solid-liquid food mixtures in cans and model validation under pasteurization conditions. *J. Food Eng.* **2010**, *97*, 449–456.

13. Dimou, A.; Yanniotis, S. 3-D numerical simulation of asparagus sterilization using computational fluid dynamics. *J. Food Eng.* **2011**, *104*, 394–403.

14. Dimou, A.; Panagou, E.; Stoforos, N.G.; Yanniotis, S. Analysis of thermal processing of table olives using Computational Fluid Dynamics. *J. Food Sci.* **2013**, *78*, E1695–E1703.

15. Lopez, A. *A Complete Course in Canning and Related Processes*, Book III, 12th ed.; The Canning Trade Inc.: Baltimore, MD, USA, 1987.

16. Bird, R.B.; Stewart, W.E.; Lightfoot, E.N. *Transport Phenomena*; John Wiley and Son: New York, NY, USA, 1960.

17. The Sugar Engineers. Material Properties. Available online: http:www.sugartech.co.za (accessed on 15 June 2012).

18. Mohos, F.A. *Confectionery and Chocolate Engineering: Principles and Applications*; Wiley-Blackwell: Ames, IA, USA, 2010.

19. Phomkong, W.; Srzedninchi, G.; Driscoll, R.H. Desorption isotherms of stone fruit. *Dry. Technol.* **2006**, *24*, 201–210.

20. Ball, C.O.; Olson, F.C.W. *Sterilization in Food Technology: Theory, Practice and Calculations*; McGraw-Hill Book Co.: New York, NY, USA, 1957.

21. Silva, F.V.; Gibbs, P. Target selection in designing pasteurization processes for shelf-stable high-acid fruit products. *Crit. Rev. Food Sci. Nutr.* **2004**, *44*, 353–360.

22. Dimou, A.; Stoforos, N.G.; Yanniotis, S. CFD simulations in still cans filled with solid food items in liquid. *Procedia Food Sci.* **2011**, *1*, 1216–1222.

23. Silva, F.V.M.; Gibbs, P.A. Principles of thermal processing: Pasteurization. In *Engineering Aspects of Thermal Food Processing*; Simpson, R., Ed.; CRC Press: Boca Raton, FL, USA, 2009; pp. 14–48.

8

Change in Color and Volatile Composition of Skim Milk Processed with Pulsed Electric Field and Microfiltration Treatments or Heat Pasteurization [†]

Anupam Chugh [1], Dipendra Khanal [1], Markus Walkling-Ribeiro [1,2,]*, Milena Corredig [1], Lisa Duizer [1] and Mansel W. Griffiths [1]

[1] Department of Food Science, University of Guelph, Guelph, ON N1G 2W1, Canada;
E-Mails: achugh@uoguelph.ca (A.C.); dkhanal@uoguelph.ca (D.K.);
mcorredi@uoguelph.ca (M.C.); lduizer@uoguelph.ca (L.D.); mgriffit@uoguelph.ca (M.W.G.)

[2] Department of Food Science, Cornell University, Ithaca, NY 14853, USA

[†] This article is an extended version of a conference paper named "Effect of non-thermal pulsed electric fields and tangential-flow microfiltration processing on flavor and color properties of skim milk", published in the proceedings of the 2013 IFT Annual Meeting and Food Expo.

* Author to whom correspondence should be addressed; E-Mail: mw757@cornell.edu

Abstract: Non-thermal processing methods, such as pulsed electric field (PEF) and tangential-flow microfiltration (TFMF), are emerging processing technologies that can minimize the deleterious effects of high temperature short time (HTST) pasteurization on quality attributes of skim milk. The present study investigates the impact of PEF and TFMF, alone or in combination, on color and volatile compounds in skim milk. PEF was applied at 28 or 40 kV/cm for 1122 to 2805 μs, while microfiltration (MF) was conducted using membranes with three pore sizes (lab-scale 0.65 and 1.2 μm TFMF, and pilot-scale 1.4 μm MF). HTST control treatments were applied at 75 or 95 °C for 20 and 45 s, respectively. Noticeable color changes were observed with the 0.65 μm TFMF treatment. No significant color changes were observed in PEF-treated, 1.2 μm TFMF-treated, HTST-treated, and 1.4 μm MF-treated skim milk ($p \geq 0.05$) but the total color difference indicated better color retention with non-thermal preservation. The latter did not affect raw skim milk volatiles significantly after single or combined processing ($p \geq 0.05$), but HTST caused considerable changes in their composition, including ketones, free fatty acids, hydrocarbons, and sulfur compounds ($p < 0.05$). The findings indicate that for the particular thermal and non-thermal treatments selected for this study, better retention of skim milk color and flavor components were obtained for the non-thermal treatments.

Keywords: pulsed electric field (PEF); microfiltration (MF); thermal pasteurization; skim milk; volatile compounds; color degradation; non-thermal processing; hurdle technology

---

1. Introduction

Milk is a widely consumed beverage due to its nutritional importance, a pleasant aroma and mouth-feel, and a slightly sweet taste. Color and flavor are important sensorial attributes of milk that are influenced by several factors, such as milk composition, cow's feed and metabolism, environmental factors and processing conditions [1,2]. Few studies are available on the impact of conventional thermal treatments and emerging processing technologies like pulsed electric field (PEF), high hydrostatic pressure (HHP) and ultra high pressure (UHP) on volatile compounds in milk or milk based beverages [3–6]. Pulsed electric field (PEF) and microfiltration (MF) are emerging innovative technologies that could meet the increasing consumer demand for "fresh-like" minimally processed foods. The mode of action that underlies PEF is based on short electric pulses of high voltage applied to a product that is placed between a pair of electrodes, thereby, bringing about electroporation of the bacterial cell wall and its subsequent breakdown [7]. In addition to the type of microorganism and its innate resistance [8], the efficacy of PEF for food preservation depends on processing factors such as electric field strength, number of pulses applied and the treatment time [9], as well as product parameters, including electrical conductivity, viscosity, and pH [10].

MF is a membrane-driven separation process typically employing membranes with a pore size of 1.2 to 1.4 μm that allows removal of bacteria in vegetative and spore forms from milk [11,12]. Commercially available thermal processing methods promise a high degree of microbial safety but can adversely affect other food properties such as color and volatile compounds. Several studies have demonstrated a comparable impact of PEF and MF processes on microbial inactivation in foods [13–17]. Milk has its natural color due to the reflectance of light by dispersed milk fat globules, proteins, and natural milk pigments like riboflavin and carotenoids [18,19]. However, milk color is altered at high processing temperatures as a result of Maillard browning or a temporary increase in lightness due to denaturation of soluble whey proteins [20,21]. The changes in milk compounds that are responsible for change in milk color may, at the same time, affect the perception of flavor in milk due to intensification of some volatile components. It is generally accepted that prolonged or severe heat treatments cause the degradation of the volatile profile of milk, with the increase in concentration of volatile compounds being positively correlated to intensity of the heating [22,23]. The development of such volatile compounds may alter consumer acceptance as suggested by Gandy and others [24]. In contrast, non-thermal processing may have a minimal effect on the concentration of volatiles in foods due to shorter processing times at temperatures below those used for pasteurization. Zhang *et al.* [5] observed increased levels of aldehydes and ketones in pasteurized milk as compared to raw and PEF processed milk. PEF-treated and HHP-treated orange juice-milk beverage showed a reduced loss of volatiles as compared to thermally processed beverage [3]. From a quality and safety perspective, it may be advantageous to use hurdle technology rather than using a single technology to achieve sufficient product safety and maximized quality while maintaining minimal processing. Microfiltration is one of the non-thermal technologies that have been commercially allowed for extending shelf life of milk, although its use as a stand-alone treatment is restricted due to regulatory requirements [25]. Based on microbiological data, previous studies have indicated that the safety of skim milk processed with a combination of PEF and MF is comparable to that achieved by HTST pasteurization [13,17], potentially representing a non-thermal alternative to the well-established heat pasteurization. However, the effect of this novel process on color and volatile compounds has not been determined. Thus, the objective of the present study was to study the effect of different PEF, MF, and hurdle (PEF/MF) processing conditions on the color and flavor profiles of skim milk. In addition, the retention of color and flavor following treatments with these non-thermal technologies was compared to that obtained by heat pasteurization.

2. Experimental Materials and Methodology

2.1. Supply, Preparation, and Storage Conditions of Skim Milk

Raw milk obtained from a local dairy processing plant was separated at the Canadian Research Institute for Food Safety at the Department of Food Science, University of Guelph, using a cream separator (STsM-100-18, Motor Sich JSC, Zaporozhye, Ukraine). Skim milk was selected for processing in this study due to its lower fat content, allowing efficient microfiltration without membrane blockage, and because it is common research model for whole milk and has grown in consumer popularity.

2.2. Pulsed Electric Field (PEF) Treatment of Skim Milk

Raw skim milk processing was carried out in a PEF treatment chamber designed at the University of Guelph using an exponential decay pulse generator (PPS 30, University of Waterloo, Waterloo, ON, Canada) to form monopolar pulses with an average width of 1.5 µs. PEF parameters were measured and monitored as described previously [17]. The single PEF chamber used consists of two co-axial stainless steel electrodes encased in an insulating plastic casing. Liquid passes between electrodes through a gap of 0.21 cm. The chamber has a hold up volume of 44 cm^3 of which 25 cm^3 is the volume between the electrodes where milk is treated by means of electric pulses. An external jacket with a volume of 790 cm^3 surrounds the outer electrode for additional temperature control, allowing the PEF chamber to be heated or cooled during operation. Skim milk was pumped through Masterflex silicone tubing (size L/S 16) using a peristaltic pump (Masterflex pump drive 7524-40 and pump head 77201-60, Cole Parmer Instrument Co., Vernon Hills, IL, USA). The product was pumped at a flow rate of 20 to 35 mL/min and the temperature at the entry and exit of the PEF chamber was recorded using T-type thermocouples (TMTSS-040G-6, Omega, Stamford, CT, USA) connected to a wireless temperature data logger (OM-SQ2020-2F8, Omega, Stamford, CT, USA). After treatment the outgoing product was cooled by flow through tubing submerged in a cooling water bath (NESLAB RTE 7, Thermo Fisher Scientific Inc., Newlington, NH, USA), set at 2.5 °C. For hurdle treatment, involving subsequent microfiltration, samples were kept in ice in a cold room at 4 °C before further processing to maintain the cold chain and thus, avoid growth of mesophilic microorganisms and potential quality degradation in line with common practice used in the dairy industry. Experiments were conducted at three PEF intensities that reflect different processing demands for skim milk with regard to food safety and energy consumption, which can vary considerably, based on production facilities and locations, and have not been investigated for comparable effects on milk quality parameters, such as color and volatile composition, based on these treatment intensities. The skim milk was processed using electric field strength and treatment time combinations of 28 kV/cm and 2805 µs for low-intensity PEF (PEF-L), 40 kV/cm and 1122 µs for moderate-intensity PEF (PEF-M), and 40 kV/cm and 1571 µs for high-intensity PEF (PEF-H). These processing conditions corresponded to energy densities of 83, 157 and 198 kJ/L, respectively. A pulse frequency of 25 Hz was applied for PEF-L, whereas 17.5 Hz were used for PEF-M and PEF-H treatments. Milk temperature at the inlet of the PEF chamber was 17 °C on average and depending on the treatment intensity applied maximum temperatures of 37 (PEF-L), 56 (PEF-M), and 65 (PEF-H) °C were obtained at the outlet. Following PEF and the cooling step the product temperature was 10 °C and samples were collected at this temperature.

2.3. Microfiltration (MF) of Skim Milk

Bearing in mind that smaller sized membrane pores allow for enhanced microbial reduction in skim milk, the latter was microfiltered at different pore size diameters to compare their effect on the product quality. Moreover, lab and pilot scale systems featuring different membrane materials and designs were used for MF to determine whether or not the processing scale and/or membrane materials and/or membrane designs affect color and volatile compound composition in skim milk. Tangential-flow microfiltration (TFMF) was performed using a laboratory scale Supor tangential-flow filtration system featuring either 0.65 µm (medium screen, 1.4 cm thick) or 1.2 µm (suspended screen, 1.8 cm thick) pore-sized polyethersulfone membranes (Pall Corporation, Port Washington, NY, USA). For laboratory-scale TFMF, a flow rate of 300 mL/min was generated using a Masterflex EW 77200-60 pump (Cole Parmer Instrument Co., Vernon Hills, IL, USA) and the skim milk was delivered by silicone tubing (Masterflex 96400-25, Cole Parmer Instrument Co., Vernon Hills, IL, USA) through either of the TFMF membrane cassettes, with trans-membrane pressures ranging from 0.3 to 0.6 kPa. The permeate flow rate was between 7 and 13 mL/min for milk passing through the TFMF 0.65 µm membrane (MF-0.65), whereas it was between 37 and 53 mL/min for the TFMF 1.2 µm (MF-1.2) membrane. In addition, skim milk processing was also carried out on a pilot scale 1.4 µm pore size, ceramic membrane, cross-flow microfiltration (CFMF) unit (MFS-1, Tetra Pak Filtration Systems, Aarhus, Denmark). The CFMF unit was used at a permeate flow rate (100 L/h) ten times that of retentate (10 L/h), using filter module Type-SCT Membralox (Tetra Pak Filtration Systems). The trans-membrane pressure of the ceramic membrane module was 106 kPa. All microfiltration treatments were carried out at a constant temperature of 35 °C after pre-heating skim milk in a water bath (Isotemp 210, Thermo Fisher Scientific Inc., Newlington, NH, USA) set to this temperature and equipped with a thermometer. In agreement with membrane and equipment manufacturer specifications milk volumes were selected for TFMF and CFMF processing that allowed proper milk circulation through the membranes and process stability before sampling. To avoid membrane clogging over time, laboratory scale membranes were cleaned by recirculating 0.5 N sodium hydroxide followed by 0.1 N nitric acid and 2% bleach at 50 °C, and a final rinse with water immediately after each run. Following the same approach but using different cleaning agents the pilot plant CFMF ceramic membrane was cleaned by recirculation of commercially available mixed acid detergent descaler (Diversey Divos2 (VM13), Diversey, Oakville, ON, Canada) and caustic detergent (Diversey Liquid Bril Tak (VC85)) followed by a final rinse with water.

2.4. PEF/MF Processing of Skim Milk

For hurdle processing, milk was processed using PEF as described above and stored at 4 °C prior to processing with MF. All three PEF treatments: PEF-L, PEF-M and PEF-H were used for hurdle processing followed by MF-0.65 μm and MF-1.2 μm membranes (PEF/MF). However, due to the difficulty in achieving the high sample volumes required for pilot scale microfiltration (MF-1.4) using the laboratory scale PEF unit, only skim milk processed using the PEF-M treatment was coupled with the MF-1.4 μm treatment.

2.5. High-Temperature Short-Time Pasteurization of Skim Milk

High-temperature short-time (HTST) processing of raw skim milk was carried out with a pilot scale, dual stage heat exchanger unit (UHT/HTST Lab-25 EDH, Microthermics Inc., Raleigh, NC, USA) at the Guelph Food Technology Centre (GFTC), Guelph, ON, Canada. Skim milk was kept at 4 °C before processing in the tubular heat exchanger. Milk was processed at two different temperatures of 75 (HTST-75) and 95 °C (HTST-95) for respective holding times of 20 and 45 s, corresponding to flow rates of 0.9 and 0.4 L/min, respectively. The higher intensity HTST pasteurization (95 °C, 45 s) for milk takes into consideration an additional safety margin and extended shelf life in part applied by the industry, but at the same time increased thermal load also renders changes in color and volatile compounds more likely. The heat exchanger pre-heated milk to 55 °C prior to reaching the respective temperatures and cooled the product to 10 °C after the treatments and prior to sampling.

2.6. Color Analysis

Color attributes were measured in Hunter Lab color space using a CM 3500-d spectrophotometer (Konica Minolta Sensing Inc., Mahwah, NJ, USA) equipped with SpectraMagic NX CM-S 100 software (Konica Minolta Sensing Inc.). Reflectance measurements were collected over a wavelength range from 400 to 700 nm. The color values were expressed in Hunter Lab color space as lightness (L), redness (a), and yellowness (b). In order to compare the total color difference (ΔE), between the color properties of untreated samples to those of values obtained after subjecting milk to different treatments, the following equation [26,27] was utilized:

$$\Delta E = \sqrt{\Delta L^2 + \Delta a^2 + \Delta b^2} \qquad (1)$$

where, $\Delta L = L_{standard} - L_{sample}$, $\Delta a = a_{standard} - a_{sample}$, $\Delta b = b_{standard} - b_{sample}$.

In addition, the whiteness was calculated for the skim milk samples by converting Hunter Lab to CIE 1931 XYZ color space values using following formulas:

$$Y = \left(\frac{L}{10}\right)^2 \tag{2}$$

$$X = \left(Y + \left(\frac{L}{10} \times \frac{a}{17.5}\right)\right)\Big/1.02 \tag{3}$$

$$Z = \left(Y - \left(\frac{b}{7}\right) \times \left(\frac{L}{10}\right)\right)\Big/0.847 \tag{4}$$

and then using American Society for Testing and Materials (ASTM) E313 standard practice for calculating Whiteness index (WI):

$$WI = (3.388 \times Z) - (3 \times Y) \tag{5}$$

For each treatment, samples were collected in triplicate. Color measurements were done in duplicate for each sample.

2.7. Analysis of Volatile Compounds

Headspace volatiles in skim milk were analyzed using SIFT-MS (Selected In Flow Tube-Mass Spectrometry), which is a direct mass spectrometer that applies chemical ionization for the analysis of volatile components, yielding product ions that are analyzed with SIFT-MS technology (Voice200®, Syft Technologies Ltd., Christchurch, New Zealand). The latter identifies and quantifies volatile compounds with a built-in scientific library, that is experimentally determined with product ions and reaction rate coefficients, in combination with the flow tube geometry, the ionic reaction time, the measured flow rates and the pressure applied during sample analysis and requires no chemical standards for calibration. Skim milk samples (35 mL) were placed in 300 mL glass jars with lids closed and then warmed to room temperature before analysis. An empty jar was used as a blank. Twenty-one flavor compounds associated with milk were quantified using selected ion mode with 100 ms time limit, count limit of 10,000, 30 s background scan time and a product scan time of 60 s. Helium was applied as the carrier gas (250 kPa) and the scan was performed following charge transfer from three positively-charged reagent ions (H_3O^+, NO^+, and O_2^+) to the analyte in the flow tube. The temperature of the inlet arm extension was 120 °C and flow tube pressure was 113.6 ± 1.7 mTorr.

Samples were collected in triplicate for each treatment with a duplicate analysis for each sample.

2.8. Statistical Data Analysis

Color and flavor data were statistically analyzed with Sigmaplot version 11.0 (Systat Software Inc., London, UK). One-way analysis of variance (ANOVA) was carried out for three different batches of skim milk ($n = 3$), analyzing all treatments at a confidence interval of 95%. When statistical differences between treatment means were observed, the Holm-Sidak method for multiple pairwise comparisons was used. For adequate reproducibility each treatment was repeated three times and analysis was performed in duplicate.

3. Results

3.1. Color Attributes of Non-Thermally and Thermally Treated Skim Milk

No significant difference in color was found in untreated skim milk and when processed using PEF-L, PEF-M, PEF-H, MF-1.2 and MF-1.4 and different combinations of PEF with MF treatments. The values for the color attributes L, a, and b following various treatments are provided in Table 1.

Table 1. Comparison of Hunter color space attributes L (lightness), a (redness) and b (yellowness) measured in raw skim milk and after processing with high-temperature short-time pasteurization at 75 (HTST-75) and 95 (HTST-95) °C, pulsed electric field at low (PEF-L), moderate (PEF-M) and high (PEF-H) intensities, 0.65 and 1.2 μm pore size tangential flow microfiltration (0.65 TFMF and 1.2 TFMF, respectively), 1.4 μm-pore size cross-flow microfiltration (1.4 CFMF), and different pulsed electric field-based combinations with the latter three membrane filtration treatments.

Treatments	L	a	b	ΔE
Raw Skim Milk	64.50 (±0.87) [a]	−3.92 (±0.08) [a]	−0.15 (±0.15) [a]	-
HTST-75	65.00 (±0.61) [a]	−4.01 (±0.10) [a]	−0.08 (±0.02) [a]	0.51
HTST-95	65.21 (±0.10) [a]	−4.03 (±0.01) [a]	0.19 (±0.04) [a]	0.79
0.65 TFMF	53.90 (±1.25) [b]	−3.92 (±0.03) [a]	−3.21 (±0.28) [b]	11.03
1.2 TFMF	64.22 (±0.05) [a]	−3.94 (±0.05) [a]	−0.28 (±0.17) [a]	0.31
1.4 CFMF	64.71 (±0.08) [a]	−3.89 (±0.03) [a]	−0.33 (±0.04) [a]	0.10
PEF-L	64.26 (±0.68) [a]	−3.88 (±0.02) [a]	−0.20 (±0.17) [a]	0.25
PEF-M	64.58 (±0.51) [a]	−3.88 (±0.01) [a]	−0.16 (±0.26) [a]	0.10
PEF-H	64.85 (±0.60) [a]	−3.96 (±0.05) [a]	−0.07 (±0.16) [a]	0.37
PEF-L/0.65 TFMF	53.07 (±0.65) [b]	−3.99 (±0.04) [a]	−3.12 (±0.11) [b]	11.81
PEF-M/0.65 TFMF	53.37 (±1.23) [b]	−3.99 (±0.04) [a]	−3.52 (±0.03) [b]	11.63
PEF-H/0.65 TFMF	53.44 (±1.38) [b]	−4.02 (±0.04) [a]	−3.28 (±0.24) [b]	11.49
PEF-L/1.2 TFMF	64.51 (±0.50) [a]	−3.92 (±0.02) [a]	−0.15 (±0.22) [a]	0.02
PEF-M/1.2 TFMF	64.36 (±0.42) [a]	−3.92 (±0.01) [a]	−0.23 (±0.03) [a]	0.15
PEF-H/1.2 TFMF	64.65 (±0.32) [a]	−4.00 (±0.02) [a]	−0.20 (±0.06) [a]	0.18
PEF-M/1.4 CFMF	64.57 (±0.18) [a]	−3.91 (±0.02) [a]	−0.34 (±0.07) [a]	0.20
P [1]	*	NS	*	-
SEM [2]	0.539	0.010	0.148	0.711

[a, b] Different superscripted letters in the same column indicate statistical significance between the two means. Values in parentheses following the mean values of the color attributes indicate the standard deviations.
[1] P stands for the statistical probability. * Refers to a statistical significance of $p < 0.05$. NS indicates no statistical significance. [2] SEM abbreviates the standard error of the mean.

Total color difference, ΔE, representing an overall color difference of treated samples from the untreated skim milk was calculated using equation 1 and presented in Table 1. Color difference classification was adopted from Cserhalmi and others [27]. Based on this classification system ΔE can be categorized as: 0 to 0.5 = "not noticeable", 0.5 to 1.5 = "slightly noticeable" and >1.5 = "noticeable". Slightly noticeable color differences of 0.51 and 0.79 were observed in skim milk treated with HTST treatment at 75 °C or 95 °C, respectively. However, a noticeable ΔE of 11.03 was obtained in skim milk processed with MF-0.65 ($p < 0.05$). Similar values ($p \geq 0.05$) were obtained when combination treatments were applied consisting of PEF (PEF-L, PEF-M, and PEF-H) and subsequent MF-0.65 for hurdle treatment of skim milk, yielding ΔEs of 11.83, 11.63 and 11.49, respectively. Overall, noticeably better color retention was achieved with PEF-L, PEF-M and PEF-H treatments; 1.2 μm TFMF and 1.4 μm CFMF-based processing, achieving lower ΔE values (in the range between 0.10 and 0.37 ($p < 0.05$)) than by the HTST and 0.65 μm TFMF treatments.

While skim milk pasteurized at 95 °C showed a positive yellowness value, no significant difference in any of the color attributes was observed ($p \geq 0.05$) between the heat-treated skim milk compared to that processed by PEF, MF-1.2 and MF-1.4, and their combinations. Lightness was 16%–18% lower for all treatments that involved skim milk processing through 0.65 μm MF and the corresponding samples also exhibited a higher blueness as indicated by a lower b value ($p < 0.05$). As a result significantly lower whiteness index values were obtained for MF-0.65 and hurdle treatments combining PEF of different processing intensities with MF-0.65 as shown in Figure 1 ($p < 0.05$).

Figure 1. Whiteness obtained in untreated raw skim milk and by processing the latter with low-intensity (PEF-L), medium-intensity (PEF-M) and high-intensity (PEF-H) pulsed electric field, 0.65 (MF-0.65), 1.2 (MF-1.2), and 1.4 (MF-1.4) μm microfiltration, high-temperature short-time (HTST) pasteurization treatments and different pulsed electric field-based combinations with microfiltration treatments (after conversion from Hunter Lab color space data to CIE 1931 XYZ color space values). Different letters (*i.e.*, *a*, *b*) above bars indicate significant differences between treatment means ($p < 0.05$).

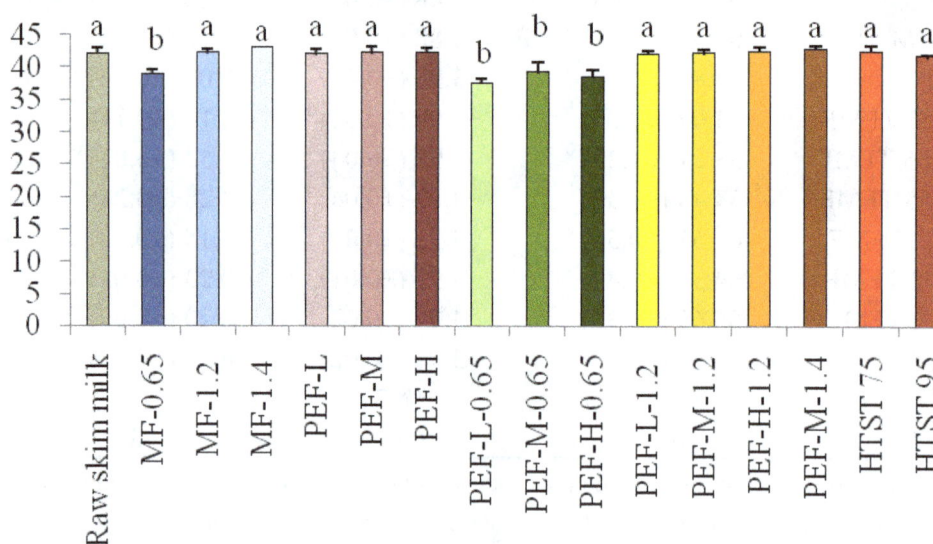

For the remaining non-thermal or thermal treatments no significant differences in whiteness were determined when compared among one another or with raw skim milk ($p \geq 0.05$). Overall, non-thermal treatments using MF-1.2 MF-1.4 and PEF-L, PEF-M, PEF-H, and heat pasteurization did not have a significant impact on the single color properties of skim milk but, with the exception of MF-0.65-based treatments, thermally pasteurized skim milk was in a higher ΔE category than non-thermally treated skim milk.

3.2. Flavor Attributes of Non-Thermally and Thermally Treated Skim Milk

The effect of heat, pulsed electric field and microfiltration treatments on the volatile profile of skim milk was measured using SIFT-MS. Volatile compounds generated as a result of heat treatment and non-enzymatic browning reactions in skim milk were of main interest and, therefore, 21 compounds were selected for this study. Quantities of acetic acid and acetaldehyde were not detected by the mass spectrometer, so that data for 19 compounds (Tables 2–4) was obtained. Concentrations of ketones were found to be significantly higher in skim milk treated with HTST at 95 °C for a holding time of 45 s when compared to other treatments ($p < 0.05$). The major ketones contributing to this were 2-hexanone, 2-heptanone, 2-octanone, acetone, and butanone. However, there was no marked increase in concentration of these ketones in skim milk pasteurized at 75 °C for 20 s, nor for PEF-treated and MF-treated skim milk ($p \geq 0.05$).

The aldehydes in skim milk were altered in a different way as compared to ketones in skim milk. Thermal treatments had a significant impact on 3-methylbutanal levels. The concentration of 3-methylbutanal was increased significantly upon heat treatment at 95 °C ($p < 0.05$). By contrast, microfiltration did not have an impact on aldehyde concentrations in skim milk. HTST treatments at 75 and 95 °C, PEF and microfiltration treatments did not have a significant impact on the concentration of heptanal or decanal ($p \geq 0.05$).

Short chain fatty acids were the other group of volatiles analyzed in skim milk. Of these, hexanoic acid and butanoic acid were detected in raw skim milk, heat-treated, PEF-treated and microfiltered skim milk. Increased intensity of thermal treatments resulted in an increased concentration of hexanoic acid and the values were significantly higher in skim milk subjected to both HTST-75 and HTST-95 ($p < 0.05$) than in raw skim milk. Pasteurization at the higher temperature (HTST-95) led to a significant increase ($p < 0.05$) in the concentration of butanoic acid, whereas no impact ($p \geq 0.05$) on its concentration was found for skim milk pasteurized at 75 °C. Moreover, no significant difference ($p \geq 0.05$) in hexanoic and butanoic acid concentrations was observed between raw, PEF, MF, and hurdle treated skim milk.

The concentrations of both ethanol and propanol increased in skim milk following heat pasteurization at 95 °C ($p < 0.05$). Contrasting results ($p \geq 0.05$) were obtained for PEF, MF, hurdle processing and HTST-75 treatments, indicating that their concentrations did not change in skim milk subjected to these processing conditions. Heat treatment at 95 °C resulted in increased levels of toluene ($p < 0.05$), while there was no significant impact by HTST-75, PEF, and MF treatments on this volatile compound ($p \geq 0.05$).

The methyl acetate concentration was significantly higher ($p < 0.05$) in skim milk processed at 95 °C. None of the other processes affected the concentration of this volatile component ($p \geq 0.05$). Furthermore, two sulfur compounds, hydrogen sulfide and dimethyl sulfide, were detected by the SIFT-MS equipment. Hydrogen sulfide concentrations were found to be at significantly higher levels in HTST-75 and HTST-95 treated skim milk, reaching the highest levels with the latter treatment ($p < 0.05$). The highest concentration of dimethyl sulfide occurred in skim milk that underwent the HTST-95 treatment. PEF processing and microfiltration did not increase the content of sulfur compounds.

Table 2. Retention of in 3-methyl butanal, heptanal, decanal, butanoic acid, hexanoic acid, dimethyl sulfide, and hydrogen sulfide obtained in skim milk after high-temperature short-time pasteurization (HTST$_{75}$ = 75 °C for 20 s; HTST$_{95}$ = 95 °C for 45 s) and lab-scale tangential-flow microfiltration (TFMF$_{0.65}$ = 0.65 μm; TFMF$_{1.2}$ = 1.2 μm), pilot scale cross-flow microfiltration (CFMF$_{1.4}$ = 1.4 μm), pulsed electric field at low (PEF-L: 28 kV/cm; 2805 μs), moderate (PEF-M: 40 kV/cm, 1122 μs), and high (PEF-H: 40 kV/cm, 1571 μs) processing intensities and selected non-thermal combination treatments using PEF and subsequent microfiltration.

Treatments	3-Methyl butanal [ppb]	Heptanal [ppb]	Decanal [ppb]	Butanoic acid [ppb]	Hexanoic acid [ppb]	Dimethyl sulfide [ppb]	Hydrogen sulfide [ppb]
Raw Skim Milk	3.6 (±1.1) [a]	15 (±6.2) [a]	2.7 (±0.5) [a]	14 (±6.0) [a]	2.3 (±0.8) [a]	12 (±5.9) [a]	1.4 (±0.7) [a,b]
HTST$_{75}$	3.3 (±0.1) [a]	15 (±1.9) [a]	3.3 (±0.0) [a]	13 (±1.4) [a]	9.7 (±0.3) [d]	17 (±2.2) [a,b]	4.2 (±0.1) [c]
HTST$_{95}$	6.3 (±0.7) [b]	24 (±5.2) [a]	4.8 (±1.6) [a]	44 (±6.0) [b]	4.8 (±0.2) [c]	25 (±6.1) [b]	5.5 (±0.2) [d]
TFMF$_{0.65}$	3.9 (±0.0) [a]	14 (±10.3) [a]	2.2 (±0.3) [a]	15 (±5.8) [a,b]	1.1 (±0.3) [a]	11 (±0.7) [a,b]	n.d.
TFMF$_{1.2}$	4.5 (±1.7) [a]	17 (±7.3) [a]	3.4 (±1.4) [a]	13 (±1.8) [a]	2.4 (±0.2) [a,b]	12 (±0.9) [a,b]	1.1 (±0.5) [a,b]
CFMF$_{1.4}$	3.5 (±1.5) [a]	11 (±1.8) [a]	2.3 (±0.8) [a]	11 (±3.1) [a]	2.5 (±0.3) [a,b]	11 (±0.5) [a,b]	1.1 (±0.4) [a,b]
PEF-L	2.4 (±0.4) [a]	10 (±0.7) [a]	2.5 (±0.7) [a]	12 (±3.3) [a]	1.6 (±0.3) [a,b]	9.1 (±3.4) [a]	0.8 (±0.1) [a]
PEF-M	3.0 (±1.7) [a]	17 (±2.9) [a]	3.4 (±0.8) [a]	16 (±7.2) [a,b]	1.8 (±0.5) [a,b]	15 (±7.0) [a,b]	1.3 (±0.2) [a, b]
PEF-H	3.9 (±2.8) [a]	14 (±2.2) [a]	2.6 (±1.1) [a]	15 (±5.1) [a,b]	2.7 (±0.1) [b]	18 (±2.1) [a,b]	2.1 (±0.1) [b]
PEF-L/TFMF$_{0.65}$	2.5 (±0.4) [a]	20 (±0.7) [a]	3.9 (±2.3) [a]	10 (±3.4) [a]	1.7 (±0.2) [a,b]	5.0 (±0.8) [a]	1.1 (±0.3) [a,b]
PEF-M/TFMF$_{0.65}$	3.6 (±2.8) [a]	21 (±1.0) [a]	3.3 (±1.6) [a]	13 (±5.2) [a]	1.7 (±0.5) [a,b]	7.5 (±1.3) [a]	0.9 (±0.3) [a,b]
PEF-H/TFMF$_{0.65}$	4.8 (±2.5) [a]	16 (±2.1) [a]	3.2 (±0.5) [a]	9.1 (±2.3) [a]	1.5 (±0.5) [a,b]	7.9 (±0.7) [a]	1.1 (±0.2) [a,b]
PEF-L/TFMF$_{1.2}$	2.1 (±0.5) [a]	14 (±2.7) [a]	2.5 (±0.1) [a]	8.6 (±1.9) [a]	1.7 (±0.1) [a,b]	9.2 (±1.9) [a]	0.9 (±0.2) [a,b]
PEF-M/TFMF$_{1.2}$	5.6 (±0.2) [a]	20 (±0.4) [a]	1.5 (±1.2) [a]	16 (±0.3) [a,b]	2.5 (±0.1) [a,b]	17 (±2.7) [a,b]	1.6 (±0.2) [a,b]
PEF-H/TFMF$_{1.2}$	2.7 (±1.3) [a]	13 (±3.0) [a]	1.8 (±0.2) [a]	8.9 (±3.2) [a]	2.1 (±0.5) [a,b]	12 (±2.4) [a,b]	1.2 (±0.3) [a,b]
PEF-M/CFMF$_{1.4}$	2.0 (±0.2) [a]	17 (±1.2) [a]	3.5 (±0.1) [a]	9.1 (±0.7) [a]	1.4 (±0.4) [a,b]	14 (±2.3) [a,b]	1.0 (±0.0) [a,b]
P [1]	*	NS	NS	*	*	*	*
SEM [2]	0.19	0.64	0.12	1.53	0.20	0.87	0.11

[a, b, c] Different superscripted letters in the same column indicate statistical significance between the two means. Values in parentheses following the mean values of compound indicate the standard deviations. [1] P stands for the statistical probability. * Refers to a statistical significance of $p < 0.05$. NS indicates no statistical significance. [2] SEM abbreviates the standard error of the mean.

Table 3. Retention of in ethanol, 1-propanol, 1-octene, toluene, p-xylene, and methyl acetate obtained in skim milk after high-temperature short-time pasteurization (HTST$_{75}$ = 75 °C for 20 s; HTST$_{95}$ = 95 °C for 45 s) and lab-scale tangential-flow microfiltration (TFMF$_{0.65}$ = 0.65 μm; TFMF$_{1.2}$ = 1.2 μm), pilot scale cross-flow microfiltration (CFMF$_{1.4}$ = 1.4 μm), pulsed electric field at low (PEF-L: 28 kV/cm; 2805 μs), moderate (PEF-M: 40 kV/cm, 1122 μs), and high (PEF-H: 40 kV/cm, 1571 μs) processing intensities and selected non-thermal combination treatments using PEF and subsequent microfiltration.

Treatments	Ethanol [ppb]	1-Propanol [ppb]	1-Octene [ppb]	Toluene [ppb]	p-Xylene [ppb]	Methyl acetate [ppb]
Raw Skim Milk	65 (±18) [a]	10 (±2.0) [a]	34 (±8.3) [a]	4.2 (±0.8) [a]	1.6 (±0.9) [a]	6.7 (±1.3) [a]
HTST$_{75}$	57 (±9.9) [a,b]	11 (±1.8) [a,b]	31 (±2.1) [a]	4.6 (±0.8) [a]	1.5 (±0.3) [a]	10 (±3.0) [a,b]
HTST$_{95}$	131 (±1.8) [c]	16 (±2.1) [b]	38 (±1.5) [a]	6.9 (±1.6) [b]	2.0 (±0.8) [a]	19 (±4.1) [b]
TFMF$_{0.65}$	51 (±6.0) [a,b]	11 (±0.8) [a,b]	15 (±1.7) [a]	4.0 (±0.7) [a]	1.6 (±1.1) [a]	6.7 (±0.8) [a,b]
TFMF$_{1.2}$	60 (±0.9) [a,b]	10 (±1.3) [a,b]	33 (±8.3) [a]	4.6 (±0.6) [a]	1.3 (±0.3) [a]	6.3 (±0.5) [a,b]
CFMF$_{1.4}$	61 (±27) [a,b]	9.2 (±3.3) [a,b]	22 (±3.7) [a]	4.5 (±1.8) [a]	2.7 (±0.5) [a]	5.6 (±2.3) [a,b]
PEF-L	39 (±2.0) [a]	7.1 (±1.8) [a,b]	28 (±7.8) [a]	2.8 (±0.2) [a]	0.9 (±0.5) [a]	3.5 (±0.1) [a]
PEF-M	64 (±16) [a,b]	9.3 (±2.8) [a,b]	39 (±8.0) [a]	5.5 (±0.2) [a]	1.4 (±0.4) [a]	7.8 (±3.1) [a,b]
PEF-H	61 (±7.1) [a,b]	9.6 (±4.3) [a,b]	29 (±4.7) [a]	2.4 (±0.1) [a]	2.5 (±0.2) [a]	7.5 (±0.7) [a,b]
PEF-L/TFMF$_{0.65}$	18 (±13) [a]	4.8 (±0.2) [a]	31 (±11) [a]	2.6 (±0.0) [a]	1.9 (±0.0) [a]	5.4 (±0.4) [a]
PEF-M/TFMF$_{0.65}$	39 (±36) [a]	6.1 (±1.7) [a,b]	26 (±5.7) [a]	3.9 (±0.3) [a]	2.1 (±0.6) [a]	3.8 (±0.4) [a]
PEF-H/TFMF$_{0.65}$	44 (±2.9) [a]	5.2 (±2.2) [a]	24 (±3.4) [a]	2.2 (±0.2) [a]	1.2 (±0.5) [a]	5.2 (±0.5) [a]
PEF-L/TFMF$_{1.2}$	58 (±2.6) [a,b]	5.2 (±0.1) [a]	28 (±7.7) [a]	2.0 (±0.4) [a]	0.9 (±0.3) [a]	4.1 (±0.6) [a]
PEF-M/TFMF$_{1.2}$	84 (±5.3) [b,c]	12 (±0.8) [a,b]	37 (±1.8) [a]	5.6 (±0.7) [a]	3.0 (±1.1) [a]	6.4 (±0.5) [a,b]
PEF-H/TFMF$_{1.2}$	52 (±3.4) [a,b]	7.4 (±3.2) [a]	29 (±2.2) [a]	3.0 (±0.9) [a]	1.1 (±0.6) [a]	3.8 (±2.2) [a]
PEF-M/CFMF$_{1.4}$	42 (±12) [a,b]	6.6 (±1.5) [a,b]	25 (±5.8) [a]	2.2 (±0.4) [a]	1.5 (±0.1) [a]	4.1 (±0.1) [a]
P [1]	*	*	N.S.	*	N.S.	*
SEM [2]	3.06	0.34	0.95	0.17	0.13	0.37

[a, b, c] Different superscripted letters in the same column indicate statistical significance between the two means. Values in parentheses following the mean values of compound indicate the standard deviations. [1] P stands for the statistical probability. * Refers to a statistical significance of $p < 0.05$. NS indicates no statistical significance. [2] SEM abbreviates the standard error of the mean.

Table 4. Retention of in acetone, butanone, 2-hexanone, 2-heptanone, 2-octanone and 2-nonanone obtained in skim milk after high temperature short time pasteurization (HTST$_{75}$ = 75 °C for 20 s; HTST$_{95}$ = 95 °C for 45 s) and lab-scale tangential-flow microfiltration (TFMF$_{0.65}$ = 0.65 μm; TFMF$_{1.2}$ = 1.2 μm), pilot scale cross-flow microfiltration (CFMF$_{1.4}$ = 1.4 μm), pulsed electric field at low (PEF-L: 28 kV/cm; 2805 μs), moderate (PEF-M: 40 kV/cm, 1122 μs), and high (PEF-H: 40 kV/cm, 1571 μs) processing intensities and selected non-thermal combination treatments using PEF and subsequent the microfiltration.

Treatments	Acetone [ppb]	Butanone [ppm]	2-Hexanone [ppb]	2-Heptanone [ppb]	2-Octanone [ppb]	2-Nonanone [ppb]
Raw Skim Milk	2.7 (± 0.8) [a]	174 (±51) [a]	6.8 (±3.1) [a]	4.9 (±2.5) [a]	9.7 (±6.5) [a]	4.2 (±1.1) [a]
HTST$_{75}$	4.0 (±0.4) [a, b]	225 (±23) [a, b]	9.0 (±1.1) [a, b]	4.8 (±1.7) [a, b]	11 (±1.8) [a, b]	5.5 (±0.8) [a]
HTST$_{95}$	6.2 (±1.0) [b]	344 (±0.7) [b]	9.3 (±0.3) [b]	12 (±2.6) [b]	16 (±3.6) [b]	5.9 (±0.3) [a]
TFMF$_{0.65}$	2.4 (±0.4) [a]	169 (±23) [a]	2.8 (±1.6) [a]	4.6 (±1.1) [a, b]	8.9 (±1.4) [a, b]	3.3 (±0.1) [a]
TFMF$_{1.2}$	2.6 (±0.2) [a, b]	188 (±56) [a]	3.4 (±1.8) [a, b]	4.9 (±4.0) [a, b]	9.5 (±3.5) [a, b]	3.7 (±2.7) [a]
CFMF$_{1.4}$	2.3 (±0.9) [a]	140 (±51) [a]	7.0 (±1.6) [a, b]	3.7 (±0.3) [a, b]	6.5 (±0.6) [a]	5.0 (±2.4) [a]
PEF-L	1.8 (±1.0) [a]	120 (±32) [a]	5.7 (±1.5) [a, b]	3.9 (±0.2) [a, b]	7.2 (±1.4) [a, b]	2.6 (±0.4) [a]
PEF-M	2.6 (±1.9) [a, b]	135 (±74) [a]	7.0 (±0.7) [a, b]	5.1 (±2.8) [a, b]	8.8 (±3.5) [a, b]	3.5 (±1.9) [a]
PEF-H	3.1 (±2.2) [a, b]	147 (±58) [a]	7.8 (±0.3) [a, b]	4.8 (±2.2) [a, b]	7.9 (±4.2) [a, b]	5.4 (±0.9) [a]
PEF-L/TFMF$_{0.65}$	1.5 (±0.9) [a]	54 (±8.0) [a]	5.0 (±0.1) [a, b]	6.0 (±1.2) [a, b]	8.7 (±1.8) [a, b]	5.2 (±3.8) [a]
PEF-M/TFMF$_{0.65}$	1.9 (±1.1) [a]	75 (±40) [a]	5.7 (±0.2) [a, b]	6.5 (±1.4) [a, b]	8.8 (±1.8) [a, b]	5.4 (±3.6) [a]
PEF-H/TFMF$_{0.65}$	1.9 (±1.1) [a]	80 (±41) [a]	6.9 (±0.6) [a, b]	4.7 (±3.1) [a, b]	7.4 (±2.8) [a, b]	4.5 (±0.9) [a]
PEF-L/TFMF$_{1.2}$	1.5 (±0.4) [a]	56 (±18) [a]	5.0 (±0.1) [a, b]	3.5 (±0.3) [a]	7.5 (±0.3) [a, b]	1.9 (±0.5) [a]
PEF-M/TFMF$_{1.2}$	3.0 (±0.2) [a, b]	74 (±26) [a]	7.6 (±0.4) [a, b]	6.6 (±1.0) [a, b]	12 (±0.0) [a, b]	5.7 (±0.4) [a]
PEF-H/TFMF$_{1.2}$	2.2 (±0.6) [a]	84 (±60) [a]	5.4 (±1.3) [a, b]	4.8 (±1.2) [a, b]	8.3 (±0.9) [a, b]	2.9 (±0.1) [a]
PEF-M/CFMF$_{1.4}$	2.1 (±0.4) [a]	81 (±11) [a]	8.9 (±0.2) [b]	3.4 (±0.3) [a]	8.8 (±2.4) [a]	4.0 (±0.7) [a]
P [1]	*	*	*	*	*	NS
SEM [2]	0.12	8.01	0.38	0.36	0.54	0.23

[a, b, c] Different superscripted letters in the same column indicate statistical significance between the two means. Values in parentheses following the mean values of compound indicate the standard deviations. [1] P stands for the statistical probability. * Refers to a statistical significance of $p < 0.05$. NS indicates no statistical significance. [2] SEM abbreviates the standard error of the mean.

4. Discussion

For Hunter *L*, *a*, *b* color space values, no significant differences were observed in lightness, redness, and yellowness for skim milk treated with PEF, HTST and, MF-1.2 and MF-1.4 ($p \geq 0.05$). A significant reduction in *L*, *b* and whiteness index values for skim milk that was microfiltered through a 0.65 μm pore size membrane indicates that higher cut-off membranes partially retain suspended milk components [28,29]. Pafylias *et al.* [30] suggested the use of larger pore size membranes of about 1.4 μm to maintain a balance between reduction of bacterial cells and retention of milk components. However, we did not see any significant differences in color attributes for skim milk processed through 1.2 and 1.4 μm pore size MF. Silva *et al.* [31] compared instrumental color differences between microfiltered and pasteurized skim milk and found smaller changes in color coordinates of microfiltered skim milk due to a lack of reactions caused by heating skim milk to pasteurization temperatures. HTST treatments caused a small increase in lightness values for all the samples, which may be due to an increase in number of dispersed components as a result of denaturation of milk proteins at high temperatures [20]. However, the increase in lightness was not significant ($p \geq 0.05$). The Hunter-*b* value was positive for HTST-95 treatment indicating yellowness in skim milk. This might be attributed to heat induced browning reactions in milk between lactose and amino acids [32–34]. However, PEF and MF-treated skim milk retained milk color better than heat treatments as indicated by "not noticeable" total color difference values compared to raw skim milk. As indicated by ΔE values, there was a lower impact on color of skim milk produced with the non-thermal technologies, with the exception of the smallest pore size MF that produced "unacceptable" color values. However, our results are not in agreement with results of Bermúdez-Aguirre *et al.* [35] who observed significant changes in *L*, *a*, *b* values of PEF-treated skim milk. The authors, however, did not observe any trends in color change and attributed this change partly to the wearing down of the electrode due to arcing.

Among the 21 volatile compounds analyzed, ketones represented a major class based on their expected contribution to flavor. Ketones in milk can originate from different sources. Some of them may be present in raw milk as a result of feed, e.g., acetone and 2-butanone, while others may be generated or accumulated as a result of β-oxidation of saturated fatty acids followed by decarboxylation or by β-ketoacid decarboxylation [36,37]. A significant increase in concentration of all methyl ketones, except 2-nonanone, was observed for samples subjected to heat treatment ($p < 0.05$). Increased levels of methyl ketones in milk due to thermal treatments have also been reported by other researchers [23,38]. The methyl ketone content of skim milk treated with either PEF or MF alone was similar to that of raw skim milk ($p \geq 0.05$), indicating that they have a minimum effect on the concentration of lipids in skim milk. Comparable results ($p \geq 0.05$) were obtained by hurdle processing of skim milk using both PEF and MF, suggesting a non-significant impact of combined processes on ketones in skim milk.

Another important class of compounds identified and relevant to this study were the sulfur compounds: dimethyl and hydrogen sulfide. These compounds are formed in milk as a result of thermal denaturation of milk whey proteins during processing and are responsible for cooked flavors in milk. Hydrogen sulfide is produced mainly from Strecker degradation of sulfhydryl groups of sulfur-containing amino acids (*i.e.*, cysteine and methionine) in whey proteins. Its production is also attributed to denatured proteins associated with the milk fat globule membrane and the rearrangement of a thiazole group during thermal degradation of thiamine [36,39,40]. In our research, the concentration of hydrogen sulfide in raw skim milk was found to be 1.39 ± 0.68 µg/L. Its concentration was higher when skim milk was pasteurized at 95 °C for 45 s (5.53 µg/L) as compared to skim milk subjected to 75 °C for 20 s (4.21 µg/L). Thus, the hydrogen sulfide concentration was found to increase with the severity of heat treatment. In contrast, PEF treatment applied at the highest intensity, which is known to generate increased temperature in the treatment chamber to about 65 °C by means of Joule heating, however, did not cause any significant increase in hydrogen sulfide concentration (2.05 µg/L) ($p \geq 0.05$). The PEF-L treatment that resulted in the longest exposure (2805 µs) of skim milk to electric pulses also did not cause any significant changes in hydrogen sulfide concentration (0.82 µg/L) when compared to raw skim milk ($p \geq 0.05$). The increase in concentration of sulfur compounds has been associated with growing denaturation of sulfur containing polypeptides affected to a large degree by increasing exposure to heat [39,41]. Dimethyl sulfide has been reported to occur naturally in cow's milk as a result of diet and at low levels it contributes to milk flavor but at high concentrations it may impart an off-flavor to the milk [42,43]. It can also be formed as a result of thermal denaturation of sulfur containing amino acids in milk. Increased dimethyl sulfide content in milk subjected to heat treatment at pasteurization temperatures and ultra-high-temperatures has been reported by different research groups [22,44,45]. In this study, dimethyl sulfide concentration was observed to be significantly higher in skim milk processed at 95 °C for 45 s (23.6 µg/L) as compared to raw skim milk (6.17 to 17.95 µg/L). However, no significant differences in concentrations of dimethyl sulfide from those in raw skim milk were observed for milk that underwent HTST-75, PEF, and MF treatments ($p \geq 0.05$). Hurdle processing of PEF with MF also did not cause any significant changes in concentrations of sulfur compounds in comparison to raw skim milk ($p \geq 0.05$).

Among aldehydes, acetaldehyde concentrations were below the detection limit of SIFT-MS for raw as well as treated skim milk. Aliphatic aldehydes are produced as a result of auto-oxidation of unsaturated fatty acids as well as by breakdown of amino acids in milk [37,40]. No significant differences were found in heptanal and decanal concentrations among raw skim milk, HTST-treated, PEF-treated, and MF-treated skim milk ($p \geq 0.05$). In contrast, Zhang and others [5] observed significantly increased heptanal and decanal levels in pasteurized milk in comparison to raw milk while no significant differences were found in PEF-treated skim milk. However, the increase in aldehyde concentrations in skim milk upon exposure to heat treatment at 95 °C, though not significant ($p \geq 0.05$) could contribute to significant changes in aroma of skim milk due to lower threshold values for some aldehydes [44]. 3-Methyl butanal is a product of Strecker degradation of leucine during the non-enzymatic browning reaction, which occurs during heat treatment of milk [36,37]. No significant increase in its concentration was found after non-thermal PEF and MF treatments ($p \geq 0.05$). Its presence in skim milk treated at 95 °C was found to be significantly higher ($p < 0.05$) than in raw skim milk and in skim milk processed by all other treatments tested. Contarini and Povolo [23] also observed a positive correlation between increased levels of 3-methyl butanal and severity of heat treatment, while no significant differences were found in heptanal concentrations.

Short-chain and medium-chain fatty acids (C-4 to C-14 and some C-16) comprise 45% of total fatty acids in milk fat and are produced in the mammary gland from acetate and β-hydroxybutyrate, the products of bacterial fermentation in the rumen. Activation of acetyl to acetyl CoA and its carboxylation to malonyl CoA results in a stepwise addition of two CH_2 groups and hence, increase in fatty acid chain length from short- to medium-chain fatty acids [46]. Out of these straight-chained, even numbered fatty acids, butanoic and hexanoic acids were quantified in this study. An increase in concentration of these two short-chain fatty acids as a result of heat exposure has been reported by Gandy and others [24]. Significantly higher amounts of both butanoic acid and hexanoic acid were observed in skim milk treated at 95 °C. However, HTST pasteurization at 75 °C showed a greater increase in hexanoic acid and, interestingly, the increase was twice the increase observed for the 95 °C treatment. Increased levels of butanoic and hexanoic acid in heat treated milk may impart cheesy, sour, and cream notes to milk as classified by Zhang and others [5] under the category of high olfactometric intensity odorants.

Formation of alcohols by reduction of corresponding carbonyl compounds [29] was observed for milk stored at refrigeration temperatures, where alcohol concentration increased as result of reduction of carbonyl compounds [30,47]. HTST-75, PEF, and MF treatments did not have any significant impact on the concentration of ethanol and 2-propanol in skim milk. Similarly, Zhang et al. [5] reported no marked change in concentrations of alcohols subsequent to PEF and HTST processing. However, at a processing temperature of 95 °C, a marked increase in concentration of both these alcohols was observed, which indicates some reducing reactions take place at this temperature. Toluene and *p*-xylene are considered as products of thermal degradation of β-carotene [48] and these compounds have been associated with heat treated milk [24,38,49]. Esters are formed by reaction of carboxylic acids with alcohols. Methyl acetate was the only ester quantified in this study and its concentration increased three-fold in HTST-95 treated skim milk. As observed in similar studies [3,50,51] non-thermal processing had a lesser impact on the concentration of volatile compounds in skim milk than thermal processing and indicates that skim milk of better quality results from non-thermal processing. Last but not least, it is worthy of mention that different scales (lab, pilot plant) of MF processing equipments (1.2 μm TFMF and 1.4 μm CFMF, respectively) used in this study indicated comparable effects on color and composition of volatiles in skim milk, thereby, also substantiating that different designs and materials of the MF membrane modules had no significant effect on these organoleptical quality parameters.

5. Conclusions

MF (1.2 and 1.4 μm) and PEF alone and in combination (hurdle technology) caused no significant changes to skim milk color and its volatile compound composition, thus, the hurdle approach can be regarded as promising for preservation of skim milk of very good sensory quality. No noticeable color differences from raw skim milk were observed by application of PEF, and MF with a pore size of 1.2 μm or above. Smaller pore size microfiltration (0.65 μm) was found to be unsatisfactory when applied alone or in combination with PEF, resulting in significant changes to skim milk color attributes. Different scale, membrane materials, and membrane designs of the MF systems showed no effect on color and volatile compounds. A significant increase in ketones, fatty acids, hydrocarbons and sulfur compounds was obtained following heat treatment of skim milk at the higher intensity treatment (95 °C, 45 s) applied in this study, while lower intensity heat treatment (75 °C, 20 s) also resulted in increased concentration of selected volatile compounds. However, increased PEF intensity did not significantly alter concentrations of volatile compounds. These findings indicate PEF/MF is an emerging technology with potential to produce skim milk of high sensory quality. Further research on this promising hurdle technology is envisaged such as the study of enzyme activity and vitamin retention in PEF/MF-treated skim milk as well as its application for processing other liquid dairy products.

Acknowledgments

The authors are thankful for the financial support by the Dairy Farmers of Ontario and the Natural Sciences and Engineering Research Council of Canada that both made the present study possible. Moreover, we are grateful for the supply of raw milk by the Gay Lea Foods Co-operative Limited and for the use of the pilot scale heat exchanger and microfiltration units provided by the Guelph Food Technology Centre.

Author Contributions

The present manuscript was written by Anupam Chugh, who carried out the experimental work together with Dipendra Khanal under the guidance and supervision of Markus Walkling-Ribeiro,

Milena Corredig, and Mansel W. Griffiths. The experimental design of the study was devised by Walkling-Ribeiro, also responsible for the article correspondence, and Griffiths, the principal investigator of the research project. Due to her expertise in sensory analyses Lisa Duizer, also a co-investigator of the research project, mentored Chugh and Khanal during the course of their volatile compound analyses.

Conflicts of Interest

The authors declare no conflict of interest.

References

1. Doan, F.J. The color of cow's milk and its value. *J. Dairy Sci.* **1924**, *7*, 147–153.
2. Thomas, E.L. Trends in milk flavors. *J. Dairy Sci.* **1981**, *64*, 1023–1027.
3. Sampedro, F.; Geveke, D.J.; Fan, X.; Zhang, H.Q. Effect of PEF, HHP and thermal treatment on PME inactivation and volatile compounds concentration of an orange juice-milk based beverage. *Innov. Food Sci. Emerg. Technol.* **2009**, *10*, 463–469.
4. Pereda, J.; Jaramillo, D.P.; Quevedo, J.M.; Ferragut, V.; Guamis, B.; Trujillo, A.J. Characterization of volatile compounds in ultra-high-pressure homogenized milk. *Int. Dairy J.* **2008**, *18*, 826–834.
5. Zhang, S.; Yang, R.; Zhao, W.; Hua, X.; Zhang, W.; Zhang, Z. Influence of pulsed electric field treatments on the volatile compounds of milk in comparison with pasteurized processing. *J. Food Sci.* **2011**, *76*, C127–C132.
6. Chugh, A.; Khanal, D.; Duizer, L.; Walkling-Ribeiro, M.; Griffiths, M.W. Effect of non-thermal pulsed electric fields and tangential-flow microfiltration processing on flavor and color properties of skim milk. In the Proceedings of the IFT Annual Meeting and Food Expo, Chicago, IL, USA, 14 July 2013; session 074-04.
7. Zimmermann, U. The effect of high intensity electric field pulses on eukaryotic cell membranes: Fundamentals and applications. In *Electromanipulation of Cells*; Zimmermann, U., Neil, G.A., Eds.; CRC Press: Boca Raton, FL, USA, 1996; pp. 1–106.
8. Waite-Cusic, J.G.; Diono, B.H.S.; Yousef, A.E. Screening for *Listeria monocytogenes* surrogate strains applicable to food processing by ultra high pressure and pulsed electric field. *J. Food Prot.* **2011**, *74*, 1655–1661.
9. Grahl, T.; Markl, H. Killing of microorganisms by pulsed electric fields. *Appl. Microbiol. Biotechnol.* **1996**, *45*, 148–157.
10. Griffiths, M.W.; Walkling-Ribeiro, M. Implementation of microbially safe foods with pulsed electric fields. In *Food Microbiology: Novel Food Preservation and Microbial Assessment Techniques*; Boziaris, I.S., Ed.; Science Publishers—Taylor & Francis CRC Press: Boca Raton, FL, USA, 2014.
11. GEA Filtration. Available online: http://www.geafiltration.com/applications/microfiltration_whole_skim_milk.asp (accessed on 1 September 2013).

12. Gésan-Guiziou, G. Removal of bacteria, spores and somatic cells from milk by centrifugation and microfiltration techniques. In *Improving the Safety and Quality of Milk, Volume 1—Milk Production and Processing*; Griffiths, M.W., Ed.; CRC Press: Boca Raton, FL, USA, 2010; pp. 356–369.

13. Rodríguez-González, O.; Walkling-Ribeiro, M.; Jayaram, S.; Griffiths, M.W. Factors affecting the inactivation of the natural microbiota of milk processed by pulsed electric fields and cross-flow microfiltration. *J. Dairy Res.* **2011**, *78*, 270–278.

14. Sobrino-López, A.; Martín-Belloso, O. Enhancing the lethal effect of high-intensity pulsed electric field in milk by antimicrobial compounds as combined hurdles. *J. Dairy Sci.* **2008**, *91*, 1759–1768.

15. Swanson, B.G.; Barbosa-Cánovas, G.V. Inactivation of *Listeria innocua* in skim milk by pulsed electric fields and nisin. *Int. J. Food Microbiol.* **1999**, *51*, 19–30.

16. Odriozola-Serrano, I.; Bendicho-Porta, S.; Martín-Belloso, O. Comparative study on shelf life of whole milk processed by high-intensity pulsed electric field or heat treatment. *J. Dairy Sci.* **2006**, *89*, 905–911.

17. Walkling-Ribeiro, M.; Rodríguez-González, O.; Jayaram, S.; Griffiths, M.W. Microbial inactivation and shelf life comparison of "cold" hurdle processing with pulsed electric fields and microfiltration, and conventional thermal pasteurisation in skim milk. *Int. J. Food Microbiol.* **2011**, *144*, 379–386.

18. Solah, V.A.; Staines, V.; Honda, S.; Limley, H.A. Measurement of milk colour and composition: Effect of dietary intervention on Western Australian Holstein-Friesian cow's milk quality. *J. Food Sci.* **2007**, *72*, S560–S566.

19. Nozière, P.; Graulet, B.; Lucas, A.; Martin, B.; Grolier, P.; Doreau, M. Carotenoids for ruminants: From forages to dairy products. *Anim. Feed Sci. Technol.* **2006**, *131*, 418–450.

20. Dunkerley, J.A.; Ganguli, N.C.; Zadow, J.G. Reversible changes in the colour of skim milk on heating suggest rapid reversible heat-induced changes in casein micelle size. *Aust. J. Dairy Technol.* **1993**, *48*, 66–70.

21. Browning, E.; Lewis, M.; Macdougall, D. Predicting safety and quality parameters for UHT-processed milks. *Int. J. Dairy Technol.* **2001**, *54*, 111–120.

22. Datta, N.; Elliott, A.J.; Perkins, M.L.; Deeth, H.C. Ultra-high-temperature (UHT) treatment of milk: Comparison of direct and indirect modes of heating. *Aust. J. Dairy Technol.* **2002**, *57*, 221–227.

23. Contarini, G.; Povolo, M. Volatile fraction of milk: Comparison between purge and trap and solid phase microextraction techniques. *J. Agric. Food Chem.* **2002**, *50*, 7350–7355.

24. Gandy, A.L.; Schilling, M.W.; Coggins, P.C.; White, C.H.; Yoon, Y.; Kamadia, V.V. The effect of pasteurisation temperature on consumer acceptability, sensory characteristics, volatile compound composition, and shelf-life of fluid milk. *J. Dairy Sci.* **2008**, *91*, 1769–1777.

25. Lewis, M. Improving pasteurized and extended shelf-life milk. In *Improving the Safety and Quality of Milk: Volume One, Milk Production and Processing*; Griffiths, M.W., Ed.; CRC Press: Boca Raton, FL, USA, 2010; pp. 277–301.

26. Walkling-Ribeiro, M.; Noci, F.; Riener, J.; Cronin, D.A.; Lyng, J.G.; Morgan, D.J. The impact of thermosonication and pulsed electric fields on *Staphylococcus aureus* inactivation and selected quality parameters in orange juice. *Food Bioprocess Technol.* **2009**, *2*, 422–430.

27. Cserhalmi, Z.; Sass-Kiss, Á.; Tóth-Markus, M.; Lechner, N. Study of pulsed electric field treated citrus juices. *Innov. Food Sci. Emerg. Technol.* **2006**, *7*, 49–54.

28. Sachdeva, S.; Buchheim, W. Separation of native casein and whey proteins using micrfiltration. *Aust. J. Dairy Technol.* **1997**, *52*, 92–97.

29. Vivekanand, V.; Kentish S.E.; Connor, O.; O'Connor, A.J.; Barber, A.R.; Stevens, G.W. Microfiltration offers environmentally friendly fractionation of milk proteins. *Aust. J. Dairy Technol.* **2004**, *59*, 186–188.

30. Pafylias, I.; Cheryan, M.; Mehaiab, M.A.; Saglam, N. Microfiltration of milk with ceramic membranes. *Food Res. Int.* **1996**, *29*, 141–146.

31. Silva, R.C.S.N.; Vasconcelos, C.M.; Suda, J.Y.; Minim, V.P.R.; Pires, A.C.S.; Carvalho, A.F. Acceptance of microfiltered milk by consumers aged from 7 to 70 years. *Rev. Inst. Adolfo Lutz* **2012**, *71*, 481–487.

32. Pagliarini, E.; Vernile, M.; Peri, C. Kinetic study on colour changes in milk due to heat. *J. Food Sci.* **1990**, *55*, 1766–1767.

33. Patton, S. Browning and associated changes in milk and its products: A review. *J. Dairy Sci.* **1955**, *38*, 457–478.

34. Boekel, M.A. Effect of heating on Maillard reactions in milk. *Food Chem.* **1998**, *62*, 403–414.

35. Bermúdez-Aguirre, D.; Fernández, S; Esquivel, H.; Dunne, P.C.; Barbosa-Cánovas, G.V. Milk processed by pulsed electric fields: Evaluation of microbial quality, physicochemical characteristics, and selected nutrients at different storage conditions. *J. Food Sci.* **2011**, *76*, S289–S299.

36. Fennema, O.R. *Food Chemistry*, 3rd ed.; Fennema, O.R., Ed.; CRC Press: New York, NY, USA, 1996; pp. 254–273.

37. Calvo, M.M.; de la Hoz, L. Flavour of heated milks. A review. *Int. Dairy J.* **1992**, *2*, 69–81.

38. Contarini, G.; Povolo, M.; Leardi, R.; Toppino, P.M.; Lattiero-Caseario, I.S.; Lombardo, V.; Salerno, V.B. Influence of heat treatment on the volatile vompounds of milk. *J. Agric. Food Chem.* **1997**, *45*, 3171–3177.

39. Hutton, J.T.; Patton, S. The origin of sulfhydrylgroups in milk proteins and their contributions to "cooked" flavour. *J. Dairy Sci.* **1952**, *35*, 699–705.

40. Zabbia, A.; Buys, E.M.; de Kock, H.L. Undesirable sulphur and carbonyl flavour compounds in UHT milk: A review. *Crit. Rev. Food Sci. Nutr.* **2012**, *52*, 21–30.

41. Al-Attabi, Z.; D'Arcy, B.R.; Deeth, H.C. Volatile sulphur compounds in UHT milk. *Crit. Rev. Food Sci. Nutr.* **2009**, *49*, 28–47.

42. Toso, B.; Procida, G.; Stefanon, B. Determination of volatile compounds in cows' milk using headspace GC-MS. *J. Dairy Res.* **2002**, *69*, 569–577.

43. Patton, S.; Forss, D.A.; Day, E.A. Methyl sulfide and the flavour of milk. *J. Dairy Sci.* **1956**, *39*, 1469–1470.

44. Vazquez-Landaverde, P.A.; Velazquez, G.; Torres, J.A; Qian, M.C. Quantitative determination of thermally derived off-flavour compounds in milk using solid-phase microextraction and gas chromatography. *J. Dairy Sci.* **2005**, *88*, 3764–3772.

45. Vazquez-Landaverde, P.A.; Torres, J.A.; Qian, M.C. Quantification of trace volatile sulfur compounds in milk by solid-phase microextraction and gas chromatography-pulsed flame photometric detection. *J. Dairy Sci.* **2006**, *89*, 2919–2927.

46. *Advanced Dairy Chemistry*, 3rd ed.; Fox, P.F., McSweeney, P.L., Eds.; Kluwer Academic/Plenum: New York, NY, USA, 2003–2006; p. 4.

47. Nursten, H.E. The flavour of milk and dairy products: I. Milk of different kinds, milk powder, butter and cream. *Int. J. Dairy Technol.* **1997**, *50*, 48–56.

48. Mader, I. Beta-Carotene: Thermal degradation. *Science* **1964**, *144*, 533–534.

49. Valero, E.; Villamiel, M.; Miralles, B.; Sanz, J.; Martínez-Castro, I. Changes in flavour and volatile components during storage of whole and skimmed UHT milk. *Food Chem.* **2001**, *72*, 51–58.

50. Vallverdú-Queralt, A.; Bendini, A.; Barbieri, S.; di Lecce, G.; Martin-Belloso, O.; Toschi, T.G. Volatile profile and sensory evaluation of tomato juices treated with pulsed electric fields. *J. Agric. Food Chem.* **2013**, *61*, 1977–1984.

51. Sampedro, F.; Geveke, D.J.; Fan, X.; Rodrigo, D.; Zhang, Q.H. Shelf-life study of an orange juice-milk based beverage after PEF and thermal processing. *J. Food Sci.* **2009**, 74, S107–S112.

9

Seafood and Water Management

Saskia M. van Ruth [1,2,*], **Erwin Brouwer** [1,†], **Alex Koot** [1,†] **and Michiel Wijtten** [1,†]

[1] RIKILT Wageningen UR, P.O. Box 230, 6700 EV Wageningen, The Netherlands;
E-Mails: erwin.brouwer@wur.nl (E.B.); alex.koot@wur.nl (A.K.); michiel.wijtten@wur.nl (M.W.)

[2] Food Quality and Design Group, Wageningen University, P.O. Box 17,
6700 AA Wageningen, The Netherlands

[†] These authors contributed equally to this work.

[*] Author to whom correspondence should be addressed; E-Mail: saskia.vanruth@wur.nl

External Editor: Christopher J. Smith

Abstract: Seafood is an important food source for many. Consumers should be entitled to an informed choice, and there is growing concern about correct composition labeling of seafood. Due to its high price, it has been shown to be vulnerable to adulteration. In the present study, we focus on moisture levels in seafood. Moisture and crude protein contents of chilled and frozen cod, pangasius, salmon, shrimp and tilapia purchased from various retail outlets in the Netherlands were examined by reference methods and the values of which were compared with the reported data from other studies in literature. Significant differences in proximate composition were determined for different species and between chilled and frozen products of the same species. Pangasius products showed the highest moisture contents in general (86.3 g/100 g), and shrimp products revealed the largest differences between chilled and frozen products. Comparison with literature values and good manufacturing practice (GMP) standards exposed that, generally, chilled pangasius, frozen pangasius and frozen shrimp products presented considerably higher moisture and lower crude protein/nitrogen contents than those found in other studies. From the GMP standards, extraneous water was estimated on average at 26 g/100 g chilled pangasius product, and 25 and 34 g/100 g product for frozen shrimp and pangasius products, respectively.

Keywords: cod; fish; crude protein; moisture; pangasius; salmon; shrimp; tilapia; water

1. Introduction

Seafood is an important source of nutrients in the human diet in various places in the world and comprises valuable proteins and lipids [1]. European food law underpins the concept of informed consumer choice in the purchase of food (Regulation (EC) No. 178/2002; [2]). The European Food Labelling Directive includes the requirement for a Quantitative Ingredients Declaration, which means that most pre-packed fish products are required to be labelled with a declaration of the amount of fish present as the percentage of the final weight of the product. There is, however, growing concern regarding the correct composition and labeling of seafood. Due to its premium price, seafood is susceptible to mislabeling, e.g., 75 fraud cases have been reported in the U.S. Pharmacopeial Convention food fraud database (http://www.foodfraud.org) to date [3]. Most often, this concerns replacement of species, replacement of wild products by farmed products, adulterated geographical origin and excessive water addition. In the European Rapid Alert System for Food and Feed [4], fish and fish products are listed 86 times in conjunction with fraud/adulteration. This system concerns mostly fraudulent health certificates and illegal imports. Food adulteration appears when a number of criteria are fulfilled. This requires an opportunity (a suitable target), motivation (economic, social and moral drivers), rationalization (justification by those involved) and lack of guardianship. In the very nature of food fraud, the actor conspires to manipulate product composition in the attempt to evade quality assurance and quality control plans implemented by manufacturers, distributors and purchasers and, therefore, requires a different approach from food safety assurance [3]. Food supply networks are optimized for rapid, low-cost production from all sources, which has consequently resulted in nontransparent, fragile systems with little guardianship when it comes to fraud. This certainly holds for seafood, which is sourced from many different parts around the globe and is often supplied through extensive intersecting networks.

Seafood proximate chemical composition is commonly categorized in water, protein, lipid and ash. Low levels of carbohydrates may be present, but the concentrations are considered negligible. The body composition varies and is associated with the fish appetite, growth and efficiency of feed utilization [5]. Water is the main component, both in volume and weight, in all seafood products. It is a determinant of the value of the products, sensory attributes and shelf-life. Commercial practices have evolved to retain and add moisture to seafood during harvest, processing and storage in order to reduce moisture or drip loss during frozen storage and thawing. However, there is a thin line between water addition to make up for moisture losses and excessive extraneous water addition for one's own economic gain. Water and water binder addition are technologies borrowed from the meat industry.

The second major seafood component is protein. The protein content of foods is commonly derived by calculation, based on the measurement of nitrogen content using the Kjeldahl method [6]. The nitrogen in fish is distributed between proteins and other nitrogen-containing compounds, which is often referred to as non-protein nitrogen (NPN). The hundreds of proteins in muscle consist of the contractile proteins, *i.e.*, the myofibrillar proteins; the sarcoplasmic proteins from the fluid within the muscle cells are mainly enzymes and are the only water-soluble proteins; and the stromal proteins, which are present in the connective tissue and hold the muscle bundles together and to the skeleton, and include proteins that are associated with cellular membranes. NPN compounds are comprised of peptides, amino acids, amines, amine oxides, guanidine compounds, quaternary ammonium compounds, purines and urea. These compounds originate mostly from the sarcoplasm. Nitrogen levels vary with the different fish species, but most finfish muscle tissue consists of 18–22 g crude protein per 100 g product; converted, this is approximately 2.9–3.5 g nitrogen per 100 g product [7].

For physiological reasons, strong relationships exist between protein and moisture levels in meat [8] and seafood [5,9]. In general, in muscle meat, the water content is close to 77% and protein to 23%, which results in a water-to-protein ratio of 3.35. This information is also used to control excessive water addition. The Poultry Meat Marketing Standards Regulation (EC) 543/2008 [10] regulates the amounts of extraneous water in poultry meat and limits the water-to-protein ratio to 3.40 for chicken breasts (non-preparations). Furthermore, the Food Industries Manual states limits for water-to-protein ratios on a fat-free basis for pork (3.40–3.50) and beef (3.60–3.70) [11]. In addition to the relationship between moisture and protein contents, correlations with lipid content may exist, as well.

Considering the vulnerability of seafood to adulteration and the simplicity of the measurements, surprisingly few reports are available surveying the actual moisture/protein contents of seafood products on the market. Therefore, the current study aims for insights into common water management practice for a variety of seafood species on the Dutch market, representing a considerable share of the seafood products consumed in the Netherlands. For this study, seafood species were selected that were reported to be among those most frequently consumed in the Netherlands in 2013 [12] and are sold chilled and frozen, *i.e.*, cod, pangasius, salmon, shrimp and tilapia products. The moisture and crude protein contents of the products obtained from supermarkets, specialty shops and open markets (mobile outlets) were examined with the official reference methods and evaluated against values in the literature.

2. Experimental Section

2.1. Sample Material

Seafood samples (110) of five species, sold as chilled and frozen (55/55), were purchased at various retail outlets (supermarkets, specialty seafood shops, open air markets) in the Netherlands in spring, 2014 (Table 1). On arrival to the laboratory, all samples were labeled and stored at −18 °C. Samples were ground prior to moisture and crude protein analysis.

Table 1. Sampling design.

Variable	Categories	Number of Samples
Seafood species	Cod (*Gadus morhua*)	20
	Pangasius (*Pangasius bocourti*)	20
	Salmon (*Salmo salar*)	26
	Shrimp (*Crangon* spp., *Penaeus* spp., *Pandalus* spp., peeled)	24
	Tilapia (*Oreochromis* spp.)	20
Sales temperature	Chilled	55
	Frozen	55
Retail outlet	Supermarket, convenience	26
	Supermarket, intermediate	42
	Supermarket, discounter	12
	Specialty shop	18
	Open air market	12

2.2. Moisture Analysis

Moisture contents were determined by the reference method, ISO 1442:1997 [13]. This method considers the loss in mass obtained after thorough mixing of the test portion with sand and drying to constant mass at 103 ± 2 °C, divided by the mass of the test portion. Moisture analyses were carried out on each individual sample in duplicate.

2.3. Crude Protein Analysis

Crude protein contents were determined by the reference method, ISO 937:1978 [14]. This Kjeldahl method involves the digesting of a test portion with concentrated sulfuric acid, using copper (II) sulfate as a catalyst, to convert organic nitrogen to ammonia ions. Then, alkalization, distillation of the liberated ammonia into an excess of boric acid solution and titration with hydrochloric acid to determine the ammonia bound by the boric acid are followed by the calculation of the nitrogen content of the sample from the amount of ammonia produced. A standard factor, termed the nitrogen factor, of 6.25 was and is typically used in the determination of crude protein in foods, based on the assumption that the average nitrogen content of proteins is 16% ($1/0.16 = 6.25$). Protein analyses were carried out on each individual sample in duplicate. Protein values reported in the publication refer to these crude protein analyses.

2.4. Statistical Analysis

Moisture and protein content data were subjected to multi-factor analysis of variance (MANOVA; factors: species and sales temperature), and Fisher's least significant difference (LSD) tests were carried out to determine significant differences among groups using XLSTAT 2014.3.02 (Addinsoft, Paris, France). MANOVA can assess two or more independent variables (in this case, species and sales temperature) for the significance of the effects on two or more metric dependents (in this case, moisture and protein contents). This allows a joint analysis of each dependent rather than performing several univariate tests, thus avoiding multiple testing risks. A significance level of $p < 0.05$ was used throughout the study.

3. Results and Discussion

3.1. Water and Protein Contents

The moisture and protein levels of five different seafood species, sold chilled and frozen, from various retail outlets were analyzed with the ISO reference methods. In the current study, the moisture content assessment was based on oven drying, but also, other technologies are generally available for moisture determinations, e.g., near-infrared, nuclear magnetic resonance and guided microwave spectroscopy [15–17]. The protein (nitrogen) levels were determined by the widely-used Kjeldahl methodology. Alternatives for determining the nitrogen content are the Dumas combustion method, as well as colorimetric, electrophoretic, chromatography, mass spectrometry and immunology-based methods [18]. The moisture and protein contents of the seafood samples analyzed in the current study are presented for the individual 110 samples in Figure 1.

Figure 1. Moisture *versus* protein contents for individual samples of five seafood species (different colors), for both chilled (filled triangle) and frozen products (open circle). Horizontal lines indicate protein levels associated with minimal nitrogen factors for GMP (good manufacturing practice) fish ingredients for white fish (orange) and washed and peeled shrimp (grey) [7].

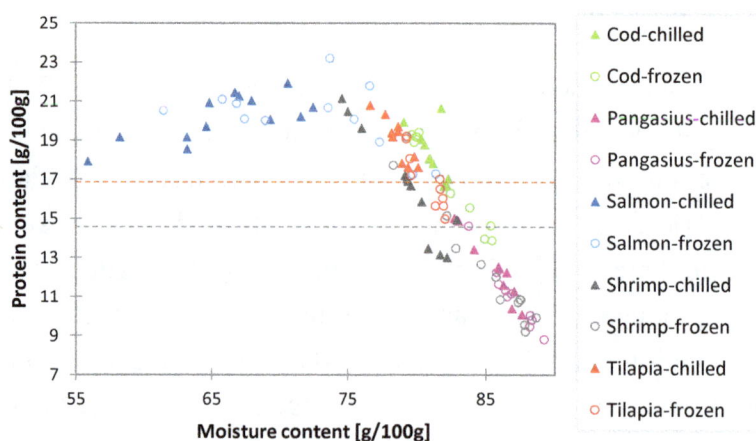

Mean values and statistical comparison of the species and chilled/frozen samples are presented in Table 2. MANOVA indicated significant differences in moisture contents between the groups ($F_{(5,109)} = 70.7$, $p < 0.0001$), to which both the factor, fish species ($F_{(4,109)} = 80.0$, $p < 0.0001$), and the factor, sales temperature ($F_{(1,109)} = 33.7$, $p < 0.0001$), contributed. Similarly MANOVA pointed out significant differences in protein contents between the groups ($F_{(5,109)} = 53.6$, $p < 0.0001$), with also the factor, fish species ($F_{(4,109)} = 60.9$, $p < 0.0001$) and the factor, sales temperature ($F_{(1,109)} = 24.4$, $p < 0.0001$), having a significant effect on the protein contents.

Salmon samples showed the lowest moisture contents, which is due to the relatively high lipid fraction in comparison to the other samples, *i.e.*, white fish and shrimps. On the other hand, the protein contents of the salmon samples were relatively high, and the moisture contents were more variable. For tilapia and cod samples, significantly higher moisture and lower protein contents were observed compared to the salmon samples, and in turn, pangasius demonstrated significantly higher moisture and lower protein contents than tilapia and cod (Fisher's LSD test, $p < 0.05$). Shrimp moisture and protein contents overlapped with those of the tilapia/cod and pangasius groups.

Table 2. Moisture contents, protein contents and water/protein ratios in various seafood samples presented for different groups (mean ± SD) [1].

Sample	Moisture Content (g/100 g)	Protein Content (g/100 g)	Water/Protein Ratio
All	79.4 ± 7.2	16.5 ± 3.7	4.8
Cod	81.6 ± 1.9 [B,C]	17.6 ± 2.0 [II]	4.6
Pangasius	86.3 ± 1.7 [A]	11.7 ± 1.8 [IV]	7.4
Salmon	69.3 ± 6.6 [D]	20.1 ± 1.4 [I]	3.4
Shrimp	82.4 ± 4.3 [B]	14.2 ± 3.5 [III]	5.8
Tilapia	79.7 ±1.6 [C]	18.0 ±1.7 [II]	4.4
Chilled	77.5 ± 7.5 [Y]	17.4 ± 3.2 [S]	4.4
Frozen	81.3 ± 6.2 [X]	15.5 ± 4.0 [T]	5.2

[1] A–D, I–IV, X and Y, S and T, values with either different letters or different Roman numbers in a column are significantly different (MANOVA, Fisher's LSD test $p < 0.05$).

The determined moisture and protein contents in the cod samples are in agreement with those reported by Krzynowek and Murphy [19], who reported a moisture content of 82.1% and a protein content of 17.4% for fresh fillets. Similarly, the salmon proximate composition is also in line with previous studies, e.g., a moisture content range of 60%–75% and a protein range of 17%–25% was reported in a study examining a large number of salmon samples [20]. Furthermore, a second large study analyzing 178 salmon samples from Ireland, Norway and Scotland (178 samples) resulted in an average moisture content of 69.1% moisture [17], which is all fairly close to the data gathered in the present study. The analyses of the pangasius, shrimp and tilapia samples revealed, on average, higher moisture and lower protein contents than reported in other studies. Rathod and Pagarkar [21] reported recently pangasius moisture and protein contents of 76.6% and 14.4%, respectively, and Karl and co-workers [22] similarly reported pangasius moisture and protein contents of 82.7% and 14.2%, respectively. For black tiger shrimp, moisture contents of 80.5% and protein contents of 17.1% have been reported, for white shrimp, 77.2% moisture and 18.8% protein [1], and for pink shrimp, moisture contents of 80.1% and protein contents of 18.1% [19]. Finally, for tilapia, moisture contents of 75.8% and protein contents of 18.8% [23], as well as a moisture range of 74.4%–77.8% have been published [24]. In conclusion, for the cod and salmon samples in the present study, moisture and protein contents are similar to those reported by other authors, whereas pangasius, shrimp and tilapia revealed, on average, higher moisture and lower protein contents.

The water and protein content of the chilled and frozen samples were statistically compared (Table 3). Both for moisture ($F (9,109) = 46.3, p < 0.0001$) and protein ($F (9,109) = 36.8, p < 0.0001$), significant differences between groups were observed (MANOVA). It is remarkable that for all species, frozen samples showed higher moisture and lower protein contents than the chilled samples. Salmon and shrimp sample groups presented significant differences in moisture content between chilled and frozen samples (Fisher's LSD test, $p < 0.05$), and the shrimp and tilapia sample groups demonstrated also significant differences in protein content between chilled and frozen samples.

Table 3. Moisture contents, protein contents and water/protein ratios in various seafood samples (mean ± SD) [1].

Seafood Species	Sales Temperature	Moisture Content (g/100 g)	Protein Content (g/100 g)	Water/Protein Ratio
Cod	Chilled ($n = 10$)	81.1 ± 1.0 [B,C]	18.3 ± 1.3 [II,III]	4.4
	Frozen ($n = 10$)	82.2 ± 2.5 [B]	17.0 ± 2.4 [III,IV]	4.8
	Δ Frozen-chilled	+1.1 (+1.4%)	−1.3 (−7.1%)	+0.4 (+10.0%)
Pangasius	Chilled ($n = 10$)	85.6 ± 1.7 [A]	12.4 ± 1.7 [V]	6.9
	Frozen ($n = 10$)	86.9 ± 1.6 [A]	11.0 ± 1.7 [V]	7.9
	Δ Frozen-chilled	+1.3 (+1.5%)	−1.4 (−11.3%)	+1.0 (+14.4%)
Salmon	Chilled ($n = 13$)	65.8 ± 4.9 [E]	20.2 ± 1.2 [I]	3.3
	Frozen ($n = 13$)	72.9 ± 6.2 [D]	20.1 ± 1.7 [I]	3.6
	Δ Frozen-chilled	+7.1 (+10.8%)	−0.1 (−0.5%)	+0.3 (+10.0%)
Shrimp	Chilled ($n = 12$)	79.2 ± 2.7 [C]	16.6 ± 2.8 [IV]	4.8
	Frozen ($n = 12$)	85.5 ± 3.1 [A]	11.9 ± 2.5 [V]	7.2
	Δ Frozen-chilled	+6.3 (+8.0%)	−4.7 (−28.3%)	+2.4 (+50.0%)
Tilapia	Chilled ($n = 10$)	78.7 ±1.0 [C]	19.0 ±1.1 [I,II]	4.1
	Frozen ($n = 10$)	80.8 ± 1.2 [B,C]	16.9 ± 1.5 [III,IV]	4.6
	Δ Frozen-chilled	+2.1 (+2.7%)	−2.1 (−11.1%)	+0.5 (+12.0%)

[1] A–D, I–IV, values with either different letters or different Roman numbers in a column are significantly different (MANOVA, Fisher's LSD test $p < 0.05$).

Mean water-to-protein ratios varied from 3.3 to 4.4 for chilled cod, salmon and tilapia. Chilled shrimp (4.8) and pangasius (6.9) demonstrate relatively high mean water-to-protein ratios. Regardless of species, frozen samples had higher water-to-protein ratios than the chilled samples, with striking mean values for the shrimp and pangasius groups, *i.e.*, 7.2 and 7.9, respectively.

3.2. Comparison of Nitrogen Factors

In the U.K., the determination of nitrogen as a quantitative marker for seafood fat-free protein is well established and is the official chemical enforcement method. It is also widely used by food producers to check the specification and added water of their seafood raw materials. A "nitrogen factor" is the average nitrogen content of seafood tissues, on a fat-free basis, unless the fat content is low, as in white fish [25]. In the U.K., Code of Practice on the declaration of fish content in fish products [7], a nitrogen factor of 2.65 has been set for white fish (reflecting a protein content of 16.5 g/100 g in the present study) and a limit of 2.33 for washed and peeled shrimp (reflecting a protein content of 14.6 in the present study). Both lines are presented in Figure 1 and show that a considerable number of samples have lower nitrogen (protein) contents than these nitrogen (protein) factors. Furthermore, other relevant nitrogen factors for good manufacturing practice (GMP) fish ingredients of various species have been published recently and concern (minced) cod (2.67), pangasius (2.66) and tilapia (2.88) [25]. Salmon values are not considered here, since its higher lipid content would need to be taken into account to make appropriate comparisons. Table 4 presents the number of cod, pangasius, shrimp and tilapia products analyzed, which shows higher or lower nitrogen content values than the nitrogen factors published. Those samples not meeting the nitrogen factors were subjected to further estimation of the extraneous water based on these nitrogen factors: [(nitrogen factor − nitrogen content measured)/nitrogen factor] × 100%.

Table 4. Comparison of the nitrogen contents of seafood samples with Kjeldahl nitrogen factors for GMP fish ingredients [7,25]: the number of samples with nitrogen contents below or over minimal nitrogen factors [1].

Fish Species	Chilled Products		Frozen Products	
	Below Minimal Nitrogen Factor	Exceeding Minimal Nitrogen Factor	Below Minimal Nitrogen Factor	Exceeding Minimal Nitrogen Factor
Cod	1	9	5	5
Pangasius	10	0	10	0
Shrimp	3	9	10	2
Tilapia	3	7	7	3
Sum	17 (40%)	25 (60%)	32 (76%)	10 (24%)

[1] The nitrogen factors considered are: cod (2.67 [25]), pangasius (2.66 [25]), shrimp (2.33 [7]) and tilapia (2.88 [25]).

Chilled cod and tilapia included some samples with nitrogen contents below the factors published, but only a marginal difference existed (0.2%–3%). Estimation of water replacement based on the nitrogen factors indicated that the chilled shrimp samples that did not meet the nitrogen factors comprised 9 g extraneous water/100 g and chilled pangasius 26 g extraneous water/100 g product. Only a single chilled product had water addition labelled. Water replacement in the frozen products with lower nitrogen contents than the nitrogen factors was estimated as well, and amounted to tilapia 10 g extraneous water/100 g product, for cod 14 g/100 g product, for shrimp 25 g/100 g product and for pangasius 34 g/100 g product, on average. Most of the frozen products not meeting the nitrogen factors had water addition labelled. The quantity indicated was usually 10% extraneous water (drip loss, including glazing), except for some shrimp products, which indicated 10%–20% added water. Nonetheless, the estimations indicated higher extraneous water levels, which implies under-labelling of the water addition, especially for frozen shrimp and pangasius products.

4. Conclusions

The present study in which 110 samples of seafood were analyzed for their moisture and protein contents showed significant differences in proximate composition for different species and products sales temperature (chilled or frozen). Chilled pangasius, frozen pangasius and frozen shrimp products revealed consistently higher moisture and lower protein/nitrogen contents compared to other studies in the literature, as well as compared to GMP standards. These results indicate water addition, which seems to occur unlabeled in certain chilled products and under-labelled in particular frozen products.

Author Contributions

Saskia van Ruth initiated the study, was in charge of the experimental design, carried out the various calculations and composed the major part of the manuscript. Erwin Brouwer, Alex Koot and Michiel Wijtten collected sample materials, carried out sample preparation, analysis and quality assurance activities, gathered the data in a database and contributed to the manuscript.

Conflicts of Interest

The authors declare no conflict of interest.

References

1. Sriket, P.; Benjakul, S.; Visessanguan, W.; Kijroongrojana, K. Comparative studies on chemical composition and thermal properties of black tiger shrimp (*Penaeus monodon*) and white shrimp (*Penaeus vannamei*) meats. *Food Chem.* **2007**, *103*, 1199–1207.
2. Regulation (EC) No. 178/2002 of the European Parliament and of the Council of 28 January 2002 Laying Down the General Principles and Requirements of Food Law, Establishing the European Food Safety Authority and Laying Down Procedures in Matters of Food Safety. Available online: http://eur-lex.europa.eu/LexUriServ/LexUriServ.do?uri=CONSLEG:2002R0178:20090807:EN:PDF (accessed on 21 July 2014).
3. Moore, J.C.; Spink, J.; Lipp, M. Development and application of a database of food ingredient fraud and economically motivated adulteration from 1980 to 2010. *J. Food Sci.* **2012**, *77*, R118–R126.
4. RASFF (The Rapid Alert System for Food and Feed)—Food and Feed Safety Alerts. Available online: http://ec.europa.eu/food/safety/rasff/index_en.htm (accessed on 21 July 2014).
5. Breck, J.E. Body composition in fishes: Body size matters. *Aquaculture* **2014**, *433*, 40–49.
6. AOAC (Association of Official Analytical Chemists). AOAC Official Method 950.46 Moisture in Meat. In *Official Methods of Analysis*, 15th ed.; AOAC International: Arlington, VA, USA, 1990.
7. UK Association of Frozen Food Producers; British Frozen Food Federation; British Retail Consortium; British Hospitality Association; Sea Fish Industry Authority; LACOTS; Association of Public Analysist. Code of Practice on the Declaration of Fish Content in Fish Products. Available online: http://www.seafish.org/media/Publications/Fish_Content_CoP.pdf (accessed on 21 July 2014).
8. De Greef, K.H.; Verstegen, M.W.A.; Kemp, B. Validation of a porcine growth model with emphasis on chemical body compostion. *Livest. Prod. Sci.* **1992**, *32*, 163–180.
9. Yeannes, M.I.; Almandos, M.E. Estimation of fish proximate composition starting from water content. *J. Food Compos. Anal.* **2003**, *16*, 81–92.
10. Regulation (EC) No. 543/2008 of 16 June 2008 Laying Down Detailed Rules for the Application of Council Regulation (EC) No 1234/2007 as Regards the Marketing Standards for Poultry Meat. Available online: http://eur-lex.europa.eu/legal-content/EN/TXT/?uri=CELEX:32008R0543 (accessed on 21 July 2014).
11. Ranken, M.D.; Kill, R.C.; Baker, C.G.J. *Food Industries Manual*, 24th ed.; Blackie Academic and Professional: London, UK, 1997; pp. 37–39.

12. Nederlands Visbureau. Visconsumptie in 2013—GfK Jaarcijfers, 2014. Available online: http://www.visspecialisten.nl/l/library/download/30769 (accessed on 21 November 2014).

13. ISO 1442:1997 Meat and Meat Products—Determination of Moisture Content (Reference Method). Available online: http://www.iso.org/iso/catalogue_detail.htm?csnumber=6037 (accessed on 21 July 2014).

14. ISO 937:1978 Meat and Meat Products—Determination of Nitrogen Content (Reference Method). Available online: http://www.iso.org/iso/catalogue_detail.htm?csnumber=5356 (accessed on 21 July 2014).

15. Honikel, K.O. Moisture and water-holding capacity. In *Handbook of Muscle Foods Analysis*, 1st ed.; Nollet, L.M.L., Toldrá, F., Eds.; CRC Press: Boca Raton, FL, USA, 2009; pp. 315–334.

16. Wold, J.P.; Isaksson, T. Non-destructive determination of fat and moisture in whole Atlantic salmon by near-infrared diffuse spectroscopy. *J. Food Sci.* **1997**, *62*, 734–736.

17. He, H.-J.; Wu, D.; Sun, D.-W. Non-destructive and rapid analysis of moisture distribution in farmed Atlantic salmon (*Salmo salar*) fillets using visible and near-infrared hyperspectral imaging. *Innov. Food Sci. Emerg. Technol.* **2013**, *18*, 237–245.

18. Greaser, M.L. Proteins. In *Handbook of Muscle Foods Analysis*, 1st ed.; Nollet, L.M.L., Toldrá, F., Eds.; CRC Press: Boca Raton, FL, USA, 2009; pp. 57–73.

19. Krzynowek, J.; Murphy, J. *Proximate Composition, Energy, Fatty Acid, Sodium, and Cholesterol Content of Finfish, Shellfish, and Their Products*; US Department of Commerce: Gloucester, MA, USA, 1987; pp. 3–42.

20. Colwell, P.; Ellison, L.R.; Walker, M.J.; Elahi, S.; Burns, D.T.; Gray, K. Nitrogen factors for Atlantic Salmon, *Salmo salar*, farmed in Scotland and in Norway and for the derived ingredient, "salmon frame mince", in fish products. *J. Assoc. Public Anal.* **2011**, *39*, 44–78.

21. Rathod, N.; Pagarkar, A. Biochemical and sensory quality changes of fish cutlets, made from pangasius fish (*Pangasianodon hypophtalmus*), during storage in refrigerated display unit at −15 to −18 °C. *Int. J. Food Agric. Vet. Sci.* **2013**, *3*, 1–8.

22. Karl, H.; Lehmann, I.; Rehbein, H.; Schubring, R. Composition and quality attributes of conventionally and organically farmed *Pangasius* fillets (*Pangasius hypophtalmus*) on the German market. *Int. J. Food Sci. Technol.* **2010**, *45*, 56–66.

23. Olagunju, A.; Muhammad, A.; Mada, S.B.; Mohammed, A.; Mohammed, H.A.; Mahmoud, K.T. Nutrient composition of *Tilapia zilli, Hemi-synodontis membranacea, Clupea harengus* and *Scomber scombrus* consumed in Zaria. *World J. Life Sci. Med. Res.* **2012**, *2*, 16–19.

24. Tan, Y.T. Proximate composition of freshwater fish—Grass carp, *Puntius gonionotus* and *Tilapia*. *Hydrobiologia* **1971**, *37*, 361–366.

25. Analytical Methods Committee. Seafood nitrogen factors. *Anal. Methods* **2014**, *6*, 4490–4492.

10

High Pressure Treatment in Foods

Edwin Fabian Torres Bello [1,*]**, Gerardo González Martínez** [2]**, Bernadette F. Klotz Ceberio** [2]**, Dolores Rodrigo** [1] **and Antonio Martínez López** [1]

[1.] Institute of Agrochemistry and Food Technology (CSIC), Avenida Agustín Escardino, 7 Parque Científico, 46980 Paterna (Valencia), Spain; E-Mails: lolesra@iata.csic.es (D.R.); amartinez@iata.csic.es (A.M.L.)

[2] Alpina Research Institute (IAI), Alpina Productos Alimenticios S.A, Edificio Corporativo Km 3 vía, Briceño-Sopó, Cundinamarca, 251001, Colombia; E-Mails: gerardo.gonzalez@alpina.com.co (G.G.M.); bernadette.klotz@alpina.com (B.F.K.C.)

* Author to whom correspondence should be addressed; E-Mail: fabiantb@iata.csic.es

Abstract: High hydrostatic pressure (HHP), a non-thermal technology, which typically uses water as a pressure transfer medium, is characterized by a minimal impact on food characteristics (sensory, nutritional, and functional). Today, this technology, present in many food companies, can effectively inactivate bacterial cells and many enzymes. All this makes HHP very attractive, with very good acceptance by consumers, who value the organoleptic characteristics of products processed by this non-thermal food preservation technology because they associate these products with fresh-like. On the other hand, this technology reduces the need for non-natural synthetic additives of low consumer acceptance.

Keywords: high pressure; microorganism; spores; protein; enzyme; packaging; cheese

1. Introduction

Currently consumers worldwide are more demanding with regard to the quality and safety of the foods they consume, especially those that produce the perception of healthy products. To meet these demands, the food industry has improved its heat preservation processes by developing continuous high temperaturc/short time (HTST) and ultra high temperature (UHT) treatments and aseptic packaging. In addition, consumption of minimally processed products has increased significantly. These products maintain a high standard of nutrition and flavor, while meeting the required safety level and achieving a long shelf life [1].

Minimally processed foods have been developed alongside the development of various emerging preservation technologies. Within this group of technologies there are the so-called "non-thermal preservation technologies," which do not use heat as the main form of microbial and enzyme inactivation. Although heat is generated by some of these processes, the temperature increase never reaches the levels of a conventional thermal process and can be suitably controlled by a cooling station. These new preservation technologies include oscillatory magnetic fields, pulsed electric fields, ultrasound, irradiation, and high hydrostatic pressure. Probably the most developed and most widely implanted technology at the industrial level is high hydrostatic pressure. This technology has demonstrated its capability of preserving sensory and nutritional qualities of foods while producing suitable levels of microbiological and enzyme inactivation.

2. High Hydrostatic Pressure Technology

The main objective of any non-thermal technology is to maximize the freshness and flavor qualities of the foodstuffs while achieving the required level of food safety. High hydrostatic pressure (HHP) meets with these requirements and today it being incorporated in many companies as an alternative to conventional heat treatment procedures. Applications include the preservation of meat products, oysters, fruit jams, fruit juices, salad dressings, fresh calamari, rice cake, duck liver, jam, guacamole, and many ready-to-eat foods. In all these cases, microbial and enzyme inactivation is achieved without altering the product quality [2]. In relation to the total percentage utilization of HHP equipment, vegetable products account for 28%, meat products for 26%, sea foods and fish for 15%, juices and beverages for 14%, and other products for 17%, generating an amount of 350,000,000 kg of processed products in 2012, according to data from Hiperbaric, S.A. [3].

All this makes HHP the most commercially developed non-thermal technology, with very good acceptance by consumers, who value the organoleptic characteristics of pressure-treated products with a quality barely affected by treatment. Currently the world market has experienced significant growth in the incorporation of equipment at industrial level (Table 1).

Table 1. Number of HHP machines around the world. Source: Hiperbaric, S.A. [3].

Time (Years)	HPP machines in industry
1990	2
1991	2
1992	3
1993	3
1994	4
1995	4
1996	4
1997	4
1998	5
1999	9
2000	14
2001	21
2002	27
2003	38
2004	52
2005	68
2006	78
2007	95
2008	109
2009	122
2010	147
2011	167

In general, microbial inactivation is achieved at pressures that vary from 100 to 800 MPa during relatively short times (from a few seconds to several minutes). Some treatments are combined with mild temperatures between 20 and 50 °C to inactivate enzymes. The processing conditions depend fundamentally on the food to be treated and the microorganisms and enzymes to be inactivated; we note that this technology at the pressure currently used in the food industry does not inactivate bacterial spores [4,5].

3. Packaging

The package is an important part in the development and industrial application of HHP as a preservation technology. It is possible to use a great variety of packages with different shapes; however, food must be packed in a flexible and resistant package, able to withstand pressure and maintain the integrity. Polyethylene (PE), polyethylene terephthalate (PET), polypropylene (PP), ethylene-vinyl alcohol (EVOH), polyamide (PA), and nylon films are some of the packaging materials currently used in industrial food processing by HHP treatments [6,7]. Juliano *et al.* [7] suggest minimize the headspace up to 30% to maximize the utilization of the vessel capacity and minimize the time needed for preheating, if the treatment requires temperature. Usually, an HPP vessel will utilize its 50%–70% volume capacity depending on the shape of the package and the vessel design [8].

4. Microbial Inactivation

The objective of any preservation process is the inactivation of microorganisms that can spoil the food and/or produce illness in the consumer (pathogenic microorganisms). The response of microorganisms to HHP has been extensively studied [9–12] varies according to the following factors: molds and yeasts are the most sensitive microorganisms; Gram-negative bacteria have medium sensitivity, whereas Gram-positive bacteria are the most resistant among vegetative cells and their spores need very high pressures to be inactivated. Regarding the action mechanisms of pressure, according to the studies carried out by Huang *et al.* [13] a pressure of 50 MPa can affect or inhibit protein synthesis and produce a reduction in the number of microbial ribosomes. A pressure of 100 MPa can cause partial denaturalization of cellular proteins; when the pressure is increased to 200 MPa it produces internal damage in the microbial structure and external damage in the cellular membrane. Pressures equal or similar to 300 MPa produce irreversible damage to the microorganism, including leakage of intracellular components to the surrounding medium, resulting finally in cellular death [14–16].

The various effects that take place in microorganisms depend on their physiological state, microorganisms in log phase being more sensitive to HHP than those in stationary phase. This behavior could be explained by the fact that in the log phase the microorganism is in the process of cellular division and the membrane is more sensitive to environmental stresses [6]. This effect was also reported by Mañas and Mackay [17] in *Escherichia coli* strain J1, in exponential and stationary phases. The cells in stationary phase showed higher resistance to HHP treatment than those in exponential phase. Some modifications were also observed (aggregation of cytoplasmic proteins, condensation of the nucleoid) after 200 MPa treatments for 8 min at 20 °C.

Temperature is a very important environmental stress in HHP treatments because the combination of the two technologies, with short times can increases significantly microbial inactivation. According to studies carried out by Chen and Hoover [18] and Ross *et al.* [19], an HHP treatment of *L. monocytogenes* at initial temperatures of 45–50 °C and 5 min produced more than 5 log decimal reductions in the initial microbial concentration in UHT whole milk. However, was necessary to increase the treatment time to 35 min to produce the same inactivation at initial temperature of 22 °C.

HHP has proved to be an effective technology for inactivating various pathogens, as reported by Jofré *et al.* [20]. The application of a treatment of 600 MPa for 6 min at 31 °C resulted in a reduction close to 3.5 decimal log for *E. coli*, *Listeria monocytogenes*, *Salmonella enterica* subsp. *enterica*, *Yersinia enterocolitica*, and *Campylobacter jejuni* in meat products.

Although there are many studies in relation to the effect of HHP on bacteria, the information that exists on molds and yeasts is relatively scarce (Table 2). In general, yeasts and molds can be inactivated at 200–400 MPa [21], but when they are in the spore or ascospore state or in a food with a very high concentration of sugar the pressure needed to inactivate them could be close to 600 MPa [22]. These microorganisms are frequently involved in spoilage of cereals derivatives (tofu, tortillas), minimally processed vegetables, and lactic derivatives such as butter, yoghurt, and soft cheese [23,24].

Table 2. High hydrostatic pressure (HHP) inactivation of molds and yeasts in different foods.

Food product	Microorganism	HHP conditions	Inactivation results	Reference
Pineapple juice	*Byssochlamys nivea*	550–600 MPa for 3–15 min at 20–80 °C	600 MPa for 15 min at 80 °C, 5.7 log reduction	Ferreira *et al.* [25]
Apple-broccoli juice	*S. cerevisiae*; *A. flavus*	250–400 MPa for 5–20 min at 21 °C	400 MPa for 10 min at 21 °C, 5 log reduction	Houška *et al.* [26]
Apple juice	*Talaromyces avellaneus*	200–600 MPa for 10–60 min at 17–60 °C	600 MPa for 50 min at 60 °C, 5 log reduction ascospores	Voldřich *et al.* [27]
Concentrated orange juice	*S. cerevisiae*	100–400 MPa for 0–120 min at 20 °C	400 MPa for 60 min at 20 °C, 3 log reduction	Basak *et al.* [28]
Cheese	*P. roqueforti*	50–800 MPa for 20 min at 10–30 °C	400 MPa for 20 min at 20 °C, 6 log reduction	O'Reilly *et al.* [29]

5. Spore Inactivation

Spores are cellular forms that some microorganisms have developed as a response to adverse environmental situations in order to survive. Spores are characterized by their high resistance to different environmental stresses and preservation treatments. The most important spore-producing genera are *Clostridium*, *Bacillus* and *Alicyclobacillus*. The initial spore load present in foods can be significantly reduced by HHP in combination with mild temperatures (Table 3). In various published studies, 3.5 decimal log reductions have been reported for *Clostridium sporogenes* and 5.7 decimal log reductions for *Bacillus coagulans* by HHP at temperatures of 60–90 °C [30,31]. Furthermore, Meyer [32] observed significant reductions in the initial spore concentration in low-acid foods after treatments ranging between 700 and 1000 MPa and a product temperature of 70 °C. With those conditions they obtained foods that were microbiologically stable at room temperature, and in many cases the quality of the products was higher than that of those processed by heat.

Table 3. HHP inactivation of spores in different foods.

Food product	Microorganism	HHP conditions	Inactivation results	Reference
Carrot juice	*B. licheniformis*	400–600 MPa for 0–40 min at 40–60 °C	241 to 465 MPa (D value range 23.3 to 31 °C)	Tola and Ramaswamy [33]
Cooked chicken	*C. botulinum*	600 MPa for 2 min at 20 °C	600 MPa for 2 min at 20 °C, 2 log reduction	Linton *et al.* [34]
Orange juice	*A. acidoterrestris*	200–600 MPa for 1–15 min at 45–65 °C	600 MPa, D55 °C = 7 min; 200 MPa, D65 °C = 5.0 min	Silva *et al.* [35]
Tomato sauce	*B. coagulans*; *A. acidoterrestris*	100–800 MPa for 10 min at 25, 40, 60 °C	700 MPa for 10 min at 60 °C, 2 log reduction	Vercammen *et al.* [36]
Tomato pulp	*B. coagulans*	300–600 MPa for 0–39 min at 50–60 °C	600 MPa for 15 at 60 °C 5.7 log reduction	Zimmermann *et al.* [31]
Orange Juice	*A. acidoterrestris*	200–600 MPa for 10 min at 20–60 °C	600 MPa for 10 min at 50 °C, 3 log reduction	Hartyáni *et al.* [37]
Milk	*B.sporothermodurans*	300–500 MPa for 10–30 min at 30–50 °C	495 MPa for 30 min at 49 °C, 5 log reduction	Aouadhi *et al.* [38]

At present, methods to germinate spores before HHP treatment are under study. Exist different methods for germination of spores such as combining extremely high pressure and temperature, methods that involves using low or medium pressure (150–300 MPa), temperature, and other factors as single amino acids, sugars, asparagine, glucose, fructose to germinate the spores and produce bacterial vegetative cells, after which the bacterial vegetative cells are inactivate using HHP [39,40]. In addition, there are other germinant agents, which include lysozyme, salts, and cationic surfactants such as dodecylamine, that can be used in combination with high pressure. It is important to point out however that the spores of proteolytic *Clostridium botulinum* and *Clostridium sporogenes* germinate in response to L-alanine but not to universal germinant AGFK (a mixture of L-asparagine, D-glucose, D-fructose, and potassium ions) or inosine [40–43]. This initial process can be followed by HHP treatment of 300–900 MPa at 30–60 °C [34,44].

According to the study carried out by Georget *et al.* [45] to germinate *Geobacillus stearothermophilus* spores under moderate high pressure in buffer N-(2-acetamido)-2-aminoethanesulfonic acid (ACES) applying a treatment of 200 MPa with temperature of 55 °C , an inactivation over 2 log10 was achieved after 5 min of treatment. A 200 MPa for 40 min at 55 °C treatment led an inactivation of 3 log reduction following the subsequent inactivation to 80 °C for 20 min. In case of the spores of *Clostridium botulinum* earlier studies in cooked chicken with 2% sodium lactate, showed that germination of spores occurred at 4 °C and a spore reduction in the initial inoculum of 1.7 log10 cfu/g with a treatment at 600 MPa for 2 min at 20 °C was achieved [38]. For the germination and inactivation of *Clostridium perfringens* spores in poultry meat, spores were incubated for 15 min at 55 °C with an addition of L-asparagine and potassium chloride, followed of a treatment of 568 MPa at 73 °C for 10 min achieving ~4 log reductions in the concentration of spores [33].

6. Effects of HHP on Proteins

HHP technology has been used fundamentally to reduce the microbial load and increase the safety and shelf life of treated foods with superior nutritional and sensory properties to those thermally treated. Nevertheless, the effect of HHP on proteins has raised interest and studies have been carried out to elucidate it. High Hydrostatic Pressure treatments affect the non-covalent links (ionic, hydrophobic, and hydrogen links) of proteins, which means that the secondary, tertiary, and quaternary structures can be unfolded and dissociated while the primary structure remains stable [46]. Messens *et al.* [47] reported that it is necessary to apply a pressure of around 150 MPa to observe changes in the quaternary structure, and it is necessary to apply more than 200 MPa to significantly modify the secondary and tertiary structures. Owing to these changes, Liu *et al.* [48] and Tabilo-Munizaga *et al.* [49] studied the application of this technology to develop industrial applications to confer unique characteristics to foods (gel formation, emulsions, foams, new flavors and textures) or to seek a fat replacement. The possibility of using these new products as fat replacements has encouraged in-depth studies of stabilizing and gelling agents, agents that are usually incorporated in foodstuffs to give stability, texture, and palatability [50,51]. It is important to note that the changes depend directly on the type of protein used (disulfide bridges, linked by hydrophobic interactions, isoelectric points) and the HHP treatment (pressure, time, and temperature) [52,53].

All these studies make HHP a promising technology for the revalorization of waste and agro-industrial by-products.

According to He *et al.* [54], when proteins isolated from peanuts were treated at pressures between 50 and 200 MPa for 5 min the isolates increased their water-holding capacity (WHC) and oil-binding capacity (OBC), producing changes of interest in relation to protein properties. Additionally, the effect of HHP on milk proteins and whey has been studied in depth under various treatment conditions. The results indicated various changes in protein structure. Casein micelles experienced significant changes at pressures between 150 and 400 MPa and at a temperature of 20 °C [55]. However, a greater denaturalization of proteins from whey β-lactoglobulin and α-lactalbumin was observed at pressures higher than 100 and 400 MPa, respectively [47,55,56].

The effect of pressure on vegetable proteins has also been studied. Protein isolates from peanuts (5% w/v of protein) produced gels at 100 MPa for 5 min at 25 °C, while isolates from soya protein (9% w/v of protein) produced gels at 600 MPa for 5 to 10 min and a temperature of 33.5 °C [57,58]. For gelification of the isolates it was necessary to add $CaCl_2$ at a concentration of 0.015–0.020 mol L^{-1}, in accordance with the work reported by Maltais *et al.* [58], who indicated that calcium concentrations are very important and determine the final characteristics of gels. At low calcium concentrations filamentous gels occurred, while at high concentrations disordered phase separation gels or aggregates appeared.

7. Effect of HHP on Enzymes

There are two important regions in an enzyme, one responsible for recognizing the substrate and the other responsible for catalyzing the reaction when joined to the substrate. Minimal conformational change in the structure may completely affect the enzyme functionality.

Enzymes can be divided into two groups according to the effect of treatment by high hydrostatic pressure. In the first group are enzymes that are activated with pressures of 100–500 MPa, an activation that occurs only in monomeric proteins [59,60]. The second group includes enzymes that are inactivated when exposed to pressures higher than 500 MPa in combination with relatively high temperatures [61–63]. The main studies conducted on the effect of HHP on enzymes are based on the enzymes that are most often present in foodstuffs and produce deterioration of it or unacceptable sensory changes (Table 4). Among them we can highlight the enzymes peroxidase (POD), pectin methylesterase (PME), lipoxygenase (LOX), and polyphenol oxidase (PPO), as shown in the study carried out by Ludikhuyze *et al.* [64]. In general, polyphenol oxidase (PPO) and peroxidase (POD) are inactivated by applying a pressure equal to or greater than 400 MPa in combination with temperatures between 20 and 90 °C. Under these conditions, enzyme activity can be reduced by up to 50%, although the percentages may vary depending on the intrinsic properties of processed foods. It should be noted that the predictive models used in thermal inactivation are often inadequate to describe inactivation by HHP treatment [65].

Table 4. HHP inactivation of enzymes in different foods.

Food product	Enzyme	HHP conditions	Inactivation achieved	Reference
Jam	Pectin methylesterase (PME); Peroxidase (POD)	550–700 MPa for 2.5–75 min at 45–75 °C	PME: 27%–40% POD: 51%–70%	Igual et al. [61]
Feijoa puree	Peroxidase (POD); Polyphenol oxidase (PPO); Pectin methylesterase (PME)	600 MPa for 5 min at 25 °C	POD: 78% PPO: 55.6% PME: 56%	Ortuño et al. [62]
Camarosa strawberry	Polyphenol oxidase (PPO)	600 MPa for 15 min at 34–62 °C	PPO: 82%	Sulaiman and Silva [65]
Fruit smoothies	Polyphenol oxidase (PPO)	600 MPa for 10 min at 20 °C	PPO: 83%	Keenan et al. [66]
Dry-cured ham	Glutathione peroxidase (GSHPx); Superoxide dismutase (SOD)	900 MPa for 5 min at 12 °C	GSHPx: 44.2% SOD: 17.6%	Clariana et al. [67]
Strawberry pulps	β-Glucosidase; Polyphenol oxidase (PPO); Peroxidase (POD)	400–600 MPa for 5–25 min at 25 °C	β-Glu: 41.4% PPO: 74.6% POD: 74.6%	Cao et al. [68]

8. Some Industrial Applications of HHP

HHP technology has become a commercially implemented technology in fruit juice processing, spreading from its origins in Japan to the USA and Europe, and now Australia, with worldwide utilization increasing almost exponentially since 2000. In the U.S., Genesis Juice Corp.® processes eight types of organic juices by HHP, including apple, carrot, apple-ginger, apple-strawberry, ginger lemonade, strawberry lemonade, a herbal tea beverage, and apple- and banana-based smoothies, other company of high interest by its increment in sales in U.S., is Suja™, situated in San Diego, CA, produces a variety of mixture vegetable and fruits juices. European companies presently employing this technology in fruit juice processing include Invo® making smoothies in Spain, UltiFruit® making orange and grapefruit juices and a mixture of strawberry-orange juice in France, Frubaça® manufacturing various fruit-based beverages in Portugal, Juicy Line-Fruity Line® in Holland, Beskyd Frycovice, a.s® manufacturing mixtures of broccoli-apple-lemon and broccoli-orange-lemon in the Czech Republic, ATA S.P.A.® manufacturing carrot and apple juices in Italy, and Puro® commercializing smoothies in the UK.

Regarding processing conditions, treatments are optimized at a pressure level of 600 MPa in combination with moderate heat. In addition, due to the special characteristics of fruit juices, (nutritional components, flavor) and the perception by the consumer as a healthy food, quantities ranging from 500 to 2000 kg/h can be produced to satisfy current consumer demand considering the current capacities of industrial equipment. Shelf lives are estimated at *ca.* 10–35 day under refrigeration conditions, depending on the type of juice. Products are sold in supermarket chains, specialty and gourmet stores, and food services providing fruit preparations and dressings. Two main packaging formats are used, a small volume containing 250 mL, corresponding to a single portion, and a larger format containing 1 L.

One application of HHP that has great appeal is the stabilization of fresh cheese due its global consumption and the increase in global production of 3.5 million tons in the last 10 years (Figure 1), the cheese belongs to the ready-to-eat (RTE) food group, this product is characterized by special physical and chemical properties such as a near neutral pH, high water activity of 0.97, and high relative humidity, and is very prone to contamination from pathogens such as *Staphylococcus aureus*, *Listeria monocytogenes*, *Salmonella* spp., *Escherichia coli* O157:H7 [69,70], and spoilage microorganisms, such as molds and yeasts. Although currently this type of fresh cheese is made from pasteurized milk [71], the microbial recontamination occurs during subsequent processes, commonly in the stages of handling and packaging [72]. That is why high hydrostatic pressure technology could be of great interest in the microbiological stabilization of this product, avoiding high annual losses from foodborne diseases in which fresh cheeses are involved and rejections due to spoilage.

Figure 1. Cheese production for selected countries (in 1000 metric tons). Source: USDA [73].

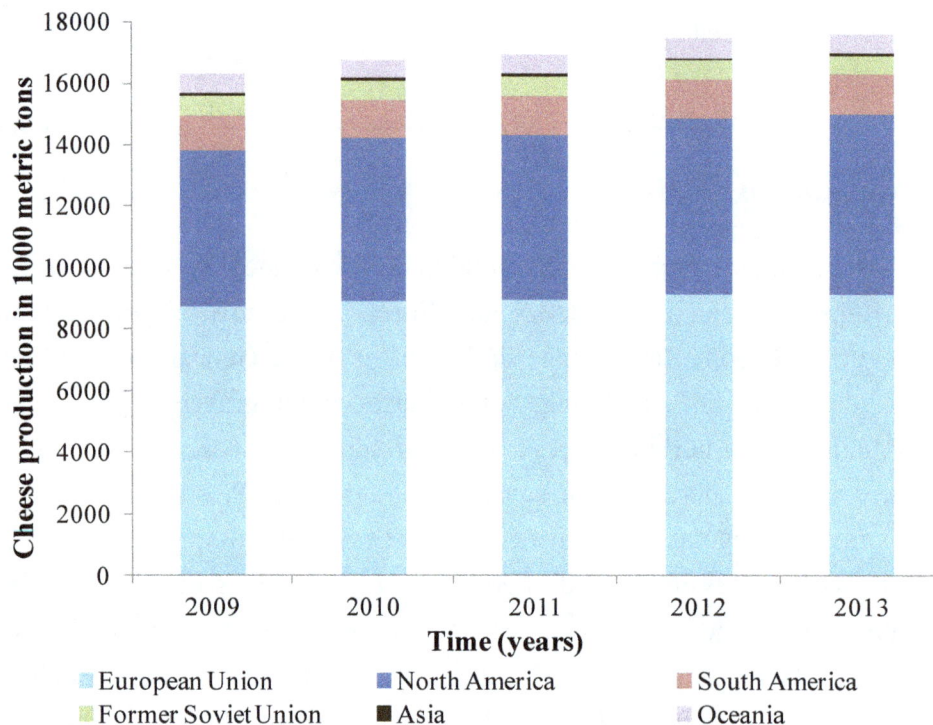

9. Conclusions

HHP treatment has proven to be an effective technology to reduce the microbial load of foods for both pathogenic and spoilage microorganisms with minimal impact on the initial quality of the foods. To apply HHP to food preservation, various parameters such as time, pressure, temperature, and pH should be considered because these parameters determine the optimum pressure intervals for microbial inactivation. Likewise, a combined treatment of moderate temperature and HHP has proven to have great potential both for the inactivation of microorganisms and enzymes and for the development of new products due to the modification of proteins of animal or vegetable origin.

Acknowledgements

We want to thank Alpina Research Institute (IAI) of Colombia for providing the necessary funds for this review and the Institute of Agrochemistry and Food Technology (IATA-CSIC) for their facilities.

Author Contributions

All authors contributed extensively to the work presented in this review, the authors analyzed the results and opined on the manuscript at all stages.

High hydrostatic pressure technology: Edwin F. Torres Bello, Bernadette F. Klotz Ceberio.

Packaging: Dolores Rodrigo.

Microbial inactivation: Edwin F. Torres Bello, Bernadette F. Klotz Ceberio.

Spore inactivation: Edwin F. Torres Bello, Antonio Martínez López.

Effects of HHP on proteins: Edwin F. Torres Bello, Gerardo González Martínez.

Effect of HHP on enzymes: Edwin F. Torres Bello, Antonio Martínez López, Dolores Rodrigo.

Some industrial applications of HHP: Edwin F. Torres Bello, Gerardo González Martínez.

Critical revision: Edwin F. Torres Bello, Antonio Martínez López, Gerardo González Martínez, Bernadette F. Klotz Ceberio, Dolores Rodrigo.

Conflict of Interest

The authors declare no conflict of interest.

References

1. Chevalier, D.; le Bail, A.; Ghoul, M. Effects of high pressure treatment (100–200 MPa) at low temperature on turbot (*Scophthalmus maximus*) muscle. *Food Res. Int.* **2001**, *34*, 425–429.

2. Polydera, A.C.; Stoforos, N.G.; Taoukis, P.S. Comparative shelf life study and vitamin C loss kinetics in pasteurised and high pressure processed reconstituted orange juice. *J. Food Eng.* **2003**, *60*, 21–29.

3. Hiperbaric, S.A. High Pressure Processing for Seafood & Meat Products. Available online: http://www.csiro.au/~/media/CSIROau/Images/Food/HPP Workshop pdf/HPP MeatSeafood Workshop May2012.pdf (accessed on 20 March 2014).

4. Ratphitagsanti, W.; Ahn, J.; Balasubramaniam, V.M.; Yousef, A.E. Influence of pressurization rate and pressure pulsing on the inactivation of *Bacillus amyloliquefaciens* spores during pressure-assisted thermal processing. *J. Food Prot.* **2009**, *72*, 775–782.

5. Barbosa-Canovas, G.V.; Gongora-Nieto, M.M.; Pothakamury, U.R.; Swanson, B.G. *Preservation of Foods with Pulsed Electric Fields*; Academic Press Ltd.: London, UK, 1999; pp. 1–9, 76–107, 108–155.

6. Ayvaz, H.; Schirmer, S.; Parulekar, Y.; Balasubramaniam, V.M.; Somerville, J.A.; Daryaei, H. Influence of selected packaging materials on some quality aspects of pressure-assisted thermally processed carrots during storage. *LWT Food Sci. Technol.* **2012**, *46*, 437–447.

7. Juliano, P.; Koutchma, T.; Sui, Q.A.; Barbosa-Canovas, G.V.; Sadler, G. Polymeric-based food packaging for high-pressure processing. *Food Eng. Rev.* **2010**, *2*, 274–297.

8. Lambert, Y.; Demazeau, G.; Largeteau, A.; Bouvier, J.M.; Laborde-Croubit, S.; Cabannes, M. New packaging solutions for high pressure treatments of food. *High Press. Res.* **2000**, *19*, 597–602.

9. Gayan, E.; Torres, J.A.; Paredes-Sabja, D. Hurdle approach to increase the microbial inactivation by high pressure processing: Effect of essential oils. *Food Eng. Rev.* **2012**, *4*, 141–148.

10. Mújica-Paz, H.; Valdez-Fragoso, A.; Tonello Samson, C.; Welti-Chanes, J.; Torres, J.A. High-pressure processing technologies for the pasteurization and sterilization of foods. *Food Bioprocess Technol.* **2011**, *4*, 969–985.

11. Norton, T.; Sun, D.-W. Recent advances in the use of high pressure as an effective processing technique in the food industry. *Food Bioprocess Technol.* **2008**, *1*, 2–34.

12. Torres, J.A.; Velázquez, G. Commercial opportunities and research challenges in the high pressure processing of foods. *J. Food Eng.* **2005**, *67*, 95–112.

13. Huang, H.W.; Lung, H.M.; Yang, B.B.; Wang, C.Y. Responses of microorganisms to high hydrostatic pressure processing. *Food Control* **2014**, *40*, 250–259.

14. Yang, B.; Shi, Y.; Xia, X.; Xi, M.; Wang, X.; Ji, B. Inactivation of foodborne pathogens in raw milk using high hydrostatic pressure. *Food Control* **2012**, *28*, 273–278.

15. Wang, C.Y.; Huang, H.W.; Hsu, C.P.; Shyu, Y.T.; Yang, B.B. Inactivation and morphological damage of *Vibrio parahaemolyticus* treated with high hydrostatic pressure. *Food Control* **2013**, *32*, 348–353.

16. Mohamed, H.M.H.; Diono, B.H.S.; Yousef, E.Y. Structural changes in *Listeria monocytogenes* treated with gamma radiation, pulsed electric field and ultra-high pressure. *J. Food Saf.* **2012**, *32*, 66–73.

17. Mañas, P.; Mackey, B.M. Morphological and physiological changes induced by high hydrostatic pressure in exponential- and stationary-phase cells of *Escherichia coli*: Relationship with cell death. *Appl. Environ. Microbiol.* **2004**, *70*, 1545–1554.

18. Chen, H.; Hoover, D.G. Modeling the combined effect of high hydrostatic pressure and mild heat on the inactivation kinetics of *Listeria monocytogenes* Scott A in whole milk. *Innov. Food Sci. Emerg. Technol.* **2003**, *4*, 25–34.

19. Ross, A.I.V.; Griffiths, M.W.; Mittal, G.S.; Deeth, H.C. Combining nonthermal technologies to control foodborne microorganisms. *Int. J. Food Microbiol.* **2003**, *89*, 125–138.

20. Jofré, A.; Aymerich, T.; Grébol, N.; Garriga, M. Efficiency of high hydrostatic pressure at 600 MPa against food-borne organisms by challenge tests on convenience meat products. *LWT Food Sci. Technol.* **2009**, *42*, 924–928.

21. Perrier-Cornet, J.M.; Hayert, M.; Gervais, P. Yeast cell mortality related to a high-pressure shift: Occurrence of cell membrane permeabilization. *J. Appl. Microbiol.* **1999**, *87*, 1–7.

22. Goh, E.L.C.; Hocking, A.D.; Stewart, C.M.; Buckle, K.A.; Fleet, G.H. Baroprotective effect of increased solute concentrations on yeast and moulds during high pressure processing. *Innov. Food Sci. Emerg. Technol.* **2007**, *8*, 535–542.

23. Evert-Arriagada, K.; Hernández-Herrero, M.M.; Juan, B.; Guamis, B.; Trujillo, A.J. Effect of high pressure on fresh cheese shelf-life. *J. Food Eng.* **2012**, *110*, 248–253.

24. Rosaria Corbo, M.; Lanciotti, R.; Albenzio, M.; Sinigaglia, M. Occurrence and characterization of yeasts isolated from milks and dairy products of Apulia region. *Int. J. Food Microbiol.* **2001**, *69*, 147–152.

25. Ferreira, E.H.D.R.; Rosenthal, A.; Calado, A.V.; Saraiva, J.; Mendo, S. *Byssochlamys nivea* inactivation in pineapple juice and nectar using high pressure cycles. *J. Food Eng.* **2009**, *95*, 664–669.

26. Houška, M.; Strohalm, J.; Kocurová, K.; Totušek, J.; Lefnerová, D.; Tříska, J. High pressure and foods—Fruit/vegetable juices. *J. Food Eng.* **2006**, *77*, 386–398.

27. Voldřich, M.; Dobiáš, J.; Tichá, L.; Čeřovský, M.; Krátká, J. Resistance of vegetative cells and ascospores of heat resistant mould *Talaromyces avellaneus* to the high pressure treatment in apple juice. *J. Food Eng.* **2004**, *61*, 541–543.

28. Basak, S.; Ramaswamy, H.S.; Piette, J.P.G. High pressure destruction kinetics of *Leuconostoc mesenteroides* and *Saccharomyces cerevisiae* in single strength and concentrated orange juice. *Innov. Food Sci. Emerg. Technol.* **2002**, *3*, 223–231.

29. O'Reilly, C.E.; O'Connor, P.M.; Kelly, A.L.; Beresford, T.P.; Murphy, P.M. Use of hydrostatic pressure for inactivation of microbial contaminants in cheese. *Appl. Environ. Microbiol.* **2000**, *66*, 4890–4896.

30. Zhu, S.; Naima, F.; Marcotte, M.; Ramaswamy, H.; Shao, Y. High-pressure destruction kinetics of *Clostridium sporogenes* spores in ground beef at elevated temperatures. *Int. Food Microbiol.* **2008**, *126*, 86–92.

31. Zimmermann, M.; Schaffner, D.W.; Aragão, G.M.F. Modeling the inactivation kinetics of *Bacillus coagulans* spores in tomato pulp from the combined effect of high pressure and moderate temperature. *LWT Food Sci. Technol.* **2013**, *53*, 107–112.

32. Meyer, R.S. Ultra High-Pressure, High Temperature Food Preservation Process. US Patent 6017572, 2000.

33. Tola, Y.B.; Ramaswamy, H.S. Combined effects of high pressure, moderate heat and pH on the inactivation kinetics of *Bacillus licheniformis* spores in carrot juice. *Food Res. Int.* **2014**, *62*, 50–58.

34. Linton, M.; Connolly, M.; Houston, L.; Patterson, M.F. The control of *Clostridium botulinum* during extended storage of pressure-treated, cooked chicken. *Food Control* **2014**, *37*, 104–108.

35. Silva, F.V.M.; Tan, E.K.; Farid, M. Bacterial spore inactivation at 45–65 °C using high pressure processing: Study of *Alicyclobacillus acidoterrestris* in orange juice. *Food Microbiol.* **2012**, *32*, 206–211.

36. Vercammen, A.; Vivijs, B.; Lurquin, I.; Michiels, C.W. Germination and inactivation of *Bacillus coagulans* and *Alicyclobacillus acidoterrestris* spores by high hydrostatic pressure treatment in buffer and tomato sauce. *Int. J. Food Microbiol.* **2012**, *152*, 162–167.

37. Hartyáni, P.; Dalmadi, I.; Knorr, D. Electronic nose investigation of *Alicyclobacillus acidoterrestris* inoculated apple and orange juice treated by high hydrostatic pressure. *Food Control* **2013**, *32*, 262–269.

38. Aouadhi, C.; Simonin, H.; Prévost, H.; Lamballerie, M.D.; Maaroufi, A.; Mejri, S. Inactivation of *Bacillus sporothermodurans* LTIS27 spores by high hydrostatic pressure and moderate heat studied by response surface methodology. *LWT Food Sci. Technol.* **2013**, *50*, 50–56.

39. Heinz, I.V.; Buckow, R. Food preservation by high pressure. *J. Verbraucherschutz Lebensmittelsicherheit* **2010**, *5*, 73–78.

40. Setlow, P. Spore germination. *Curr. Opin. Microbiol.* **2003**, *6*, 550–556.

41. Paredes-Sabja, D.; Torres, J.A.; Setlow, P.; Sarker, M.R. *Clostridium perfringens* spore germination: Characterization of germinants and their receptors. *J. Bacteriol.* **2008**, *190*, 1190–1201.

42. Clements, M.O.; Moir, A. Role of the gerI operon of *Bacillus cereus* 569 in the response of spores to germinants. *J. Bacteriol.* **1998**, *180*, 6729– 6735.

43. Moir, A.; Kemp, E.H.; Robinson, C.; Corfe, B.M. The genetic analysis of bacterial spore germination. *Soc. Appl. Bacteriol. Symp. Ser.* **1994**, *23*, 9S–16S.

44. Akhtar, S.; Paredes-Sabja, D.; Torres, J.A.; Sarker, M.R. Strategy to inactivate *Clostridium perfringens* spores in meat products. *Food Microbiol.* **2009**, *26*, 272–277.

45. Georget, E.; Kapoor, S.; Winter, R.; Reineke, K.; Songa, Y.; Callananc, M.; Anantac, E.; Heinz, V.; Mathys, A. *In situ* investigation of *Geobacillus stearothermophilus* spore germination and inactivation mechanisms under moderate high pressure. *Food Microbiol.* **2014**, *41*, 8–18.

46. Dzwolak, W.; Kato, M.; Taniguchi, Y. Fourier transform infrared spectroscopy in high-pressure studies on proteins. *Biophys. Acta* **2002**, *1595*, 131–144.

47. Messens, W.; van Camp, J.; Huyghebaert, A. The use of high pressure to modify the functionality of food proteins. *Trends Food Sci. Technol.* **1997**, *81*, 107–112.

48. Liu, R.; Zhaoa, S.M.; Xionga, S.B.; Xie, B.J.; Qinc, L.H. Role of secondary structures in the gelation of porcine myosin at different pH values. *Meat Sci.* **2008**, *80*, 632–639.

49. Tabilo-Munizaga, G.; Gordon, T.A.; Villalobos-Carvajal, R.; Moreno-Osorio, L.; Salazar, F.N.; Pérez-Won, M.; Acuña, S. Effects of high hydrostatic pressure (HHP) on the protein structure and thermal stability of Sauvignon blanc wine. *Food Chem.* **2014**, *155*, 214–220.

50. Devi, A.F.; Buckow, R.; Hemar, Y.; Kasapis, S. Structuring dairy systems through high pressure processing. *J. Food Eng.* **2013**, *114*, 106–122.

51. Chronakis, L.S.; Kasapis, S. A rheological study on the application of carbohydrate-protein incompatibility to the development of low fat commercial spreads. *Carbohydr. Polym.* **1995**, *28*, 367–373.

52. Trujillo, A.J.; Capellas, M.; Saldo, J.; Gervilla, R.; Guamis, B. Applications of high-hydrostatic pressure on milk and dairy products: A review. *Innov. Food Sci. Emerg. Technol.* **2002**, *3*, 295–307.

53. Huppertz, T.; Fox, P.F.; de Kruif, K.G.; Kelly, A.L. High pressure-induced changes in bovine milk proteins: A review. *Biochim. Biophys. Acta* **2006**, *1764*, 593–598.

54. He, X.H.; Liu, H.Z.; Liu, L.; Zhao, G.L.; Wang, Q. Effects of high pressure on the physicochemical and functional properties of peanut protein isolates. *Food Hydrocoll.* **2014**, *36*, 123–129.

55. Ye, R.; Harte, F. High pressure homogenization to improve the stability of casein-hydroxypropyl cellulose aqueous systems. *Food Hydrocoll.* **2014**, *35*, 670–677.

56. Roach, A.; Harte, F. Disruption and sedimentation of casein micelles and casein micelle isolates under high-pressure homogenization. *Innov. Food Sci. Emerg. Technol.* **2008**, *9*, 1–8.

57. Speroni, F.; Añón, M.C. Cold-set gelation of high pressure-treated soybean proteins. *Food Hydrocoll.* **2013**, *33*, 85–91.

58. Maltais, A.; Remondetto, G.E.; Gonzalez, R.; Subirade, M. Formation of soy protein isolate cold-set gels: Protein and salt effects. *J. Food Sci.* **2005**, *70*, 67–73.

59. Asaka, M.; Aoyama, Y.; Ritsuko, N.; Hayashi, R. Purification of a latent form of polyphenoloxidase from La France pear fruit and its pressure-activation. *Biosci. Biotech. Biochem.* **1994**, *58*, 1486–1489.

60. Huang, W.; Bi, X.; Zhang, X.; Liao, X.; Hu, X.; Wu, J. Comparative study of enzymes, phenolics, carotenoids and color of apricot nectars treated by high hydrostatic pressure and high temperature short time. *Innov. Food Sci. Emerg. Technol.* **2013**, *18*, 74–82.

61. Igual, M.; Sampedro, F.; Martínez-Navarrete, N.; Fan, X. Combined osmodehydration and high pressure processing on the enzyme stability and antioxidant capacity of a grapefruit jam. *J. Food Eng.* **2013**, *114*, 514–521.

62. Ortuño, C.; Duong, T.; Balaban, M.; Benedito, J. Combined high hydrostatic pressure and carbon dioxide inactivation of pectin methylesterase, polyphenol oxidase and peroxidase in feijoa puree. *J. Supercrit. Fluids* **2013**, *82*, 56–62

63. Hendrickx, M.E.; Ludikhuyze, L.R.; van den Broeck, I.; Weemaes, C.A. Effects of high-pressure on enzymes related to food quality. *Trends Food Sci. Technol.* **1998**, *9*, 197–203.

64. Ludikhuyze, L.; van den Broeck, I.; Hendrickx, M.E. High pressure processing of fruits and vegetables. In *Fruit and Vegetable Processing: Improving Quality*; Jongen, W., Ed.; CRC Press, Inc.: New York, NY, USA, 2002; pp. 346–362.

65. Sulaiman, A.; Silva, F.V.M. High pressure processing, thermal processing and freezing of 'Camarosa' strawberry for the inactivation of polyphenoloxidase and control of browning. *Food Control* **2013**, *33*, 424–428.

66. Keenan, D.F.; Rößle, C.; Gormley, R.; Butler, F.; Brunton, N.P. Effect of high hydrostatic pressure and thermal processing on the nutritional quality and enzyme activity of fruit smoothies. *LWT Food Sci. Technol.* **2012**, *45*, 50–57.

67. Clariana, M.; Guerrero, L.; Sárraga, C.; Garcia-Regueiro, J.A. Effects of high pressure application (400 and 900 MPa) and refrigerated storage time on the oxidative stability of sliced skin vacuum packed dry-cured ham. *Meat Sci.* **2012**, *90*, 323–329.

68. Cao, X.; Zhang, Y.; Zhang, F.; Wang, Y.; Yi, J.; Liao, X. Effects of high hydrostatic pressure on enzymes, phenolic compounds, anthocyanins, polymeric color and color of strawberry pulps. *J. Sci. Food Agric.* **2011**, *91*, 877–885.

69. Kousta, M.; Mataragas, M.; Skandamis, P.; Drosinos, E.H. Prevalence and sources of cheese contamination with pathogens at farm and processing levels. *Food Control* **2010**, *21*, 805–815.

70. Rosengrena, A.; Fabricius, A.; Guss, B.; Sylvén, S.; Lindqvist, R. Occurrence of foodborne pathogens and characterization of *Staphylococcus aureus* in cheese produced on farm-dairies. *Int. J. Food Microbiol.* **2010**, *144*, 263–269.

71. Devi, A.F.; Liu, L.H.; Hemar, Y.; Buckow, R.; Kasapis, S. Effect of high pressure processing on rheological and structural properties of milk-gelatin mixtures. *Food Chem.* **2013**, *141*, 1328–1334.

72. Reij, M.W.; den Aantrekker, E.D. Recontamination as a source of pathogens in processed foods. *Int. J. Food Microbiol.* **2004**, *91*, 1–11.

73. United States Department of Agriculture (USDA). Foreign Agricultural Service (FAS). Dairy: World Markets and Trade. Available online: http://www.fas.usda.gov/data/dairy-world-markets-and-trade (accessed on 20 March 2014).

11

Conventional and Innovative Processing of Milk for Yogurt Manufacture; Development of Texture and Flavor: A Review

Panagiotis Sfakianakis and Constatnina Tzia *

Laboratory of Food Chemistry and Technology, School of Chemical Engineering, National Technical University of Athens, 5 Iroon Polytechniou St., Polytechnioupoli, 15780, Zografou, Greece; E-Mail: psfakian@central.ntua.gr

* Author to whom correspondence should be addressed; E-Mail: tzia@chemeng.ntua.gr

Abstract: Milk and yogurt are important elements of the human diet, due to their high nutritional value and their appealing sensory properties. During milk processing (homogenization, pasteurization) and further yogurt manufacture (fermentation) physicochemical changes occur that affect the flavor and texture of these products while the development of standardized processes contributes to the development of desirable textural and flavor characteristics. The processes that take place during milk processing and yogurt manufacture with conventional industrial methods, as well as with innovative methods currently proposed (ultra-high pressure, ultrasound, microfluidization, pulsed electric fields), and their effect on the texture and flavor of the final conventional or probiotic/prebiotic products will be presented in this review.

Keywords: milk; yogurt; yogurt manufacture; thermal treatment; homogenization; pressure; ultra high pressure; ultrasound; microfluidization; pulsed electric fields; prebiotic; probiotic

1. Introduction—History of Yogurt

Milk and dairy products have been consumed since the domestication of mammals; yogurt and similar fermented milk products in particular are thought to originate from the Middle East. The original production of fermented milk products derived from the need to prolong the shelf life of milk instead of being disposed [1]. Yogurt manufacture was initially based on knowledge and empirical processes without standard procedures or investigation of the steps that occur during the entire process. Only after the late 20th century, when yogurt became a profitable commercial good, its manufacture became industrialized and the processes were standardized. During the last 20 years, interest in yogurt manufacture has increased tremendously for scientific and commercial reasons. Scientific findings have suggested new dairy products that benefit human health (probiotic cultures, fortification with bioactive compounds) as well as with improved sensory, especially textural characteristics. Thus, consumer demand for yogurt and similar fermented dairy products has increased.

Yogurt is defined as the product being manufactured from milk—with or without the addition of some natural derivative of milk, such as skim milk powder, whey concentrates, caseinates or cream—with a gel structure that results from the coagulation of the milk proteins, due to the lactic acid secreted by defined species of bacteria cultures. Furthermore, these bacteria must be "viable and abundant" at the time of consumption [2]. The above definition is part of the food legislation of many countries, ensuring that the essential characteristics of yogurt will be preserved, as well as that its traditional "concept" will not be compromised. The most common types of yogurt commercially available are set type yogurt and strained yogurt; though lately frozen yogurt and drinking yogurt have become quite popular as well. Set type yogurt is fermented in retail containers and no further stirring or water removal takes pace after the fermentation process. Strained (or stirred, or Greek style) yogurt is fermented in tanks under continuous mild stirring and after the completion of fermentation a portion of the whey is removed. Due to the manufacturing process, the two types develop a different texture; set type yogurt develops a continuous gel texture, whereas strained yogurt displays a viscous, creamy smooth texture [1].

2. Standardized Yogurt Manufacturing Process

Yogurt manufacture begins with the milking of the mammal, includes several processes, ending with the packaging of the final product, yogurt. The aim of this work is to present the processes that take place in the dairy industry; no further details about the milking and transportation of the milk will be included, even though these critical stages affect the quality and safety of the final product. Yogurt is mainly produced from bovine milk, although milk from other mammals is utilized for yogurt production as well. Yogurt derived from the milk of species other than bovine tends to vary in several sensory and physicochemical characteristics, due to differences per milk composition. For instance, yogurt derived from milk with high fat content (e.g., sheep, goat, and buffalo) has a more creamy texture compared to that derived from milk with lower fat content (e.g., bovine, mare, and ass). Therefore, the species of the milk-producing mammal significantly influence the characteristics of the produced yogurt [3].

2.1. Initial Treatment of Milk

Raw milk undergoes, in the dairy industry, centrifugal clarification to remove somatic cells and any other solid impurities [3]. Afterwards, a mild heating process, known as thermalization, is performed at temperature range 60–69 °C for 20–30 s, aiming at the killing of many vegetative microorganisms and the partial inactivation of some enzymes. This process causes almost no other irreversible change in the milk [4]. After thermalization, milk is cooled <5 °C or inoculated with lactic acid bacteria or other microfloras to control the growth of the psychrotrophic bacteria [3].

2.2. Standardization of Milk Components—Fat and SNF (Solid Non-Fat) Content

The standardization of milk refers to the standardization of fat and solid-non-fat content (SNF). Bovine milk fat content varies from 3.2%–4.2% w/w. The fat content of the milk is adjusted to range from <0.5%, for skim milk, to 1.5%–2%, for semi-fat milk, to 3.5% for full fat milk. As far as yogurt is concerned, the fat content ranges from 0.1%–10% according to consumer demands. In practice, to achieve the designed fat level, either the addition of skim milk or milk fat or the separation of fat from milk via centrifuge and mixing milk fat with skimmed milk is carried out [3]. The standardization process is of paramount importance, because the fat content of the milk influences the yogurt characteristics; increasing the fat content of milk results in an increase in the consistency and viscosity of yogurt [5,6]. Also, the milk fat content affects the maximum rate of pH decrease and pH lag phase during yogurt fermentation [7].

The term of standardization is also applied to the SNF content of the milk. The SNF components of milk mainly consist of lactose, protein and minerals; SNF content of milk varies from 11% to 14% of the total weight of the milk while the SNF of yogurt ranges from 9% to 16%. The SNF content of milk used for yogurt manufacture is altered, in some cases, by producers in order to attain the desired characteristics of the coagulum; the higher the SNF level, the higher the resulting yogurt's viscosity and firmness. The addition of native milk components is permitted to yogurt and fermented milk products in some countries. It is quite common in yogurt manufacturing to fortify the milk mixture with milk powder (skimmed or full fat), whey protein concentrates or casein powder, to achieve the desired SNF content and subsequently an increase in firmness and cohesiveness [5]. It must be noted that the fat and SNF content of milk has an impact on the fermentation process. In particular, the interaction of milk SNF content and fermentation temperature has a significant effect on the duration of the fermentation process; an increase of SNF increases the duration of the fermentation process [8].

2.3. Homogenization

Milk is a typical oil in water (o/w) emulsion with milk fat globules (MFG) acting as the oil droplets and the milk fat globules membrane as the emulsifier. However, because of the reaction of agglutinins and interfacial tension, the fat globules tend to collide, either by sharing the membrane or because of the Laplace principle, according to which the pressure is greater inside small globules than inside large globules and, hence, there is a tendency for large fat globules to grow at the expense of the smallest. This phenomenon, in addition to Brownian motion, forces the milk fat to rise to the surface of the milk and thus creates the undesirable effect of separation [9,10]. In order to prevent this effect, standardized milk undergoes homogenization. The basic principle of milk homogenization is to subject MFG to severe conditions in order to disrupt the membrane surrounding them and then maintain the new globules in dispersion while a new membrane is formed at the fat serum interface. The severe conditions that cause milk homogenization can be achieved by the application of pressure, high velocity flow of the milk, or high frequency vibrations (>10 kHz). The shear stress and temperature gradient developed under these conditions lead to a cavitation phenomenon that contributes to the homogenization process. Homogenization is carried out by the application of pressure. In particular, the pressure commonly applied in the dairy industry is 10–20 MPa [11,12]. The main homogenization effects are a reduction of the diameter of the MFG from 2–10 μm to 0.1–1 μm and altering the composition of the MFG membrane. According to Cano-Ruiz and Richter [13], the membrane absorbs protein molecules, mostly caseins, from the milk serum to become sufficient to emulsify the new-formed globules, since the fat surface area increases due to homogenization. Aguilera and Kessler [14] showed that a reduction of MFG size and the alterations in the MFG membrane caused by homogenization contribute to milk emulsion stability. Furthermore, homogenization affects the characteristics of acidified milk gels, like yogurt. According to Cho, et al. [15], the smaller MFG facilitate the incorporation of fat into the protein network [5], while their increased surface area favors the interactions between fat and milk proteins, casein and denatured whey, during acidification and subsequent gel formation [5,15].

2.4. Heat Treatment

Heat treatment of milk is carried out to ensure the safety of the product, whether it is milk itself or any other dairy product, and to exploit several effects that increased temperature has on certain milk components facilitating further processes for dairy products manufacture [16]. Heat treatment of milk reduces the number of pathogenic microorganisms to safe limits for the consumer's health. Various heat treatments can be applied, which are classified based on the duration and the temperature (Table 1). The most common are known as thermalization (referred in Section 2.1), low and high pasteurization, sterilization and UHT (Ultra Heat Treatment) [3,4,17]. Low pasteurization refers to heat treatment of milk at 63–65 °C for 20 min or at 72–75 °C for 15–20 s (HTST, High Temperature Short Time). During this process, most pathogens, vegetative bacteria, yeast and molds are killed. Additionally, with low temperature pasteurization, several enzymes become inactive, while the flavor of milk is hardly altered. Furthermore, little or no serum proteins are denatured, and cold agglutination and bacteriostatic properties remain virtually intact [2,4]. A more intense heat treatment is high temperature pasteurization that requires a temperature of 85 °C for 20–30 min or 90–95 °C for 5 min. During high temperature pasteurization most vegetative microorganisms are killed, except from spores; most enzymes are deactivated (except milk proteinase, plasmin in particular, some bacterial proteinases and lipases); most whey proteins are denatured, and a distinct "cooked" flavor is developed due to the formation, mostly, of ketones [4,18]; no further irreversible changes occur. Sterilization results in extermination of all microbial content of milk, including bacterial spores, and it is achieved at 110 °C for 30 min or at 130 °C for 40 s. In addition, sterilization causes inactivation of most milk enzymes (except several bacterial lipases), darkening of the milk color due to the Maillard reaction, evaporation of most flavor volatiles, thus weakening the flavor of the milk, and considerable damage to all milk proteins, even caseins. Finally, UHT is carried out at 145 °C for 1–2 s and achieves equal bacterial eradication as from sterilization, minimal flavor deterioration and causes denaturation of several whey proteins (β-lactoglobulin, serum albumin, and some immunoglobulins). UHT treatment and high pasteurization produces many volatiles in milk, such as: 2-pentanone, 2-heptanone, 2-nonanone, 2-undecanone, 2,6-dimethylpyrazine, 2-ethylpyrazine, 2-ethyl-3-methylpyrazine, methional, pentanoic acid, benzothiazole vanillin, hexanal, benzothiazole, decalactone, H_2S, methanethiol, dimethylsulphide and carboxylsulphide. These sulfur containing molecules are responsible for the "cooked" off flavor developed during UHT and high temperature pasteurization [18]. It should be mentioned that the most commonly used heat treatment in the yogurt manufacturing process is the high temperature pasteurization at 85 °C for 20 min [3,4].

Pathogens that can grow in milk, due to bad hygiene practices or hardware failure during the stages of processing, include *Mycobacterium tuberculosis, Coxiella burnetii, Staphylococcusaureus, Salmonella* species, *Listeria monocytogenes,* and *Campylobacter jejuni*. These microorganisms are killed by even mild heat treatment ensuring that processed milk is safe for consumption. The claim that milk is safe after a mild heat treatment might sound frivolous, but most high-heat resistant pathogens either do not occur in milk (e.g., *Bacillus anthracis*) or are outnumbered by other native microorganisms (e.g., *Clostridium perfringens*), or cause spoilage before their quantity is enough to cause health issues (e.g., *Bacillus cereus*).

In addition to the reduction or complete extermination of microbiological load, heat treatment causes release of CO_2 and O_2, an increase in the amount of insoluble colloidal calcium phosphate, a decrease in calcium cations, and forces lactose isomerization, degradation and Maillard reaction, thus affecting the pH of the milk and flavor. Finally for yogurt, the most important changes during heat treatment of milk concern the milk proteins; the reactions of milk proteins, during heat treatment, have a serious impact on the yogurt curd formation and will be described more thoroughly [4].

The casein molecules in milk are in the form of micelles or aggregates of submicelles which are formed from α_{s1}-, α_{s2}- and β-caseins stabilized by κ-casein molecules held together by calcium and calcium phosphate. This structure is stable and requires a high amount of energy to be disrupted. On the other hand, whey proteins in solution have a globular shape. Whey proteins, due to their structure, are fairly stable and do not interact with fatty molecules, calcium ions or caseins in their native state. However, in the case of whey proteins (β-lactoglobulin, serum albumin) over 80 °C, their peptide chains unfold, thus denaturating irreversibly. This deformation of the peptide chains exposes their thiol groups and enables them to interact with other molecules forming S–S bonds. Depending on the pH of the environment and the proximity of molecules available, whey proteins can form bonds with other whey proteins and caseins (κ- and α_{s1}- mostly) and also they can be incorporated at the MFG membrane. The denatured whey proteins, especially at pH values lower than 6.5, have the tendency to associate with casein micelles [4,11,16]. All the above phenomena are of paramount importance and are exploited during yogurt manufacture. Yogurt curd formation is based on the isoelectric precipitation of casein. However, whey proteins can be involved; if the thiol groups of the whey proteins are exposed, an interaction between casein and whey protein molecules occur, and the formation of casein-whey bonds are facilitated. Thus, whey proteins are incorporated into the curd matrix, strengthening the latter and resulting in a more firm yogurt. Therefore, the heat induced denaturation of whey protein favors the yogurt formation with high firmness and viscosity values [5,19]. Table 1 summarizes the thermal treatments utilized in dairy processing and the respective effects on milk itself and yogurt.

Table 1. Impact of different thermal treatment techniques on milk and yogurt properties affecting flavor and texture.

Milk Treatment	Treatment Description	Effect on Milk	Effect on Yogurt
Thermalisation	Heating at 60–69 °C, for 20–30 s	Death of non-heat resistance bacteria. Inactivation of several enzymes [4].	No significant effect. Characteristics affected by further processing [4].
Low Pasteurization	Heating at 63–65 °C for 20 min/at 72–75 °C for 15–20 s (HTST)	Death of most pathogens, vegetative bacteria, yeast and molds. Several enzymes denatured, denaturation of several whey proteins [4].	Slight increase in viscosity and firmness [1].
High Pasteurization	Heating at 85 °C for 20–30 min/at 90–95 °C for 5 min	Death of most vegetative microorganisms, except spores. Deactivation of most enzymes. Denaturation of most whey proteins. Development of "cooked" flavor [4,18].	Large increase in viscosity and firmness [1].
Sterilization	Heating at 110 °C for 30 min/at 130 °C for 40 s	Extermination of all microorganisms. Deactivation of most enzymes. Denaturation of whey proteins and aggregation of caseins (casein micelles) and MFG. Weakening of flavor intensity. Color darkening [4,18].	Incorporation of whey proteins into casein matrix. Very large increase in viscosity and firmness [1,4].
Ultra Heat Treatment (UHT)	Heating at 145 °C for 1–2 s	Extermination of all microorganisms. Mild flavor deterioration. Denaturation of whey proteins (β-lactoglobulin, serum albumin, several immunoglobulins) Development of off-flavors. Color darkening [4,18].	Medium increase in viscosity and firmness [1].

2.5. Fermentation Process

The fermentation process is the most important stage of yogurt manufacture. During this stage, the yogurt curd is formed, and its textural characteristics and distinct flavor are developed [3,5]. The key factor of the fermentation process is the starter culture that acts through biochemical reactions and inductively causes the formation of the curd and the development of flavor components [5]. For a fermented dairy product to be labeled as "yogurt", it should contain the two live bacterial strains of *Streptococcus salivarius* subsp. *thermophilus* and *Lactobacillus delbrueckii* subsp. *bulgaricus* in abundance. However, yogurt starter cultures may include other microorganisms as well, like *Lactobacillus acidophilus*, *Lactobacillus casei*, *Lactobacillus lactis*, *Lactobacillus jugurti*, *Lactobacillus helveticus*, *Bifidobacterium longum*, *Bifidobacterium bifidus* and *Bifidobacterium infantis*. *Streptococcus thermophilus* subsp. *thermophilus* (ST) is the only species in the streptococcus genus that is used in dairy starter cultures. ST is Gram positive and usually considered thermophilic, however, as the optimum temperature for its growth is 35–53 °C; therefore, ST can be considered as "thermotolerant". Its cells are spherical in shape, forming chains, during the early stage of their lives and as they mature develop a more rod-like morphology and favor colonial growth. *Lactobacillus delbrueckii* subsp. *bulgaricus* (LB) is rod-shaped, Gram-positive, anaerobic bacteria and its optimum growth temperature is 40–44 °C. LB can produce very high amounts of lactic acid by metabolizing lactose [5,20]. These two species display synergy in the milk environment, metabolizing lactose into lactic acid and causing reduction of milk pH. The synergism between ST and LB is based on their individual characteristics, and as a result higher lactose metabolism and lactic acid production is attained compared to each one acting individually. ST is more "aerotolerant" than LB, lacks good proteolytic ability in comparison to LB, but possesses greater peptidase activity. When grown together in milk, ST grows vigorously at first, whereas LB grows slowly. ST, because of its great proteolytic activity, creates an abundance of peptides to stimulate the growth of LB. During the early stage of fermentation, milk lactose is transported through the cell membrane of ST with the help of the enzyme galactoside permease located in the membrane. The lactose in the cell is then hydrolyzed by lactase or β-galactosidase enzyme. ST produces significant levels of lactase, which catalyzes the hydrolysis of lactose to glucose and galactose. Glucose is converted to pyruvate which is metabolized to lactic acid by the enzyme lactic dehydrogenase. Lowered oxygen tension and formate (byproduct of ST metabolism) in turn stimulate LB growth, which is further aided by the amino acids released by the active peptidases secreted by ST. Through coordinated tandem activities, both bacteria accelerate the entire fermentation, which none of them would be able to achieve individually. When the pH of the yogurt approaches 5.0, activity of ST subsides and LB gradually dominates the overall fermentation process until the target value of pH is reached and the fermentation process ceases. Normally, the fermentation period is terminated by lowering the temperature to 4 °C. At this temperature, the culture is still alive, but its activity is drastically limited to allow controlled flavor during storage and distribution [5,20].

The growth of the symbiotic culture induces changes in the native components of the milk that are responsible for the physicochemical and sensory characteristics of yogurt. During fermentation, lactose, milk proteins and microbial content, as well as several carbon compounds, suffer major changes, whereas minor changes occur for vitamins and minerals. Lactose is reduced by 30% and produces double the molar amount of lactic acid. Proteins (caseins and whey) aggregate, increasing the consistency of yogurt. Due to proteolysis caused by starter culture, amino acids (mainly proline and glycine) are released into the yogurt, even during storage at 4 °C [20]. During incubation, the starter culture growth results in an increase in the system's microbial content from 10^8 to 10^{10} CFU g^{-1}. The carbonyl compounds formed during fermentation are mostly lactic acid, acetaldehyde, dimethyl sulfide, 2,3-butanedione, 2,3-pentanedione, 2-methylthiophene, 3-methyl-2-butenal, 1-octen-3-one, dimethyl trisulfide, 1-nonen-3-one, acetic acid, methional, (*cis,cis*)-nonenal, 2-methyl tetrahydrothiophen-3-one, 2-phenyacetaldehyde, 3-methylbutyric acid, caproic acid, guaiacol and benzothiozole. These compounds contribute to the distinctive flavor of yogurt. As far as the lipids are concerned, several free fatty acid molecules are liberated due to lipase activity, mostly stearic and oleic acid. The only change in vitamin content is an increase in Vitamin B, throughout the fermentation process and storage. Finally, the quantity of minerals in yogurt remain the same as in milk, the only change is that, due to pH lowering, these minerals are in ionic rather than colloidal form [5,18].

The concentration of lactic acid in milk during fermentation increases, pH decreases, therefore the carboxyl groups dissociate, serine phosphate is ionized, and the negative charge between casein micelles is increased. However, the presence of calcium phosphate neutralizes this negative charge, keeping electrostatic repulsion down to a level where attractive forces between the protein molecules are dominant. Due to these attractive forces, the casein micelles aggregate and eventually coagulate into a network of small chains; this is responsible for the increase of viscosity and formation of the yogurt coagulum [3,19,21,22].

The milk fermentation process of yogurt can be described adequately by the evolution of pH and viscosity with respect to time; the model that expresses the evolution of pH during fermentation time is the modified Gompertz models of de Brabandere and de Baerdemaeker (1999) (Equation (1)) [23]:

$$\text{pH} = \text{pH}_0 + (\text{pH}_0 - \text{pH}_\infty) - \left\{ -\exp\left[\frac{e \cdot \mu_{\text{pH}}}{(\text{pH}_0 - \text{pH}_\infty)} \cdot (\lambda_{\text{pH}} - t) + 1 \right] \right\} \tag{1}$$

pH_0, pH_∞ = initial and end values of pH, respectively; μ_{pH} (min^{-1}) = maximum rate of pH decrease; λ_{pH} (min) = duration of pH lag phase. Furthermore, the model that describes the evolution of viscosity during fermentation is the modified Gomperz model of Soukoulis, *et al.* (2007) (Equation (2)) [7]:

$$\mu_\alpha = \mu_{\alpha 0} + (\mu_{\alpha 0} - \mu_{\alpha \infty}) - \left\{ -\exp\left[\frac{e \cdot \mu_v}{(\mu_{\alpha 0} - \mu_{\alpha \infty})} \cdot (\lambda_v - t) + 1 \right] \right\} \tag{2}$$

$\mu_{\alpha 0}$, $\mu_{\alpha \infty}$ (Pa·s) = initial and end values of viscosity respectively; μ_v (min^{-1}) = maximum rate of viscosity decrease; λ_v (min) = duration of viscosity lag phase.

2.6. Cooling

After the pH of yogurt reaches the value of 4.7–4.3, the yogurt is cooled to around 5 °C. This inhibits the growth and metabolic reaction of the starter culture and prevents the rise in acidity. Cooling of yogurt can be in one or two phases. One-phase cooling involves the rapid decrease of the coagulum temperature to less than 10 °C, where the fermentation process is inhibited leading to yogurt with low viscosity. Two-phase cooling is initiated by rapidly decreasing the temperature to less than 20 °C and then gradually reaching the storage temperature of 5 °C leading to yogurt with an increased viscosity and limited syneresis. This is quite common in the yogurt manufacture process, especially when fruits are to be added [3,24].

3. Innovative Methods for Milk and Yogurt Processing

Among the processes involved in milk and yogurt processing, the most important are homogenization, pasteurization and fermentation. Apart from conventional processes previously mentioned, new trends in milk processing that involve the utilization of ultra high pressure, ultrasound, pulsed electric field and microfluidization are presented in this section. Last, but not least, the probiotic bacteria and prebiotics are discussed that have beneficial effects on human health and have extensively been utilized in the dairy industry. Probiotics and prebiotics are worth mentioning, as dairy products are the most common medium for delivering them to the human intestine. In Table 2, the effects of conventional or innovative methods of homogenization on milk and yogurt properties are presented that generate the flavor and texture characteristics.

3.1. Ultra High Pressure Milk Treatment and Effect on Yogurt Characteristics

Ultra high pressure (UHP) involves the application of pressures from 100 to 1000 MPa. UHP utilization in food products was initiated during the early 1980s and is a non thermal pasteurization method. Studies have shown that milk treatment with pressures of 400–600 MPa for 10 min at 25 °C can achieve a similar result to low temperature pasteurization in terms of pathogenic and spoilage microorganisms inactivation [25]. Similar studies on UHP treatment of milk from Johnston, *et al.* [26] and Law, *et al.* [27] described the disintegration of the casein micelles into smaller particles and the simultaneous increase in the amount of caseins and calcium phosphate in the serum phase. Also, it was found by the same researchers that after UHP treatment, especially with pressure higher than 500 MPa, that the denaturation of several whey proteins occurs, in particular β-lactoglobulin, several immunoglobulins and α-lactalbumin [25,28]. However, Gervilla, *et al.* [29] observed a strange effect of UHP on the size and distribution of MFG. UHP up to 500 MPa at 25 and 50 °C reduced the diameter of MFG in the range of 1–2 μm, but at 4 °C the MFG displayed no tendency to shrink. This could be attributed to the fact that the MFG membrane remained unchanged.

The application of UHP in milk acid gel formation improves the texture and firmness, reduces syneresis and increases the water holding capacity in comparison to conventional yogurts [25]. The combination of UHP and thermal treatment is reported by Ferragut, *et al.* [30] to increase yogurt viscosity and lower gelation times compared to UHP treated samples.

3.2. Ultrasound Milk Treatment and Yogurt Characteristics

Ultrasound (US) is a sound wave with a frequency higher than the upper limit of human hearing, typically higher than 20 kHz. US has been utilized in the food industry since the late 1960s, for cleaning, monitoring and food component characterization. High intensity US (power level higher than 10 W), when propagated through a solution, generate immense pressure, temperature and shear gradients and thus cause cavitation [31,32]. Therefore, US is considered as an alternative method for reducing MFG size and can be effectively applied to homogenize milk. Wu, *et al.* [33] and Nguyen and Anema [34] showed that the application of US in milk reduces the MFG diameter to between 0.1 and 0.6 μm. In addition, US treatment has been referred by Krešić, *et al.* [35] and Chandrapala, *et al.* [36] to cause alterations of the MFG membrane composition and structure leading to an efficient homogenization effect compared to conventional methods. Additionally, the effect of US on milk proteins has been studied by Madadlou, *et al.* [37] and by Chandrapala, *et al.* [36]. US treatment has been shown to cause alteration in the secondary structure of the milk proteins, aggregation of protein particles as well as denaturation [36,37]. Riener, *et al.* [38] combined US with heat treatment (thermosonication) of milk and achieved a similar effect on the MFG as obtained with US treatment without heat, leading to reduction in size and changes of the membrane allowing interaction with casein micelles. Specifically, thermosonication treatment leads to an average diameter of 0.6 μm of MFG and a MFG membrane richer in casein molecules than the native [38]. Furthermore, high amplitude US has been reported to reduce the microbial content of milk [32,39]. Finally, high intensity US treatment causes the emission of volatiles from milk and formation of off flavors. Based on the study conducted by Riener, *et al.* [40], when milk is ultrasonicated, benzene, toluene, 1,3-butadiene,

5-methyl-1,3-cyclopentadiene, 1-hexene, 1-octene, 1-nonene, p-xylene, *n*-hexanal, *n*-heptanal, 2-butanone, acetone, dimethylsulfide and chloroform are emitted. The aldehydes can be produced from the breakdown of hydroperoxides generated by photo-oxidation induced by US, whereas the series of C6–C9 1-alkenes could arise from pyrolytic cleavage of fatty acid chains. The benzene formation may be attributed to cleavage of side chains of amino acids such as phenylalanine. These volatiles cause a rubbery and burned aroma [40].

The implementation of US treatment on the production of fermented dairy products has been studied with promising results. Milk gels and yogurt produced from milk treated by high intensity US have shown improved physical properties and high value of texture characteristics (firmness, cohesiveness). Increased amplitude level US treatment (20 kHz, 50–500 W, 1–10 min) significantly improved the water holding capacity of yogurt and increased viscosity and reduced syneresis; moreover, higher US intensity and a higher US exposure time of the milk resulted in increased yogurt viscosity [33]. Increased yogurt viscosity was reported by Riener, *et al.* [38] even from skim milk treated with US (22 kHz, 50 W, 0–30 min), due to high thermal denaturation of whey proteins. Milk thermosonication prior to fermentation, (25 kHz, 400 W, 45 or 75 °C for 10 min) resulted in the formation of yogurt with greater viscosity and higher water holding capacities compared to conventionally treated milk. The same treatment altered the microstructure of yogurt resulting in a honeycomb like network and exhibiting a more porous nature, whose average structural size was smaller (~2 μm) compared to conventionally heated yogurt [40]. A study by Vercet, *et al.* [41] combined thermosonication treatment of milk (40 °C, 20 kHz for 12 s) with moderate pressure (2 kg × cm^{-2}) and showed that the apparent viscosity, yield stress, and viscoelastic properties of yogurt were increased in addition to its structure being strengthened. The increased viscosity and texture of yogurt derived from ultrasonicated milk can be attributed to the denaturation of whey proteins and association of the latter with caseins. Denatured whey proteins are more susceptible to association with casein and casein micelles. Additionally, during acidification the denatured whey proteins, associated or not with casein micelles, aggregate due to the reduction of the repulsive charge. Therefore, denatured whey proteins associated with casein micelles could act as bridging material between casein micelles and as a result the bonds that form in the yogurt matrix are formed more easily, resulting in stronger yogurt coagulum [42].

3.3. Application of Microfluidization in Milk and Yogurt Manufacture

The Microfluidizer® is an apparatus that causes homogenization via shear, turbulence and cavitation. Initially, it accelerates the fluid and separates it into two microstreams that intersects in a chamber and collide. The impact causes intense turbulence and cavitation and thus the homogenization effect is achieved [43,44]. In the case of milk studied by Ciron, *et al.* [43], the microfluidization treatment reduced the diameter of the MFG to less than 2 μm [45]. The application of microfluidization in yogurt manufacture has often been used. A comparison between yogurts derived from microfluidized milk of 0% and 1.5% fat content with conventionally homogenized milk showed that non-fat yogurt from microfluidized milk displayed increased syneresis, and reduced viscosity and firmness compared to conventionally manufactured yogurt, whereas low fat yogurts from microfluidized milk had similar texture characteristics with those from conventionally homogenized milk. Microfluidization of low-fat milk resulted in yogurt with modified microstructure, giving more interconnectivity in the protein networks with embedded fat globules, but with similar texture profiles and water retention compared to yogurt prepared from conventionally homogenized milk [43]. However, this technique requires more study to assess the efficiency of yogurt manufacture [43,46].

3.4. Pulsed Electric Field Application in Milk and Yogurt Manufacture

Pulsed electric field (PEF) treatment emits intense electric pulses through a continuous medium, to inactivate microorganisms with the best results achieved in fluids. PEF has been applied in dairy systems combined with probiotic cultures [47]. The intensity of fields range between 15 and 50 kV/cm, and the treatment lasts only a few seconds [48]. The PEF principle is to destabilize the microbial cells with a high-pressure pulse. Subsequently, electroporation to the cellular membrane makes it more permeable; therefore, the cells rupture and expel their contents [49]. Additionally, Lin, *et al.* [50] combined PEF, UHP and thermal treatment in milk and noticed even more of a decrease in microbiological content. The efficiency of PEF depends on the intensity of the electric field, and the number and duration of pulses [51]. Despite its potential, PEF application requires a high tolerance to elevated electric fields, low electric conductivity and absence of bubbles [51,52].

Table 2. Impact of different homogenization techniques on milk and yogurt properties affecting flavor and texture.

Milk Treatment	Treatment Description	Effect on Milk	Effect on Yogurt
Conventional with Pressure	10–20 MPa	Decrease of MFG size. Stability of milk as an emulsion. Whey proteins absorption to the MFG membrane.	Facilitation of curd formation. Whey protein incorporation into the casein matrix. Slight increase in viscosity and firmness.
Ultra High Pressure (UHP)	100–1000 MPa	Inactivation of spoilage and pathogenic microorganism. Casein micelles disruption. Denaturation of several whey proteins. MFG size decrease with a tendency for collision and re-aggregation.	Higher value of texture characteristics. Higher viscosity. Lower syneresis. Increased water holding capacity.
High Intensity Ultrasonication	Higher than 20 kHz, amplitude higher than 100 W	MFG size reduction. Stability of milk as an emulsion. Interaction of whey proteins with casein micelles and MFG. Reduction of microbial content. Development of off-flavor volatiles.	Higher value of texture characteristics. Higher viscosity. Lower syneresis. Increased water holding capacity.
Microfluidization	Separation of milk into two steams, moving at high velocity with subsequent collision.	MFG size reduction.	Non-fat yogurt: increased syneresis and lower viscosity. Low fat yogurt: similar texture characteristics as for conventionally manufactured yogurt.

Table 2. *Cont.*

Milk Treatment	Treatment Description	Effect on Milk	Effect on Yogurt
Pulsed Electric Field (PEF)	Application of electric pulses through milk. Intensity: 1–50 kV/cm for a few seconds.	Microbial content reduction.	Similar texture and water holding capacity as for conventionally manufactured yogurt.

3.5. Probiotic and Dairy Products

Modern nutritional probiotic products are classified as those that contain probiotic microorganisms. Probiotic microorganisms are defined as live microbes, which when ingested benefit the health of the host through their effect on the intestinal microflora. In addition, probiotic cultures must be able to survive throughout the intestinal tract, resist acidic conditions during gastric passage and bile digestion. To have their beneficial effect on the health of the host, probiotic strains must, at least temporarily, establish themselves among the natural microflora of the intestine. The initial mention about microorganisms that benefit the health of host being, while delivered via food consumption is attributed to Élie Metchnikoff in the early years of the 20th century [53]. The terminology "probiotic" is attributed to Lilley and Stillwell [54] in order to differentiate them from antibiotics. These authors defined probiotics as a substance produced by one microorganism stimulating the growth of another microorganism [54]. Later on, in 1971, Sperti gave this name to tissue extracts promoting the growth of microbes [55]. The definition by Parker in 1974 is closer to the commonly accepted definition: organisms and substances which contribute to the intestinal microbial balance [56]. The accepted definition of probiotics is a slightly improved definition by Fuller in 1989 "live microbial feed supplement which beneficially affects the host by improving its intestinal microbial balance" [57]. Probiotics are added to food as adjunct cultures in concentrations of 10^7–10^8 CFU/g or mL, if not participating in the fermentative process, and if participating, they can reach a concentration between 10^8 and 10^9 CFU/g or mL after the fermentative process [58]. The most common probiotic bacteria are strains and species of Lactobacilli, Bifidobacteria, Enterococci and Lactococci. The species most thoroughly been studied are *Lactobacillus acidophilus*, *Lactobacillus casei*, *Lactobacillus lactis*, *Lactobacillus helveticus*, *Bifidobacterium longum*, *Bifidobacterium lactis*, *Bifidobacterium animalis* ssp. *lactis* and *Bifidobacterium bifidum*, *Bifidobacterium longum*, and *Bifidobacterium bifidus* [59].

Fermented dairy products, and especially yogurt, are ideal carriers for probiotic cultures to enter the human digestive system and ensure their survivability through the stomach [60]; therefore, the subject of probiotic yogurt is thoroughly studied from its medical and dairy science perspective. For the purpose of this work, only the effect of the probiotic cultures on the yogurt manufacture and characteristics is mentioned.

Most probiotic bacteria have no significant effect on the fermentation process or on the yogurt sensory properties [61]. Based on the studies of Allgeyer, *et al.* [61] and Atunes, *et al.* [62], the addition of *Bifidobacterium lactis* and *Lactobacillus acidophilus* resulted in no difference in the sensory evaluation of low and full fat probiotic and non-probiotic yogurt. The same results were noted by other researchers as well [63,64]. However, according to Akalin, *et al.* [65] the combination of probiotic culture (*Bifidobacterium animalis* ssp. *Lactis*) and fortification of WPC (whey protein concentrate) resulted in a stronger coagulum and increased firmness and adhesiveness values for the yogurt. This was attributed mostly to the effect of the fortification with WPC and not to the presence of the probiotic culture.

Overall, the incorporation of probiotic bacteria in yogurt manufacture is viable and this claim is shown by the amount of probiotic yogurt products available in the market; however, the actual challenge is to ensure that the probiotic culture reaches the intestines of the consumers alive and is able to establish itself among the native microflora [66].

3.6. Prebiotics and Dairy Products

According to Gibson and Roberfroid [67], prebiotics are classified as certain food ingredients that beneficially affect the host in a very specific way. Prebiotics are food components, non-digestible by humans, that selectively stimulate the growth and activity of certain bacterial species already existing in the human colon, and inductively improve the health of the host. Most prebiotics are oligosaccharides in general, fructooligosaccharides in particular. Most common oligosaccharides with prebiotic character are inulin, *trans*-galactooligosaccharide, lactulose, isomalt and oligofructose. They have been shown to stimulate the growth of endogenous Bifidobacteria, and make them the predominant species in human feces [68]. Based on the study of Cruza, *et al.* [69] who added oligofructose to plain yogurt, there was no influence on the pH, the proteolytic ability, or on the viability of *Streptococcus thermophilus* or *Lactobacillus bulgaricus*. However, the end product of this endeavor was characterized as a weak gel with thixotropic and pseudoplastic behavior. Finally, the oligofructose-fortified yogurts had a fairly high acceptance by consumers. Another study conducted by Pimentel [70] suggested that the addition of long chain inulin, another known prebiotic, in low fat yogurt can lead to interesting results in sensory properties and especially in texture characteristics. In particular, the replacement of native milk fats with long chain inulin created equally acceptable firmness and color as with yogurt containing native milk fats. Most studies on the addition of prebiotic oligosaccharides in yogurt agree that most of the characteristics of the final product and the process remain fairly close to the values of the originals. Several short chain prebiotics have a slightly negative effect on the firmness and creaminess of the yogurt whereas long chain prebiotics increase those values. Overall, the final choice remains with the consumer and their preference for texture.

4. Conclusions

Dairy processing and yogurt manufacture utilize several scientifically interesting procedures such as centrifugation, homogenization, heat treatment, and in the case of yogurt, manufacture and fermentation. Each procedure significantly affects the quality and sensory characteristics of the final product, whether it is milk or yogurt. Conventional heat treatment includes thermalization, low and high temperature pasteurization, ultra heat treatment and sterilization. The application of heat treatment in milk affects the flavor, the microbial content and the milk proteins. The more intense the heat treatment is, the more radical the changes that occur. Heat treatment also affects the texture of the produced yogurt, increasing the value of its texture characteristics (firmness, cohesiveness) and viscosity. Homogenization, typically used in dairy processing, is through application of pressure, reducing the milk fat globule size and preventing fat separation from the milk. Other treatments that cause the same homogenization effect with pressure in milk are ultra high pressure, ultrasound, microfluidization and pulsed electric fields. Each type of homogenization causes additional effects on milk and on the produced yogurt. Finally, conventional fermentation process, in yogurt manufacturing, included the utilization of the species *Streptococcus salivarius* subsp. *thermophilus* and *Lactobacillus delbrueckii* subsp. *bulgaricus*. Modern dairy science and nutrition have suggested the involvement of probiotic cultures and prebiotic ingredients in order to increase the nutritional value of dairy products, while minimizing detrimental effects on the sensory characteristics.

Conflicts of Interest

The authors declare no conflict of interest.

References

1. Tamime, A.Y.; Robisons, R.K. Chapter 1 Historical background. In *Tamime and Robinson's Yogurt: Science and Technology*, 3rd ed.; Woodhead Publishing LTD: Cambridge, UK, 2007; pp. 1–10.

2. Chandan, R.C. Chapter 1 History and consumption trends. In *Manufacturing Yogurt and Fermented Milks*; Chandan, R.C., Ed.; Blackwell Publishing: Ames, IA, USA, 2006; pp. 3–17.

3. Tamime, A.Y.; Robisons, R.K. Chapter 2 Backround to manufacturing practice. In *Tamime and Robinson's Yogurt: Science and Technology*, 3rd ed.; Woodhead Publishing LTD: Cambridge, UK, 2007; pp. 11–118.

4. Walstra, P.; Wouters, J.T.M.; Geurts, T.J. Chapter 7 Heat treatment. In *Dairy Science and Technology*; Taylor & Francis Group, LLC: Boca Raton, FL, USA, 2006; pp. 225–272.

5. Walstra, P.; Wouters, J.T.M.; Geurts, T.J. Chapter 22 Fermented milks. In *Dairy Science and Technology*; Taylor & Francis Group, LLC: Boca Raton, FL, USA, 2006; pp. 551–573.

6. Shaker, R.R.; Jumah, R.Y.; Abu-Jdayil, B. Rheological properties of plain yogurt during coagulation process: Impact of fat content and preheat treatment of milk. *J. Food Eng.* **2000**, *44*, 175–180.

7. Soukoulis, C.; Panagiotidis, P.; Koureli, R.; Tzia, C. Industrial yogurt manufacture: Monitoring of fermentation process and improvement of final product quality. *J. Dairy Sci.* **2007**, *90*, 2641–2654.

8. Kristo, E.; Biliaderis, C.G.; Tzanetakis, N. Modelling of the acidification process and rheological properties of milk fermented with a yogurt starter culture using response surface methodology. *Food Chem.* **2003**, *83*, 437–446.

9. Ion-Titapiccolo, G.; Alexander, M.; Corredig, M. Heating of milk before or after homogenization changes its coagulation behavior during acidification. *Food Biophys.* **2013**, *8*, 81–89.

10. Fox, P.P. Fat globules in milk. In *Encyclopedia of Dairy Sciences*, 2nd ed.; Elsevier Ltd., Academic Press: London, UK, 2011; pp. 1564–1548.

11. Walstra, P.; Wouters, J.T.M.; Geurts, T.J. Chapter 2 Milk components. In *Dairy Science and Technology*; Taylor & Francis Group, LLC: Boca Raton, FL, USA, 2006; pp. 17–108.

12. Walstra, P.; Wouters, J.T.M.; Geurts, T.J. Chapter 9 Homogenization. In *Dairy Science and Technology*; Taylor & Francis Group, LLC: Boca Raton, FL, USA, 2006; pp. 276–279.

13. Cano-Ruiz, M.E.; Richter, R.L. Effect of homogenization pressure on the milk fat globule membrane proteins. *J. Dairy Sci.* **1997**, *11*, 2732–2739.

14. Aguilera, J.M.; Kessler, H.G. Physicochemical and rheological properties of milk-fat globules with modified membranes. *Milchwissenschaft* **1988**, *43*, 411–415.

15. Cho, Y.H.; Lucey, J.A.; Singh, H. Rheological properties of acid milk gels as affected by the nature of the fat globule surface material and heat treatment of milk. *Int. Dairy J.* **1999**, *9*, 537–545.

16. Kilara, A. Chapter 5 Basic dairy processing principles. In *Manufacturing Yogurt and Fermented Milks*; Chandan, R.C., Ed.; Blackwell Publishing: Ames, IA, USA, 2006; pp. 73–89.

17. Lewis, M.J. Chapter 5 Improvements in the pasteurisation and sterilisation of milk. In *Dairy Processing*; Smith, G., Ed.; Woodhead Publishing LTD: Cambridge, UK, 2003; pp. 79–102.

18. Boelrijk, A.E.M.; de Jong, C.; Smit, G. Chapter 7 Flavour generation in dairy products. In *Dairy Processing*; Smith, G., Ed.; Woodhead Publishing LTD: Cambridge, UK, 2003; pp. 128–153.

19. Jaros, D.; Rohm, H. Chapter 8 Controlling the texture of fermented dairy products: The case of yoghurt. In *Dairy Processing*; Smith, G., Ed.; Woodhead Publishing LTD: Cambridge, UK, 2003.

20. Vedamuthu, E.R. Chapter 6 Starter cultures for yogurt and fermented milks. In *Manufacturing Yogurt and Fermented Milks*; Chandan, R.C., Ed.; Blackwell Publishing: Ames, IA, USA, 2006; pp. 89–117.

21. Horne, D.S. Formation and structure of acidified milk gels. *Int. Dairy J.* **1999**, *9*, 261–268.

22. Chandan, R.C.; O'Rell, K.R. Chapter 12 Principles of yogurt processing. In *Manufacturing Yogurt and Fermented Milks*; Chandan, R.C., Ed.; Blackwell Publishing: Ames, IA, USA, 2006; pp. 195–211.

23. De Brabandere, A.G.; de Baerdemaeker, J.G. Effects of process conditions on the pH development during yogurt fermentation. *J. Food Eng.* **1999**, *41*, 221–227.

24. Walstra, P.; Wouters, J.T.M.; Geurts, T.J. Chapter 11 Cooling and freezing. In *Dairy Science and Technology*; Taylor & Francis Group, LLC: Boca Raton, FL, USA, 2006; pp. 297–307.

25. Trujillo, A.J.; Capellas, M.; Saldo, J.; Gervilla, R.; Guamis, B. Applications of high-hydrostatic pressure on milk and dairy products: A review. *Innov. Food Sci. Emerg. Technol.* **2002**, *4*, 295–307.

26. Johnston, D.E.; Austin, B.A.; Murphy, R.J. Effects of high hydrostatic pressure on milk. *Milchwissenschaft* **1992**, *47*, 760–763.

27. Law, A.J.R.; Leaver, J.; Felipe, X.; Ferragut, V.; Pla, R.; Guamis, B. Comparison of the effects of high pressure and thermal treatments on the casein micelles in goat's milk. *J. Agric. Food Chem.* **1998**, *46*, 2523 –2530.

28. Felipe, X.; Capellas, M.; Law, A.R. Comparison of the effects of high-pressure treatments and heat pasteurisation on the whey proteins in goat's milk. *J. Agric. Food Chem.* **1997**, *45*, 627–631.

29. Gervilla, R.; Ferragut, V.; Guamis, B. High hydrostatic pressure effects on colour and milk-fat globule of ewe's milk. *J. Food Sci.* **2001**, *66*, 880–885.

30. Ferragut, V.; Martınez, V.M.; Trujillo, A.J.; Guamis, B. Properties of yoghurts made from whole ewe's milk treated by high hydrostatic pressure. *Milchwissenschaft* **2000**, *55*, 267–269.

31. Demirdöven, A.; Baysal, T. The use of ultrasound and combined technologies in food preservation. *Food Rev. Int.* **2009**, *25*, 1–11.

32. Dolatowski, Z.J.; Stadnik, J.; Stasiak, D. Applications of ultrasound in food technology. *Acta Sci. Pol. Technol. Aliment.* **2007**, *6*, 89–99.

33. Wu, H.; Hulbert, G.J.; Mount, J.R. Effects of ultrasound on milk homogenization and fermentation with yogurt starter. *Innov. Food Sci. Emerg. Technol.* **2009**, *3*, 211–218.

34. Nguyen, H.A.; Anema, S.G.; Effect of ultrasonication on the properties of skim milk used in the formation of acid gels. *Innov. Food Sci. Emerg. Technol.* **2010**, *11*, 616–622.

35. Krešić, G.; Lelas, L.; Jambrak, A.R.; Herceg, Z.; Brnčić, S.R.; Influence of novel food processing technologies on the rheological and thermophysical properties of whey proteins. *J. Food Eng.* **2008**, *1*, 64–73.

36. Chandrapala, J.; Zisu, B.; Palmer, M.; Kentish, S.; Ashokkumar, M. Effects of ultrasound on the thermal and structural characteristics of proteins in reconstituted whey protein concentrate. *Ultrason. Sonochem.* **2011**, *18*, 951–957.

37. Madadlou, A.; Mousavi, M.E.; Emam-Djomeh, Z. Comparison of pH-dependent sonodisruption of re-assembled casein micelles by 35 and 130 kHz ultrasounds. *J. Food Eng.* **2009**, *95*, 505–509.

38. Riener, J.; Noci, F.; Cronin, D.A. The effect of thermosonication of milk on selected physicochemical and microstructural properties of yoghurt gels during fermentation. *Food Chem.* **2009**, *114*, 905–911.

39. Bermúdez-Aguirre, D.; Corradini, M.G.; Mawson, R.; Barbosa-Canova, G.V. Modeling the inactivation of *Listeria innocua* in raw whole milk treated under thermo-sonication. *Innov. Food Sci. Emerg. Technol.* **2008**, *10*, 172–178.

40. Riener, J.; Noci, F.; Cronin, D.A.; Morgan, D.; Lyng, J.G. Characterisation of volatile compounds generated in milk by high intensity ultrasound. *Int. Dairy J.* **2009**, *19*, 269–272.

41. Vercet, A.; Oria, R.; Marquina, P.; Crelier, S.; López-Buesa, P. Rheological properties of yoghurt made with milk submitted to manothermosonication. *J. Agric. Food Chem.* **2002**, *50*, 6165–6171.

42. Morand, M.; Guyomarc'h, F.; Famelart, M.H. How to tailor heat-induced whey protein/κ-casein complexes as a means to investigate the acid gelation of milk—A review. *Dairy Sci. Technol.* **2011**, *91*, 97–126.

43. Ciron, C.I.E.; Gee, V.L.; Kelly, A.L.; Auty, M.A.E. Comparison of the effects of high-pressure microfluidization and conventional homogenization of milk on particle size, water retention and texture of non-fat and low-fat yoghurts. *Int. Dairy J.* **2010**, *20*, 314–320.

44. Kasaai, M.R.; Charlet, G.; Paquin, P.; Arul, J. Fragmentation of chitosan by microfluidization process. *Innov. Food Sci. Emerg. Technol.* **2003**, *4*, 403–413.

45. Skurtys, O.; Aguilera, J.M. Applications of microfluidic devices in food engineering. *Food Biophys.* **2008**, *3*, 1–15.

46. Ronkart, S.N.; Paquot, M.; Deroanne, C.; Fougnies, C.; Besbes, S.; Blecker, C.S. Development of gelling properties of inulin by microfluidization. *Food Hydrocoll.* **2010**, *24*, 318–324.

47. Da Cruz, A.G; de Assis Fonseca Faria, J.; Saad, S.M.I.; Bolini, H.M.A.; Sant'Ana, A.S.; Cristianini, M. High pressure processing and pulsed electric fields: Potential use in probiotic dairy foods processing. *Trends Food Sci. Technol.* **2010**, *21*, 483–493.

48. Ravishankar, S.; Zhang, H.; Kempkes, M.L. Pulsed electric fields. *Food Sci. Technol. Int.* **2008**, *14*, 429–432.

49. Wouters, P.C.; Bos, A.P.; Ueckert, J. Membrane permeabilization in relation to inactivation kinetics of *Lactobacillus* species due to pulsed electric fields. *Appl. Environ. Microbiol.* **2001**, *67*, 3092–3101.

50. Lin, S.; Clark, S.; Powers, J.R.; Luedecke, L.O.; Swanson, B.G. Thermal, ultra high pressure, and pulsed electric field attenuation of *Lactobacillus*: Part 2. *Agro Food Ind. Hi-Tech* **2002**, *13*, 6–11.

51. Calderon-Miranda, M.L.; Barbosa-Canovas, G.V.; Swanson, B.G. Transmission electron microscopy of *Listeria innocua* treated by pulsed electric fields and niasin in skimmed milk. *Int. J. Food Microbiol.* **1999**, *51*, 31–38.

52. Kelly, A.L.; Zeece, M. Applications of novel technologies in processing of functional foods. *Aust. J. Dairy Technol.* **2009**, *64*, 12–16.

53. Metchnikoff, E. *Essais Optimistes. Paris. The Prolongation of Life. Optimistic Studies*; Mitchell, P.C., Ed.; Heinemann: London, UK, 1907.

54. Lilly, D.M.; Stillwell, R.H. Probiotics: Growth-promoting factors produced by microorganisms. *Science* **1965**, *147*, 747–748.

55. Sperti, G.S. *Probiotics*; Avi Publishing Co.: Westport, CT, USA, 1971.

56. Parker, R.B. Probiotics: The other half of the antibiotic story. *Anim. Nutr. Health* **1974**, *29*, 4–8.

57. Kneifel, W.; Mattila-Sandholm, T.; von Wright, A. Probiotic Bacteria detection and estimation in fermented and non-fermeneted dairy products. *Encycl. Food Microbiol.* **1999**, *3*, 1783–1789.

58. Vinderola, G.; Binetti, A.; Burns, P.; Reinheimer, J. Cell viability and functionality of probiotic bacteria in dairy products. *Front. Microbiol.* **2011**, *2*, 70.

59. Salminen, S.; Kenifel, W.; Ouwhand, A.C. Bacteria, beneficial. Probiotics, applications in dairy products. In *Encyclopedia of Dairy Sciences*, 2nd ed.; Elsevier Ltd., Academic Press: London, UK, 2011; pp. 412–419.

60. Elizaquível, P.; Sánchez, G.; Salvador, A.; Fiszman, S.; Dueñas, M.T.; López, P.; Fernández de Palencia, P.; Aznar, R. Evaluation of yogurt and various beverages as carriers of lactic acid bacteria producing 2-branched (1,3)-β-D-glucan. *J. Dairy Sci.* **2011**, *94*, 3271–3278.

61. Allgeyer, L.C.; Miller, M.J.; Lee, S.-Y. Sensory and microbiological quality of yogurt drinks with prebiotics and probiotics. *J. Dairy Sci.* **2010**, *93*, 4471–4479.

62. Atunes, A.E.; Cazetto, T.F.; Bolini, H.M. Viability of probiotic microorganisms during storage, post-acidification and sensory analysis of fat-free yogurts with added whey protein concentrate. *Int. J. Dairy Technol.* **2005**, *58*, 169–173.

63. Cruz, A.G.; Cadena, R.S.; Castro, W.F.; Esmerino, E.A.; Rodrigues, J.B.; Gaze, L.; Faria, J.A.F.; Freitas, M.Q.; Deliza, R.; Bolini, H.M.A. Consumer perception of probiotic yogurt: Performance of check all that apply (CATA), projective mapping, sorting and intensity scale. *Food Res. Int.* **2013**, *54*, 601–610.

64. Illupapalayam, V.V.; Smith, S.C.; Gamlath, S. Consumer acceptability and antioxidant potential of probiotic-yogurt with spices. *LWT Food Sci. Technol.* **2014**, *55*, 255–262.

65. Akalin, A.S.; Unal, G.; Dinkci, N.; Hayaloglu, A.A. Microstructural, textural, and sensory characteristics of probiotic yogurts fortified with sodium calcium caseinate or whey protein concentrate. *J. Dairy Sci.* **2012**, *95*, 3617–3628.

66. Hekmat, S.; Reid, G. Sensory properties of probiotic yogurt is comparable to standard yogurt. *Nutr. Res.* **2006**, *26*, 163–166.

67. Gibson, G.R.; Roberfroid, M.B. Dietary modulation of the human colonic microbiota: Introducing the concept of prebiotics. *J. Nutr.* **1995**, *125*, 1401–1412.

68. Roberfroid, M.B. Prebiotics: The concept revisited. *J. Nutr.* **2007**, *137*, 830–837.

69. Cruz, A.G.; Cavalcantia, R.N.; Guerreiroa, L.M.R; Sant'Anac, A.S.; Nogueirab, L.C.; Oliveirad, C.A.F.; Delizae, R.; Cunhaa, R.L.; Fariaa, J.A.F.; Bolinia, H.M.A. Developing a prebiotic yogurt: Rheological, physico-chemical and microbiological aspects and adequacy of survival analysis methodology. *J. Food Eng.* **2013**, *114*, 323–330.

70. Pimentel, T.C.; Cruz, A.G.; Prudencio, S.H. Short communication: Influence of long-chain inulin and *Lactobacillus paracasei* subspecies paracasei on the sensory profile and acceptance of a traditional yogurt. *J. Dairy Sci.* **2013**, *96*, 6233–6241.

12

Optimization of Multistage Extraction of Olive Leaves for Recovery of Phenolic Compounds at Moderated Temperatures and Short Extraction Times

Konstantinos Stamatopoulos [1,†], **Archontoula Chatzilazarou** [2,†] **and Evangelos Katsoyannos** [3,*]

[1] Department of Food Technology, Faculty of Food Technology and Nutrition, Technological Educational Institute of Athens, 12 Ag. Spyridonos St., Egaleo, Athens 122 10, Greece; E-Mail: stamato_k@hotmail.com

[2] Department of Oenology and Beverage Technology, Faculty of Food Technology and Nutrition, Technological Educational Institute of Athens, 12 Ag. Spyridonos St., Egaleo, Athens 122 10, Greece; E-Mail: arhchatzi@yahoo.gr

[3] Department of Food Technology, Faculty of Food Technology and Nutrition, Technological Educational Institute of Athens, 12 Ag. Spyridonos St., Egaleo, Athens 122 10, Greece

[†] These authors contributed equally to this work.

[*] Author to whom correspondence should be addressed; E-Mail: ekatso@teiath.gr

Abstract: The aim of the present study was to improve the recovery of polyphenols from olive leaves (OL) by optimizing a multistage extraction scheme; provided that the olive leaves have been previously steam blanched. The maximum total phenol content expressed in ppm caffeic acid equivalents was obtained at pH 2, particle size 0.315 mm, solid-liquid ratio 1:7 and aqueous ethanol concentration 70% (v/v). The optimum duration time of each extraction stage and the operation temperature, were chosen based on qualitative and quantitative analysis of oleuropein (OLE), verbascoside, luteolin-7-O-glucoside and apigenin-7-O-glucoside performed by high performance liquid chromatography with diode array detector (HPLC-DAD). The optimum conditions for multistage extraction were 30 min total extraction time (10 min × 3 stages) at 85 °C. The 80% of the total yield of polyphenols was obtained at the 1st stage of the extraction. The total extraction yield of oleuropein was found 23 times higher (103.1 mg OLE/g dry weight (d.w.) OL) compared to the yield (4.6 mg OLE/g d.w. OL) obtained by the conventional extraction method

(40 °C, 48 h). However, from an energetic and hence from an economical point of view it is preferable to work at 40 °C, since the total extraction yield of polyphenolic compounds was only 17% higher for a double increase in the operating temperature (*i.e.*, 85 °C).

Keywords: polyphenols; antioxidant activity; multistage extraction; steam blanching; olive leaf

1. Introduction

By-products and wastes from plant food processing, which represent a major environmental problem in Mediterranean countries, are sources of added value bioactive compounds called phytochemicals or secondary metabolites [1]. Thus, several studies have focused on the composition of the extracts and improvement processes for maximum recovery of such antioxidant substances [2,3].

Among the different parts of the olive tree, olive leaves possess the highest oleuropein content, within a range of 1%–14% compared to olive oil (0.005%–0.12%) and olive mill wastewater (0.87%) [4]. Several studies have revealed that the health-promoting properties of virgin olive oil are mainly due to the presence of polyphenolic compounds [5]. Oleuropein and related phenolic compounds (e.g., luteolin-7-glucoside, apigenin-7-glucoside, hydroxytyrosol and rutin) have shown cardiovascular protective effects [6,7]. A possible explanation for these effects of olive leaves polyphenols has been attributed to the synergistic phenomena among the phenolic compounds [8].

Several conventional extraction techniques have been reported for the recovery of target compounds from raw materials. Hot water technology is the main and most common extraction method for flavone glycosides and other antioxidants [9–12]. Methanol-water mixtures [13] have been used in conventional solid-liquid extraction methods for the isolation of phytochemicals. However, the use of non-toxic solvents is preferable for natural extracts production since it leads to the development of functional foods with health-promoting properties.

Numerous extraction-assisted techniques have been developed, such as extraction with superheated liquids [4] ultrasound-assisted extraction [14] and microwave-assisted extraction [15]. Alternatives to those methods have been reported, like supercritical fluid extraction, with CO_2 as supercritical fluid and ethanol or methanol as modifier [16]. However, it should be pointed out that most of these techniques suffer from high energy costs as they operate under high pressure [17].

Most of the studies, dealing with the optimization of extraction of phenolic compounds from plant sources, have approached this issue mainly from a scientific rather than an industrial point of view. Thus, a single-stage extraction process is the main protocol that is followed by several authors which requires long extraction times (5–24 h) for complete recovery of phenolic compounds [18,19] or sophisticate techniques to improve its efficiency at shorter times. Production of natural antioxidant compounds at industrial scale requires accelerated extraction processes, small extractant volumes, non-toxic solvents, low energy consumption and technical simplicity. Moreover, a new extraction process should be easily adapted on the existed production line of industries.

The optimization of the extraction of phenolic compounds has been focused mainly on single-stage scheme rather than on multistage one. Thus, the authors propose low extraction temperatures (e.g., 40 °C)

for high extraction times. However, from an economical point of view it would be advisable to work, e.g., at 60 °C for shorter times [20].

To fulfill the requirements of industry for mass production of natural antioxidant extracts, multi-stage extraction has been proven to be more effective than single step extraction when using equal solvent volume [21]. In this sense, it is preferable to split the total extractant volume to several portions, by carrying out three or more extraction steps [21]. Moreover, Stamatopoulos, Katsoyannos, Chatzilazarou and Konteles [22] showed that steam blanching of olive leaves as pre-treatment of olive leaves extraction significantly improves the extractability of phenolic compounds about 25–35 times and increased the antioxidant activity about 4–13 times of olive leaf extract.

In the literature there is no study, to our knowledge, approaching the improvement of extraction of olive leaves by optimizing the multistage extraction of thermal treated olive leaves. Thus, the aim of this study was to associate steam blanching of olive leaves and multistage extraction scheme for satisfactory recovery of polyphenolic compounds, operating at moderated temperatures and short extraction times.

2. Experimental Section

2.1. Chemicals

Methanol, acetic acid and acetonitrile were purchased from Merck (Darmstadt, Germany), tyrosol, caffeic acid and 2,2-diphenyl-1-picrylhydrazyl reagent (DPPH) were purchased from Sigma-Aldrich (Hohenbrunn, Germany). Oleuropein and hydroxytyrosol were purchased from Extrasyntese (Genay, France). Sodium acetate trihydrate was purchased from Carlo Erba Reactivs SDS (Val de Reuil, France), rutin from Sigma (St. Louis, MO, USA) while Folin-Ciocalteau reagent, disodium hydrogen phosphate and potassium chloride from Merck (Darmstadt, Germany) and ethanol absolute from Sigma-Aldrich (St. Louis, MO, USA). Apigenin-7-*O*-glucoside, luteolin-7-*O*-glucoside and verbascoside were obtained from ExtraSynthese (Genay, France).

2.2. Steam Blanching of Olive Leaves

Prior to multistage extraction process, olive leaves had been processed according to Stamatopoulos, Katsoyannos, Chatzilazarou, and Konteles [22]. Briefly, olive leaves were steam blanched for 10 min in a household steam cooker at atmospheric pressure and immediately they were cooled down by cold water at 17 °C. The excess water was removed by an absorbent paper and subsequently the olive leaves were dried in an air oven for 4 h at 60 °C.

2.3. Optimization of Phenolic Compounds Extraction from Olive Leaves

Optimization of extraction of polyphenols from olive leaves was performed in term of particle size, pH, composition of aqueous ethanol solution (v/v) and solid-to-liquid ration, on steam blanched olive leaves. Briefly, fresh olive leaves were dried with an air tray oven at 60 °C for 4 h.

2.3.1. Effect of Particle Size

The dried olive leaves were ground and sieved. Thus, different particle size fractions were obtained, 0.05, 0.1, 0.2, 0.315 and 1.0 mm. Each fraction was extracted separately for 2 h under stirring (400 rpm) at the following conditions: 20% aqueous ethanol solution, solid-solvent ratio 1:10, pH 3 at 40 °C temperature. Then, the samples were centrifuged at 6000 rpm for 5 min. All ethanolic extracts were filtered through 0.45-μm syringe filters and were analyzed for total phenol content by (TPC) Folin-Ciocalteau assay.

2.3.2. Effect of pH

Dried olive leaves were extracted for 2 h at 40 °C, particle size 1 mm and solid-solvent 1:10 with 20% aqueous ethanol solution adjusted at pH 1.3, 2.0 and 3.0 with 0.1 M KCl/HCL buffer solution, at pH 4.2, 5.2 and 6.5 with 0.1 M sodium acetate/acetic acid buffer solution and at pH 8.3 with 0.1 M disodium hydrogen phosphate/HCl buffer system. Each sample was centrifuged at 6000 rpm for 5 min. All ethanolic extracts were filtered through 0.45-μm syringe filters and were analyzed for total phenol content by Folin-Ciocalteau assay.

2.3.3. Effect of Solid-to-Liquid Ratio (S/L)

Dried olive leaves were extracted for 2 h at 40 °C, particle size 1 mm, pH 3 and 20% aqueous ethanol solution in different solid-liquid ratios 1:5, 1:6, 1:7, 1:8 and 1:10. Each sample was centrifuged at 6000 rpm for 5 min. All ethanolic extracts were filtered through 0.45-μm syringe filters and they were analyzed for total phenol content by Folin-Ciocalteau assay. The particle size of 1 mm was preferred because, based on our experience using lower particle size causes problems at low solid-to-liquid ratio (*i.e.*, <1:7), since the mixture of olive leave and solvents became like sludge and hence mixing with magnetic stirrer was problematic.

2.3.4. Effect of Ethanol Concentration (% EtOH, v/v)

Dried olive leaves were extracted for 2 h at different aqueous ethanol concentration of 20%, 40%, 55%, 70%, 80% and 90% (v/v) with the rest parameters to be as followed: 40 °C, 1:10 solid-solvent ratio, 1 mm particle size and pH 3. Each sample was centrifuged at 6000 rpm for 5 min. All ethanolic extracts were filtered through 0.45-μm syringe filters and they were analyzed for total phenol content by Folin-Ciocalteau assay.

2.4. Multistage Extraction Scheme of Olive Leaves

After choosing the optimum conditions of extraction in terms of particle size, pH, S/L and % EtOH, a multistage extraction scheme was set up for maximum recovery of polyphenols at moderated to high temperatures and short extraction times. In order to illustrate the significance of combining steam blanching and multistage extraction process, steamed and non-steamed olive leaves were treated, based on multistage extraction protocol. Multistage extraction was performed at different temperatures, 40 °C (mainly recommended in literature), 60, 65, 70 and 85 °C; the rest parameters (pH, S/L, % EtOH) were set up based on the results of the optimization of the single stage extraction experiments (Section 2.3). Moreover, three extraction stages were performed with 60 min duration time per stage. At each stage, sample was collected every 10 min. At the end of every stage, the sample was centrifuged (6000 rpm for 5 min) and the residue was re-extracted with renewal solvent. All the process was repeated three times. Finally, non-steam blanched olive leaves were extracted at 40 °C followed a multistage extraction and the results were compared with the corresponding of steam blanched olive leaves obtained at the same temperature. All ethanolic extracts were filtered through 0.45-μm syringe filters and they were analyzed by HPLC-DAD for qualitative and quantitative analysis of oleuropein, verbascoside, luteolin-7-O-glucoside and apigenin-7-O-glucoside.

2.5. Conventional Extraction Protocol

A procedure was developed using the same extractant (70% v/v aqueous ethanol) and sample-extractant volume ratio (1:7), as used in the multistage extraction scheme, for a time enough to ensure complete extraction of the target analytes and at mild temperature to avoid potential degradation. Dried and milled steamed and non-steamed leaves were immersed separately in 70% EtOH at 1:7 ratio and were placed in a beaker and subjected to stirring at 40 °C for 48 h. The process was repeated three times. Then, all ethanolic extracts were filtered through 0.45-μm syringe filters and they were analyzed by HPLC-DAD for qualitative and quantitative analysis of oleuropein, verbascoside, luteolin-7-O-glucoside and apigenin-7-O-glucoside.

2.6. Chromatographic Conditions

The equipment utilized was a HITACHI (Tokyo, Japan) coupled to an auto-sampler L-2200, pump L-2130, column oven L-2300 and diode array detector L-2455 and controlled by MerckAgilent EZChrom Elite software (Agilent Technologies, Santa Clara, CA, USA). The column was a Pinnacle II RP C18, 3 μm, 150 mm × 4.6 mm (Restek, Bellefonte, PA, USA), protected by a Kromasil 100-5 C18 guard cartridge starter kit (Sigma-Aldrich, Seelze, Germany) for 3.0/4.6 mm i.d. column oven was set at 40 °C. Eluent (A) and (B) were 0.02 M sodium acetate adjusted to pH = 3.2 with acetic acid and HPLC grade acetonitrile, respectively. The flow rate was 1 mL/min and the injection volume 20 μL. The elution gradient profile was as follows: started (A) 90%; 2 min, 85%; 9 min, 75%; 12 min, 65%; 15 min, 55%; 18 min, 40%; 20 min, 90%. Eluting was monitored at 280 nm for oleuropein, hydroxytyrosol and tyrosol and at 355 nm for flavonols.

2.7. Determination of Antioxidant Activity

Antiradical activity measurement was performed by using 2,2,-diphenyl-2-picryl-hydrazyl (DPPH) assay according to Braca [23], with some modifications. 2.5 mg of DPPH powder was diluted in 100 mL pure methanol with absorption 0.7 (±0.03) at 517 nm. A sample of every stage at each temperature was diluted 50 times and then 33 μL of the diluted sample was added to an aliquot of 1 mL of 0.004% DPPH solution. To control, 33 μL of pure ethanol were added instead of olive leaf extract. The reaction mixture was vortex-mixed and was allowed to stand in the darkness at room temperature for 30 min before measuring the decrease in absorbance at 517 nm. The spectrophotometer

(SHIMADZU mini 1240 UV-Vis, Shimadzu, Columbia, MD, USA) was calibrated with pure methanol. Antioxidant activity (*AA*) was expressed in percentage inhibition of DPPH radical and was calculated with the following equation:

$$AA(\%) = \left[\frac{A_0 - A_i}{A_0} \right] \times 100 \qquad (1)$$

A_0 and A_i are the absorbance of the control sample and the sample containing olive leave extract respectively. The procedure was repeated three times for each sample.

2.8. Determination of Total Phenol Content (TPC)

TPC was determined according to the Folin-Ciocalteau assay method [24] with some modifications. Briefly, all the samples were diluted with ethanol (1:1). Then, 0.1 mL of the diluted sample was put into a 20 mL test tube, into which distilled H_2O was added to have a final volume of 6.75 mL. Folin-Ciocalteau phenol reagent (0.25 mL) was added to the mixture and shaken vigorously. After 3 min, 3 mL of 35% Na_2CO_3 solution was added with mixing. Ethanol (0.1 mL) was used as blank instead of diluted sample. The solution was allowed to stand for 60 min. After this time, the absorbance was measured at 750 nm in comparison the prepared blank. The TPC of sample was expressed as mg caffeic acid equivalents (CAE)/g d.w. OL. The procedure was repeated three times for each sample.

3. Results and Discussion

This study attempted to approach the procedure of extraction of olive leaves considering the basic requirements of the mass production of nutraceuticals in industrial scale. Thus, a combination of the widely used thermal treatment (steam blanching) in food industry with a multistage extraction scheme was applied. In preliminary experiments the total phenol content (TPC) of ethanolic extracts of olive leaves was determined with Folin-Ciocalteau assay and HPLC-DAD. It should be noted that when the extraction temperature was ≤40 °C there were no qualitative changes of the phenolic profile of the extracts and hence Folin-Ciocalteau assay was preferred as an easy and fast procedure for the optimization of extraction in terms of pH, particle size, solid-to-liquid ratio (S/L) and ethanol concentration (% EtOH, v/v). However, above that temperature and particular ≥60 °C, the phenolic profile was changed and high amounts of hydroxytyrosol were appeared in the chromatograms. Moreover, the pH (except the sample of 1.3), particle size, solid-to-liquid ratio (S/L) and ethanol concentration (% EtOH, v/v) did not affect significantly the phenolic profile when the extraction temperature was below or equal to 40 °C. Thus, HPLC-DAD was preferred for the optimization of multistage extraction in terms of operation temperature and duration time while the rest parameters were kept constant.

In conclusion, the total phenol content of the extracts (TPC) obtained with Folin-Ciocalteau assay was chosen as a criterion for selecting the optimum conditions of the extraction of olive leaves in terms of pH, particle size, solid-to-liquid ratio (S/L) and ethanol concentration (% EtOH, v/v). Whereas, qualitative and quantitative analyses of oleuropein, verbascoside, luteolin-7-*O*-glucoside and apigenin-7-*O*-glucoside were performed with HPLC in order to evaluate the performance of the multistage extraction *versus* the conventional method (one stage, 48 h, 40 °C) as well as the impact of the applied temperature on the degradation of phenolics during the multistage extraction.

3.1. Effect of Particle Size

In this study, the effect of particle size was examined within a range of 0.05–1.0 mm. It was expected that the smallest particle size of olive leaves leads to highest extraction yield. However, for particle size below 0.2 mm the extraction yield is decreased (Figure 1a). This is due to the fact that the particle size has to be limited because exceedingly small particles tend to agglomerate, leading to a decrease of solvent penetration in the solid matrix and, therefore, negatively affecting the mass transfer process [25]. Thus, a typical value of 0.3 mm is accordingly recommended.

Figure 1. Total phenols content expressed as ppm caffeic acid equivalents, in ethanolic extract of steamed olive leaves as a function of (**a**) particle size of olive leaves (OL); (**b**) solvent composition (v/v); (**c**) solid-to-liquid ratio; and (**d**) pH. The extraction was performed at 50 °C for 2 h under stirring ($n = 400$ rpm).

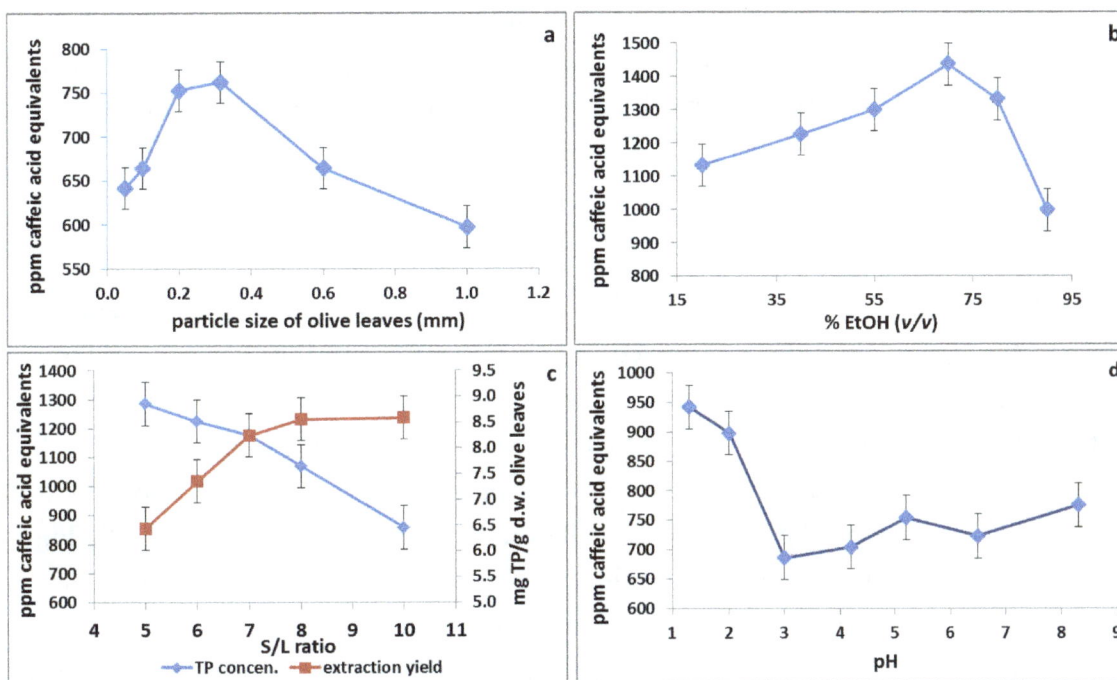

3.2. Effect of pH

Concerning the recovery of polyphenols, pH can impact according to different mechanisms on the recovery of polyphenols such as increase of solubility of the solute and alternation of the interactions of antioxidants with other constituents of the plant material [25]. Based on our results, the highest extraction yield was obtained at pH 1.3 (Figure 1d). Nevertheless, HPLC-DAD analysis of the current sample showed increased values of hydroxytyrosol due to hydrolysis of oleuropein [26]. Hence, the total phenol content seems higher, although, this is not due to the performance of the extraction at that pH value but due to higher reactivity of the hydroxytyrosol with Folin-Ciocalteau assay. Thus, pH 2.0 seems preferable since at that pH value no hydrolysis of oleuropein was observed. The proposed pH is similar to the value that Mylonaki, Kiassos, Makris, and Kefalas [19] suggest for the extraction of olive leaves. According to Japón-Luján, Luque-Rodríguez, and Luque de Castro [15], the largest amount of oleuropein extracted from olive leaves was found at pH 7; however, the authors did not work at pH values below 3. In conclusion, multistage extraction experiments should be conducted at pH 2 in order to avoid chemical hydrolysis of polyphenols at higher temperatures.

3.3. Effect of Solid-to-Liquid Ratio (S/L)

The solvent-to-solid ratio on the recovery of phenolic compounds should be carefully analysed and optimized as the solvent consumption exerts a direct influence on the extraction process cost.

Figure 1c illustrates the effect of S/L on the concentration of total phenolics in the extract (ppm CAE) and the extraction yield (mg CAE/g d.w. OL). The results show that the concentration of total phenols decreases as the S/L increases, which means that the polyphenols extracted from olive leaves were diluted since higher solvent volume was used. Moreover, the extraction yield (mg TP/g d.w. olive leaves) increased up to 1/8 S/L ratio and remains constant till 1/10 S/L. Thus, from an economical point of view a compromise has to be met between a relatively high extraction yield and a relatively high polyphenols concentration in the extract. The optimum condition that fulfils this requirement is at that S/L value in which the curves of extraction yield and concentration of total phenols are crossed each other. Thus, a solid-liquid ratio 1:7 is accordingly recommended. This ratio is similar to the value that Bilek [27] suggests for optimum extraction conditions for total phenolic compounds from dried olive leaves based on the response surface methodology analysis.

3.4. Effect of Ethanol Concentration

Several solvent systems have been used to recover phenolic compounds from plant matrices. Ethanol has been classified as generally recognized as safe (GRAS) as well as has been reported to be an effective solvent for the recovery of phenolic compounds and thus, is usually used to recover this group of phytochemicals, especially when it comes to the production of nutraceuticals [28]. Some authors reported that the effectiveness of the phenolic compound recovery through solvent extraction with ethanol can be increased by the addition of different proportions of water [29,30]. Figure 1b demonstrates the effect of ethanol-water composition on total phenols content with the highest value being obtained at 70% (v/v) aqueous ethanol concentration. Studies dealing with the extraction of polyphenols from olive leaves have recommended different proportions of ethanol-water varied at range of 40%–80% v/v [14,15,19]. This is due to the fact that there is not a single ethanol-water portion that can extract effectively all the phenolic compounds present in different plant matrices [31] and hence it needs to be carried out on a case-by-case basis.

3.5. Multistage Extraction of Olive Leaves

Most of the studies dealing with the optimization of the extraction of polyphenols have either followed protocols with long extraction times (5–48 h), often at elevated temperatures (>40 °C), or have used sophisticated techniques which may have given higher extraction yields and shorter extraction times but are still suffering from high energy consumption as they operate, e.g., under high pressures. Thus, a new procedure is required which lies between the technical simplicity of the traditional extraction methods and the high efficiency of the advanced techniques. Following this principle, thermal pre-treatment (blanching) of olive leaves and multistage extraction were combined from moderate to high temperatures. In many studies, the extraction efficiency has been mainly evaluated with determination of total phenol content using Folin-Ciocalteau assay. However, the use of this reagent for quantification of total phenols has some drawbacks. The current reagent interacts with the free hydroxyl groups of phenolic compounds via electron transfer mechanism. These hydroxyl groups can be formed either by cleaving high molecular weight polyphenols (e.g., oleuropein) to simple ones (e.g., hydroxytyrosol) or by cleaving the sugar moiety resulting in the formation of aglycones of phenolic compounds. Moreover, the phenolic compounds have different reactivity with Folin-Ciocalteau reagent [32] which can be reduced by many non-phenolic compounds; e.g., vitamin C, Cu(I), [33] resulting in overestimations or underestimations in the total phenol content. Thus, optimization of extraction of phenolic compounds from olive leaves, applying multistage extraction scheme, has to be carried out based on HPLC-DAD analysis (Figure 2) of the main phenolic compounds present in olive leaf extract at each stage of extraction.

Figure 2. Chromatographic analysis (HPLC-DAD) of ethanolic olive leaf extract at 280 nm. (**a**) first stage; (**b**) second stage; and (**c**) third stage. **1**: luteolin-7-*O*-glucoside; **2**: verbascoside; **3**: apigenin-7-*O*-glucoside; and **4**: oleuropein.

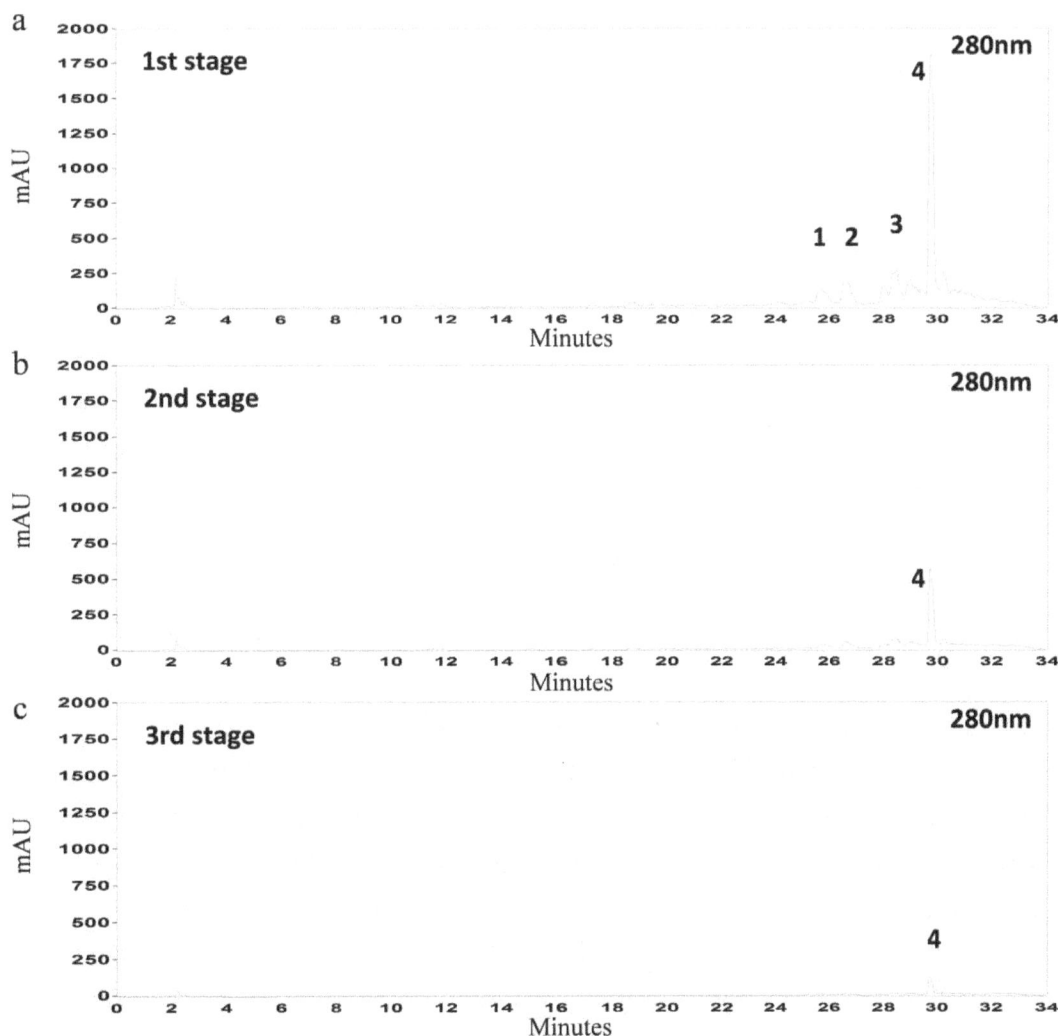

Figure 3 illustrates the yields (mg/g d.w. OL) of oleuropein (Figure 3a), luteolin-7-O-glucoside (Figure 3b), apigenin-7-O-glucoside (Figure 3c) and verbascoside (Figure 3d) at each temperature for every stage of the extraction of steamed olive leaves. The first observation is that the amount of polyphenols that has been extracted at the first 10 min does not differ significantly from the amount extracted at the last 60 min of each stage. This principle applies for every stage at each temperature and for all phenolic compounds. Hence, the duration time of each stage does not need to be more than 10 min. Consequently, the total duration time which is enough for sufficient recovery of polyphenols, should not be more than 30 min (10 min × 3 stages); provided that the olive leaves have been previously steam blanched. Regarding to the performance of the multistage extraction, at the 1st stage 75.5% ± 2.8% of the oleuropein, 80.1% ± 3.2% of the luteolin-7-O-glucoside, 84.4% ± 3.9% of the verbascoside and 80.4% ± 2.2% of the apigenin-7-O-glucoside had been extracted from olive leaves.

Figure 3. Mean values (n = 3) of extraction yields (mg/g d.w. OL) of (**a**) oleuropein; (**b**) luteolin-7-O-glucoside; (**c**) apigenin-7-O-glucoside; and (**d**) verbascoside which are obtained after multistage extraction (60 min/stage) (sampling was performed every 10 min) at 40, 60, 65, 70 and 85 °C. The extraction was performed at pH 2.5 with solid-to-liquid (S/L) ratio 1:7, 70% aqueous ethanol solution and 0.5 mm particle size of OL.

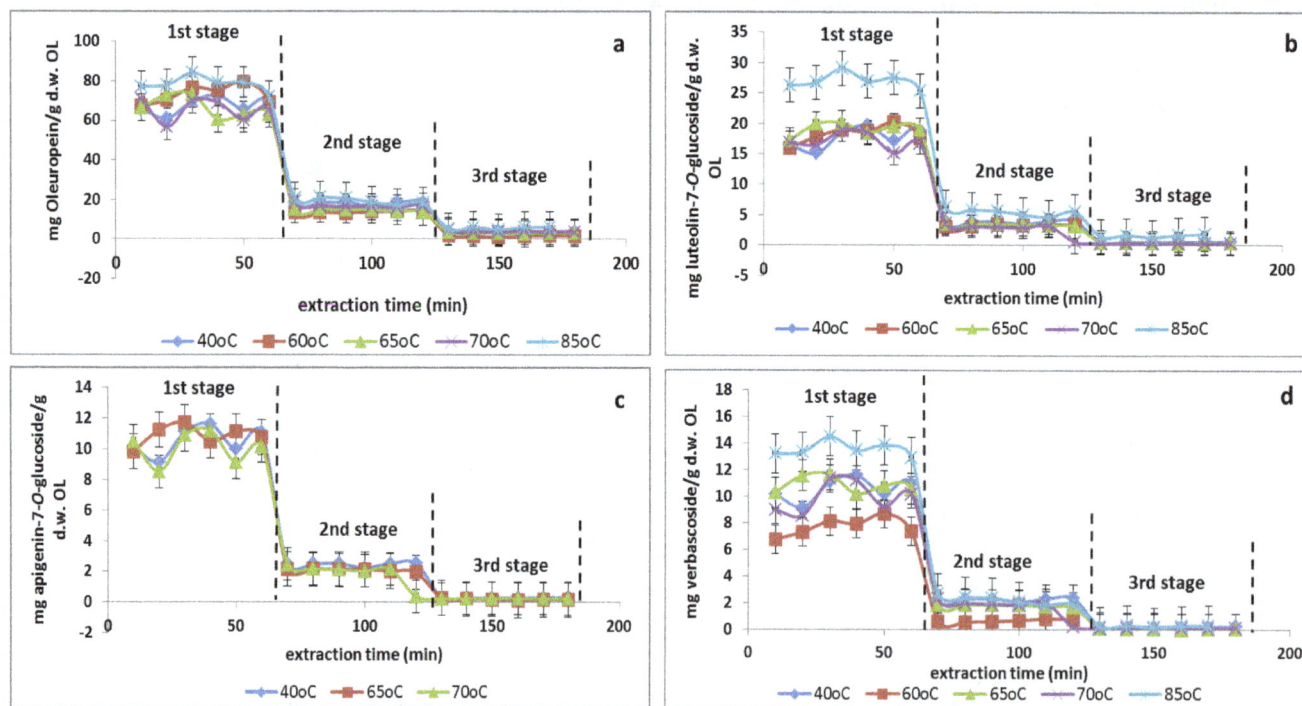

The highest total extraction yield (166.6 ± 0.9 mg/g d.w. OL) was obtained at 85 °C compared to 70 °C (140.3 ± 0.9 mg/g d.w. OL), 65 °C (133.0 ± 0.8 mg/g d.w. OL), 60 °C (133.5 ± 1.0 mg/g d.w. OL) and 40 °C (137.5 ± 0.7 mg/g d.w. OL), as presented in Table 1. Based on the results, there was only an 18% increase of the extraction yield when the temperature was raised by more than 2-fold (40–85 °C). Therefore, from energetic and hence from economical point of view, this relative improvement on the extraction yield does not justify the operation of extraction at high temperatures (>40 °C). Consequently, the combination of thermal pre-treatment and multistage extraction provides moderate temperatures and short extraction times for a satisfactory recovery of phenolic compounds from olive leaves. However, the behavior of each polyphenol was different regarding temperature fluctuations. It is observed that there was an impact on the extraction yield of luteolin-7-O-glucosed (Figure 3b) and verbascoside (Figure 3d) when multistage extraction was operated at different temperatures compared to oleuropein (Figure 3a) and apigenin-7-O-glucoside (Figure 3c). The total yield of luteolin-7-O-glucoseide and verbascoside was 33% and 18% higher (Table 1) respectively, when a 48% increase in operation temperature was applied. Nevertheless, these differences in extraction yields of each polyphenol contribute only 18% on the overall extraction yield increment.

In the case of multistage extraction at 40 °C of non-steamed olive leaves, the total yield of oleuropein was 6.5 mg/g d.w. OL compared to 4.6 mg/g d.w. OL when single extraction step was applied. Moreover, the yields of oleuropein obtained by multistage (3 stages) and single step extraction at 40 °C of steamed olive leaves, were 89.5 mg/g d.w. OL and 47.7 mg/g d.w. OL, respectively. Thus, the combination of steam blanching and multistage extraction can significantly increase the yield of about 48% at short extraction times, i.e., 30 min instead of 48 h. The yield of oleuropein was further increased at 85 °C (103.5 mg/g d.w. OL) resulting in an additional 6% increase. As mentioned before, this small increase does not justify the need for operating at such of high temperatures which consequently leads to an energy-intensive extraction process.

Examining the effect of steam blanching on the extractability of phenolic compounds during single step extraction, a highly significant increase of oleuropein (10-fold) and verbascoside (14-fold) yields was observed, compared to luteolin-7-O-glucoside (2-fold) and apigenin-7-O-glucoside (1.3-fold) (Figure 4). These dissimilarities between the polyphenols yields could be due to their different location within plant tissues [34] as well as their interaction with other plant constituents [31]. Thus, the effectiveness of the extraction seems to depend mainly on the diffusibility of the organic solvent within plant tissue when the composition of the solvent remains constant. This could be supported by the fact that the extraction yield of oleuropein from steamed olive leaves was further increased 2-fold when multistage extraction was applied, i.e., improvement of the mass transfer coefficient, without changing solvent composition and temperature.

The total amount of phenolic compounds which has been obtained from single extraction of steam blanched olive leaves was 77.92 mg/d.w. OL whereas the total amount of phenolic compounds from non-steam blanched olive leaves was 19.23 mg/g d.w. OL (Figure 4) which means a 75% increase in extraction yield.

Table 1. Polyphenols composition in olive leaf extract at each extraction stage performed at different temperatures. The values in parenthesis represent the corresponding percentage of phenolic compound in the olive leaf extract at every stage. * Mean values (mg/g d.w. OL) of each phenolic compound at every stage ($n = 6$, ±SD).

Temperature	Extraction Stage	Total	Oleuropein *	Luteolin-7-O-glucoside *	Verbascoside *	Apigenin-7-O-glucoside *
85 °C	1st	129.8 ± 1.8	78.2 ± 2.9 (60.2)	27.0 ± 1.2 (20.7)	13.5 ± 0.5 (10.3)	11.1 ± 0.9 (8.5)
	2nd	29.3 ± 0.9	19.4 ± 1.2 (66.1)	5.5 ± 0.6 (18.8)	2.2 ± 0.3 (7.5)	2.2 ± 0.5 (7.5)
	3rd	7.5 ± 0.2	5.5 ± 0.3 (72.6)	1.2 ± 0.2 (16.7)	0.3 ± 0.1 (3.9)	0.5 ± 0.1 (6.6)
	Total	166.6 ± 0.9	103.1 ± 1.5	33.7 ± 0.6	16 ± 0.9	13.8 ± 0.5
70 °C	1st	103.5 ± 2.0	65.4 ± 4.3 (63.1)	17.1 ± 1.4 (16.5)	9.9 ± 1.2 (9.5)	11.1 ± 1.0 (10.7)
	2nd	29.3 ± 0.7	16.9 ± 0.4 (57.6)	3.1 ± 1.1 (10.8)	2.0 ± 0.7 (6.8)	2.2 ± 0.7 (7.5)
	3rd	7.5 ± 0.1	3.9 ± 0.1 (51.5)	0.9 ± 0.1 (4.7)	0.7 ± 0.1 (1.8)	0.5 ± 0.1 (6.6)
	Total	140.3 ± 0.9	86.2 ± 1.6	21.4 ± 0.8	12.6 ± 0.6	13.8 ± 0.6
65 °C	1st	107.4 ± 2.0	66.5 ± 5.5 (61.9)	19.1 ± 1.1 (17.7)	10.9 ± 0.6 (10.1)	10.9 ± 0.6 (10.1)
	2nd	21.9 ± 0.2	14.6 ± 0.5 (66.6)	3.4 ± 0.1 (15.5)	1.8 ± 0.1 (8.2)	2.1 ± 0.1 (9.5)
	3rd	3.7 ± 0.1	3.04 ± 0.2 (81.9)	0.4 ± 0.1 (11.0)	0.6 ± 0.1 (1.6)	0.4 ± 0.1 (5.3)
	Total	133.0 ± 0.8	84.1 ± 2.1	23.3 ± 0.4	13.3 ± 0.3	4.5 ± 0.3
60 °C	1st	109.9 ± 2.6	73.1 ± 4.8 (66.5)	18.2 ± 4.2 (16.5)	7.7 ± 0.7 (7.0)	10.9 ± 0.8 (9.9)
	2nd	20.9 ± 0.2	13.7 ± 0.3 (65.5)	3.3 ± 0.1 (15.7)	1.8 ± 0.1 (8.6)	2.1 ± 0.2 (10.0)
	3rd	2.7 ± 0.1	1.5 ± 0.4 (55.6)	0.5 ± 0.1 (19.7)	0.5 ± 0.1 (18.2)	0.2 ± 0.1 (7.2)
	Total	133.5 ± 1.0	29.4 ± 1.8	22 ± 1.5	10 ± 0.3	13.2 ± 0.4
40 °C	1st	105.9 ± 1.9	67.3 ± 4.1 (63.5)	17.6 ± 1.6 (16.6)	10.5 ± 0.8 (9.9)	10.5 ± 0.9 (9.9)
	2nd	27.0 ± 0.2	18.3 ± 0.7 (67.7)	3.9 ± 0.2 (14.4)	2.3 ± 0.1 (8.5)	2.5 ± 0.1 (9.2)
	3rd	4.6 ± 0.1	3.9 ± 0.1 (84.7)	0.2 ± 0.1 (4.3)	0.2 ± 0.1 (4.3)	0.3 ± 0.1 (6.5)
	Total	137.5 ± 0.7	89.5 ± 1.6	21.7 ± 0.6	13.0 ± 0.3	13.3 ± 0.4

Figure 4. Total extraction yields of the main polyphenols which were determined in OL extract of steamed and non-steamed OL with HPLC-DAD after conventional extraction (70% ethanol, 48 h under stirring 400 rpm at 40 °C).

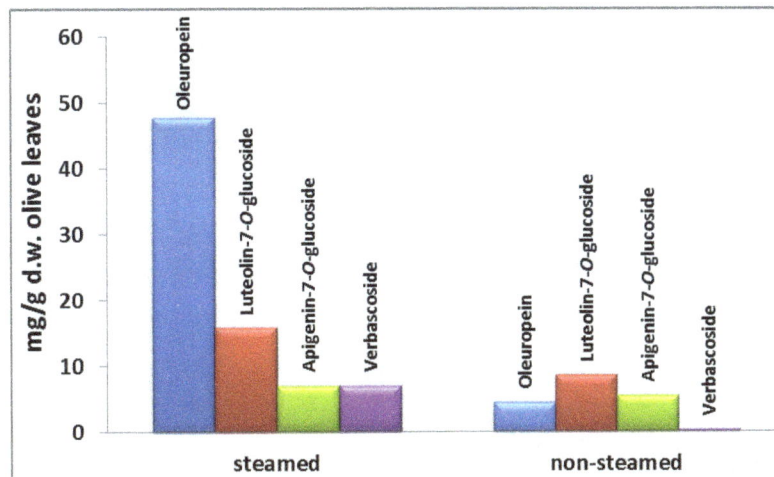

3.6. Antioxidant Activity of Ethanolic Extracts of Olive Leaves

The antioxidant activity of ethanolic extracts of olive leaves was followed a linear relationship with polyphenols concentration (Figure 5). Nevertheless, there were variations between the samples of different experimental conditions. Particularly, the sample of the 1st stage at 85 °C with concentration 32 ppm gave higher inhibition value (48%) compared to 42% inhibition value of the sample (41 ppm) which was collected from the 1st stage at 40 °C (Figure 5). The low correlation of polyphenols and antioxidant activity has been related to the different phenolic composition of ethanolic extracts (Table 1; values in parenthesis) [22] and to the antagonistic and synergistic phenomena in mixtures of pure polyphenols and OL extract [35]. The antioxidant activity of ethanolic extracts obtained after multistage extraction of steamed olive leaves was 10 times higher than the ethanolic extracts obtained with the conventional extraction method.

Figure 5. Antioxidant activity of ethanolic extracts of olive leaves. * Sum of the oleuropein, verbascoside, luteolin-7-O-glucoside and apigenin-7-O-glucoside concentrations at the end of each stage of the multistage extraction which had been performed at 40, 65, 70 and 85 °C and dissolved prior to the addition in DPPH (2,2,-diphenyl-2-picryl-hydrazyl) solution. Mean values ($n = 3$, ±SD).

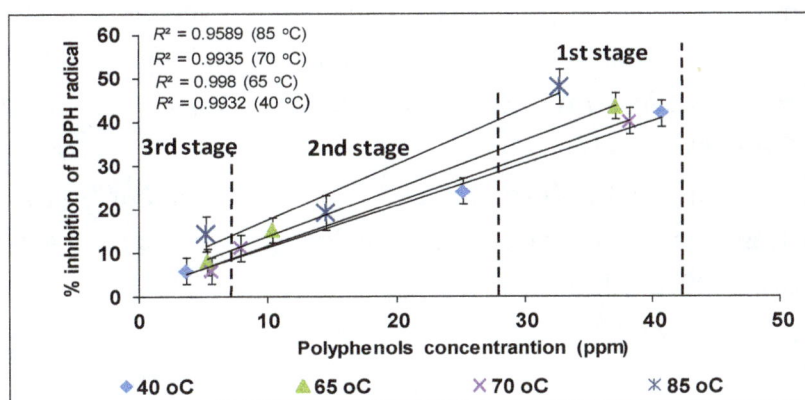

4. Conclusions

The most important highlights of the present study are as follows:

The current study showed that the combination of steam blanching process and multistage extraction (after optimization) is advantageous since it provides short extraction times (≤30 min) at moderate operation temperatures (40 °C).

The improvement of extraction yield that can be achieved is more than 50% compared to the conventional extraction method.

The antioxidant activity of ethanolic extract of steamed olive leaves which had been extracted by the multistage process was 10 times higher than that obtained by the conventional extraction method.

The current extraction scheme does not require sophisticated or complicated techniques.

Conflicts of Interest

The authors declare no conflict of interest.

References

1.   Schieber, A.; Stintzing, F.C.; Carle, R. By-products of plant food processing as a source of functional compounds-recent developments. *Trends Food Sci. Technol.* **2001**, *12*, 401–413.

2.   Moure, A.; Cruz, J.M.; Franco, D.; Domínguez, J.M.; Sineiro, J.; Domínguez, H. Natural antioxidants from residual sources. *Food Chem.* **2001**, *72*, 145–171.

3.   Shi, J.; Nawaz, H.; Pohorly, J.; Mittal, G.; Kakuda, Y.; Jiang, Y. Extraction of polyphenolics from plant material for functional foods—Engineering and technology. *Food Rev. Int.* **2005**, *21*, 139–166.

4.   Japón-Luján, R.; Luque de Castro, M.D. Superheated liquid extraction of oleuropein and related biophenols from olive leaves. *J. Chromatogr. A* **2006**, *1136*, 185–191.

5.   Servili, M.; Montedoro, G.F. Contribution of phenolic compounds to virgin olive oil quality. *Eur. J. Lipid Sci. Technol.* **2002**, *104*, 602–613.

6.   Kim, T.J.; Kim, J.H.; Jin, Y.R.; Yun, Y.P. The inhibitory effect and mechanism of luteolin 7-glucoside on rat aortic vascular smooth muscle cell proliferation. *Arch. Pharm. Res.* **2006**, *29*, 67–72.

7.   Omar, S.H. Cardioprotective and neuroprotective roles of oleuropein in olive. *Saudi Pharm. J.* **2010**, *18*, 111–121.

8.   Benavente-Garcia, O.; Castillo, J.; Lorente, J.; Ortuno, A.; del Rio, J.A. Antioxidant activity of phenolics extracted from *Olea europaea* L. leaves. *Food Chem.* **2000**, *68*, 457–462.

9.   Bendjeddou, D.; Lalaoui, K.; Satta, D. Immunostimulating activity of the hot water-soluble polysaccharide extracts of *Anacyclus pyrethrum*, *Alpinia galanga* and *Citrullus colocynthis*. *J. Ethnopharmacol.* **2003**, *88*, 155–160.

10. Dong, C.H.; Xie, X.Q.; Wang, X.L.; Zhan, Y.; Yao, Y.J. Application of Box-Behnken design in optimization for polysaccharides extraction from cultured mycelium of *Cordyceps sinensis*. *Food Bioprod. Process.* **2009**, *87*, 139–144.

11. Lee, W.C.; Yusof, S.; Hamid, N.S.A.; Baharin, B.S. Optimizing conditions for hot water extraction of banana juice using response surface methodology (RSM). *J. Food Eng.* **2006**, *75*, 473–479.

12. Corsano, G.; Montagna, J.M.; Aguirre, P.A. Design and planning optimization of multiplant complexes in the food industry. *Food Bioprod. Process.* **2007**, *85*, 381–388.

13. Savournin, C.; Elias, B.R.; Dargouth-Kesraoui, F.; Boukef, K.; Balansard, G. Rapid high-performance liquid chromatography analysis for the quantitative determination of oleuropein in *Olea europaea* leaves. *J. Agric. Food Chem.* **2001**, *49*, 618–621.

14. Japón-Luján, R.; Luque-Rodríguez, J.M.; Luque de Castro, M.D. Dynamic ultrasound-assisted extraction of oleuropein and related biophenols from olive leaves. *J. Chromatogr. A* **2006**, *1108*, 76–82.

15. Japón-Luján, R.; Luque-Rodríguez, J.M.; Luque de Castro, M.D. Multivariate optimisation of the microwave-assisted extraction of oleuropein and related biophenols from olive leaves. *Anal. Bioanal. Chem.* **2006**, *385*, 753–759.

16. Tabera, J.; Guinda, A.; Ruiz Rodríguez, A.; Señoráns, F.J.; Ibáñez, E.; Albi, T. Counter-current supercritical fluid extraction and fractionation of high-added-value compounds from a hexane extract of olive leaves. *J. Agric. Food Chem.* **2004**, *52*, 4774–4779.

17. Tsochatzidis, N.A.; Guiraud, P.; Wilhelm, A.M.; Delmas, H. Determination of velocity, size and concentration of ultrasonic cavitation bubbles by the phase-Doppler technique. *Chem. Eng. Sci.* **2001**, *56*, 1831–1840.

18. Pekić, B.; Kovač, V.; Alonso, E.; Revilla, E. Study of the extraction of proanthocyanidins from grape seeds. *Food Chem.* **1998**, *61*, 201–206.

19. Mylonaki, S.; Kiassos, E.; Makris, D.P.; Kefalas, P. Optimisation of the extraction of olive (*Olea europaea*) leaf phenolics using water/ethanol-based solvent systems and response surface methodology. *Anal. Bioanal. Chem.* **2008**, *392*, 977–985.

20. Spigno, G.; Tramelli, L.; de Faveri, D.M. Effects of extraction time, temperature and solvent on concentration and antioxidant activity of grape marc phenolics. *J. Food Eng.* **2007**, *81*, 200–208.

21. Williamson, G. Polyphenols Extraction from Foods. In *Methods in Polyphenol Analysis*, 1st ed.; Escribano-Balón, M.T., Santos-Buelga, C., Eds.; Royal Society of Chemistry: Cambridge, UK, 2003.

22. Stamatopoulos, K.; Katsoyannos, E.; Chatzilazarou, A.; Konteles, S.J. Improvement of oleuropein extractability by optimising steam blanching process as pre-treatment of olive leaf extraction via response surface methodology. *Food Chem.* **2012**, *133*, 344–351.

23. Braca, A.; de Tommasi, N.; di Bari, L.; Pizza, C.; Politi, M.; Morelli, I. Antioxidant principles from *Bauhinia tarapotensis*. *J. Nat. Prod.* **2001**, *64*, 892–895.

24. Chun, O.K.; Kim, D.O.; Smith, N.; Schroeder, D.; Han, J.T.; Lee, C.Y. Daily consumption of phenolics and total antioxidant capacity from fruit and vegetables in the American diet. *J. Sci. Food Agric.* **2005**, *85*, 1715–1724.

25. Meireles, M.A.A. *Extracting Bioactive Compounds for Food Products Theory and Applications*; CRC Press: Boca Raton, FL, USA, 2009.

26. Gikas, E.; Papadopoulos, N.; Tsarbopoulos, A. Kinetic study of the acid hydrolysis of oleuropein, the major bioactive metabolite of olive oil. *J. Liquid Chromatogr. Relat. Technol.* **2006**, *29*, 497–508.

27. Bilek, S.E. The effects of time, temperature, solvent:solid ratio and solvent composition on extraction of total phenolic compound from dried olive (*Olea europaea* L.) leaves. *GIDA J. Food* **2010**, *35*, 411–416.

28. Naczk, M.; Shahidi, F. Phenolics in cereals, fruits and vegetables: Occurrence, extraction and analysis. *J. Pharm. Biomed. Anal.* **2006**, *41*, 1523–1542.

29. Alonso, E.; Bourzeix, M.; Revilla, E. Suitabilitiy of water-ethanol mixtures for the extraction of catechins and proanthocyanidins from *Vitis vinifera* seeds contained in a winery by-product. *Seed Sci. Technol.* **1991**, *19*, 545–552.

30. Yilmaz, Y.; Toledo, R.T. Oxygen radical absorbance capacities of grape/wine industry byproducts and effect of solvent type on extraction of grape seed polyphenols. *J. Food Compos. Anal.* **2006**, *19*, 41–48.

31. Naczk, M.; Shahidi, F. Extraction and analysis of phenolics in food. *J. Chromatogr. A* **2004**, *1054*, 95–111.

32. Karadag, A.; Ozcelik, B.; Saner, S. Review of methods to determine antioxidant capacities. *Food Anal. Methods* **2009**, *2*, 41–60.

33. Huang, D.; Ou, B.; Prior, R.L. The chemistry behind antioxidant capacity assays. *J. Agric. Food Chem.* **2005**, *53*, 1841–1856.

34. Hutzler, P.; Fischbach, R.; Heller, W.; Jungblut, T.P.; Reuber, S.; Schmitz, R. Tissue localization of phenolic compounds in plants by confocal laser scanning microscopy. *J. Exp. Bot.* **1998**, *49*, 953–965.

35. Bouaziz, M.; Grayer, R.J.; Simmonds, M.S.J.; Damak, M.; Sayadi, S. Identification and antioxidant potential of flavonoids and low molecular weight phenols in olive cultivar chemlali growing in Tunisia. *J. Agric. Food Chem.* **2005**, *53*, 236–241.

Permissions

All chapters in this book were first published in Foods, by MDPI; hereby published with permission under the Creative Commons Attribution License or equivalent. Every chapter published in this book has been scrutinized by our experts. Their significance has been extensively debated. The topics covered herein carry significant findings which will fuel the growth of the discipline. They may even be implemented as practical applications or may be referred to as a beginning point for another development.

The contributors of this book come from diverse backgrounds, making this book a truly international effort. This book will bring forth new frontiers with its revolutionizing research information and detailed analysis of the nascent developments around the world.

We would like to thank all the contributing authors for lending their expertise to make the book truly unique. They have played a crucial role in the development of this book. Without their invaluable contributions this book wouldn't have been possible. They have made vital efforts to compile up to date information on the varied aspects of this subject to make this book a valuable addition to the collection of many professionals and students.

This book was conceptualized with the vision of imparting up-to-date information and advanced data in this field. To ensure the same, a matchless editorial board was set up. Every individual on the board went through rigorous rounds of assessment to prove their worth. After which they invested a large part of their time researching and compiling the most relevant data for our readers.

The editorial board has been involved in producing this book since its inception. They have spent rigorous hours researching and exploring the diverse topics which have resulted in the successful publishing of this book. They have passed on their knowledge of decades through this book. To expedite this challenging task, the publisher supported the team at every step. A small team of assistant editors was also appointed to further simplify the editing procedure and attain best results for the readers.

Apart from the editorial board, the designing team has also invested a significant amount of their time in understanding the subject and creating the most relevant covers. They scrutinized every image to scout for the most suitable representation of the subject and create an appropriate cover for the book.

The publishing team has been an ardent support to the editorial, designing and production team. Their endless efforts to recruit the best for this project, has resulted in the accomplishment of this book. They are a veteran in the field of academics and their pool of knowledge is as vast as their experience in printing. Their expertise and guidance has proved useful at every step. Their uncompromising quality standards have made this book an exceptional effort. Their encouragement from time to time has been an inspiration for everyone.

The publisher and the editorial board hope that this book will prove to be a valuable piece of knowledge for researchers, students, practitioners and scholars across the globe.

List of Contributors

Caroline Siefarth
Department of Chemistry and Pharmacy, Emil Fischer Centre, Friedrich-Alexander University of Erlangen-Nürnberg, Schuhstr. 19, Erlangen 91052, Germany
Fraunhofer Institute for Process Engineering and Packaging (IVV), Giggenhauser Str. 35, Freising 85354, Germany

Yvonne Serfert
Department of Food Technology, University of Kiel, Heinrich-Hecht-Platz 10, Kiel 24118, Germany

Stephan Drusch
Department of Food Technology and Food Material Science, Institute of Food Technology and Food Chemistry, Technical University of Berlin, Königin-Luise-Str. 22, Berlin 14195, Germany

Andrea Buettner
Department of Chemistry and Pharmacy, Emil Fischer Centre, Friedrich-Alexander University of Erlangen-Nürnberg, Schuhstr. 19, Erlangen 91052, Germany
Fraunhofer Institute for Process Engineering and Packaging (IVV), Giggenhauser Str. 35, Freising 85354, Germany

Muna Ilowefah
UPM-BERNAS Research Laboratory, Faculty of Food Science and Technology, Universiti Putra Malaysia, 43400 UPM, Serdang, Selangor, Malaysia

Chiemela Chinma
UPM-BERNAS Research Laboratory, Faculty of Food Science and Technology, Universiti Putra Malaysia, 43400 UPM, Serdang, Selangor, Malaysia

Jamilah Bakar
UPM-BERNAS Research Laboratory, Faculty of Food Science and Technology, Universiti Putra Malaysia, 43400 UPM, Serdang, Selangor, Malaysia

Hasanah M. Ghazali
UPM-BERNAS Research Laboratory, Faculty of Food Science and Technology, Universiti Putra Malaysia, 43400 UPM, Serdang, Selangor, Malaysia

Kharidah Muhammad
UPM-BERNAS Research Laboratory, Faculty of Food Science and Technology, Universiti Putra Malaysia, 43400 UPM, Serdang, Selangor, Malaysia

Mohammad Makeri
UPM-BERNAS Research Laboratory, Faculty of Food Science and Technology, Universiti Putra Malaysia, 43400 UPM, Serdang, Selangor, Malaysia

Laura Del Coco
Department of Biological and Environmental Sciences and Technologies (Di.S.Te.B.A.),
University of Salento, Prov.le Lecce-Monteroni, Lecce 73100, Italy

Sandra Angelica De Pascali
Department of Biological and Environmental Sciences and Technologies (Di.S.Te.B.A.),
University of Salento, Prov.le Lecce-Monteroni, Lecce 73100, Italy

Francesco Paolo Fanizzi
Department of Biological and Environmental Sciences and Technologies (Di.S.Te.B.A.),
University of Salento, Prov.le Lecce-Monteroni, Lecce 73100, Italy

Kadri Koppel
The Sensory Analysis Center, Kansas State University, 1310 Research Park Drive, Manhattan, KS 66502, USA

Suntaree Suwonsichon
Kasetsart University Sensory and Consumer Research Center, Department of Product Development, Kasetsart University, 50 Paholyothin Road, Jatujak, Bangkok 10900, Thailand

Uma Chitra
Department of Nutrition and Dietetics, Kasturba Gandhi College for Women, Secunderabad 500026, India

Jeehyun Lee
Department of Food Science and Nutrition, College of Human Ecology, Pusan National University, 30 Jangjeon-Dong, Geumjeoung-Ku, Busan 609 735, Korea

Edgar Chambers IV
The Sensory Analysis Center, Kansas State University, 1310 Research Park Drive, Manhattan, KS 66502, USA

Catherine C. Adley
Microbiology Laboratory, Department of Chemical and Environmental Sciences, University of Limerick, Limerick, Ireland

José M. Alvarez-Suarez
Department of Odontostomatologic and Specialized Clinical Sciences, Faculty of Medicine and Surgery, Polytechnic University of Marche, Avenue Ranieri 65, Ancona 60100, Italy
Department of Nutrition and Health, International Iberoamerican University (UNINI), Avenue Adolfo Ruiz Cortines 112, Torres de Cristal L 101 A-3, Campeche 24040, Me xico

Massimiliano Gasparrini
Department of Odontostomatologic and Specialized Clinical Sciences, Faculty of Medicine and Surgery, Polytechnic University of Marche, Avenue Ranieri 65, Ancona 60100, Italy

Tamara Y. Forbes-Hernández
Department of Odontostomatologic and Specialized Clinical Sciences, Faculty of Medicine and Surgery, Polytechnic University of Marche, Avenue Ranieri 65, Ancona 60100, Italy
Department of Nutrition and Health, International Iberoamerican University (UNINI), Avenue Adolfo Ruiz Cortines 112, Torres de Cristal L 101 A-3, Campeche 24040, Me xico

Luca Mazzoni
Department of Odontostomatologic and Specialized Clinical Sciences, Faculty of Medicine and Surgery, Polytechnic University of Marche, Avenue Ranieri 65, Ancona 60100, Italy

Francesca Giampieri
Department of Agricultural, Food and Environmental Sciences, Polytechnic University of Marche, Via Ranieri 65, Ancona 60100, Italy

Adreas Dimou
Department of Food Science and Human Nutrition, Agricultural University of Athens, Athens 11855, Greece

Nikolaos G. Stoforos
Department of Food Science and Human Nutrition, Agricultural University of Athens, Athens 11855, Greece

Stavros Yanniotis
Department of Food Science and Human Nutrition, Agricultural University of Athens, Athens 11855, Greece

Anupam Chugh
Department of Food Science, University of Guelph, Guelph, ON N1G 2W1, Canada

Dipendra Khanal
Department of Food Science, University of Guelph, Guelph, ON N1G 2W1, Canada

Markus Walkling-Ribeiro
Department of Food Science, University of Guelph, Guelph, ON N1G 2W1, Canada
Department of Food Science, Cornell University, Ithaca, NY 14853, USA

Milena Corredig
Department of Food Science, University of Guelph, Guelph, ON N1G 2W1, Canada

Lisa Duizer
Department of Food Science, University of Guelph, Guelph, ON N1G 2W1, Canada

Mansel W. Griffiths
Department of Food Science, University of Guelph, Guelph, ON N1G 2W1, Canada

Saskia M. van Ruth
RIKILT Wageningen UR, P.O. Box 230, 6700 EV Wageningen, The Netherlands
Food Quality and Design Group, Wageningen University, P.O. Box 17, 6700 AA Wageningen, The Netherlands

Erwin Brouwer
RIKILT Wageningen UR, P.O. Box 230, 6700 EV Wageningen, The Netherlands

Alex Koot
RIKILT Wageningen UR, P.O. Box 230, 6700 EV Wageningen, The Netherlands

Michiel Wijtten
RIKILT Wageningen UR, P.O. Box 230, 6700 EV Wageningen, The Netherlands

Edwin Fabian Torres Bello
Institute of Agrochemistry and Food Technology (CSIC), Avenida Agustín Escardino, 7 Parque Científico, 46980 Paterna (Valencia), Spain

Gerardo González Martínez
Alpina Research Institute (IAI), Alpina Productos Alimenticios S.A, Edificio Corporativo Km 3 vía, Briceño-Sopó, Cundinamarca, 251001, Colombia

Bernadette F. Klotz Ceberio
Alpina Research Institute (IAI), Alpina Productos Alimenticios S.A, Edificio Corporativo Km 3 vía, Briceño-Sopó, Cundinamarca, 251001, Colombia

Dolores Rodrigo
Institute of Agrochemistry and Food Technology (CSIC), Avenida Agustín Escardino, 7 Parque Científico, 46980 Paterna (Valencia), Spain

Antonio Martínez López
Institute of Agrochemistry and Food Technology (CSIC), Avenida Agustín Escardino, 7 Parque Científico, 46980 Paterna (Valencia), Spain

Panagiotis Sfakianakis
Laboratory of Food Chemistry and Technology, School of Chemical Engineering, National Technical University of Athens, 5 Iroon Polytechniou St., Polytechnioupoli, 15780, Zografou, Greece

Constatnina Tzia
Laboratory of Food Chemistry and Technology, School of Chemical Engineering, National Technical University of Athens, 5 Iroon Polytechniou St., Polytechnioupoli, 15780, Zografou, Greece

Konstantinos Stamatopoulos
Department of Food Technology, Faculty of Food Technology and Nutrition, Technological Educational Institute of Athens, 12 Ag. Spyridonos St., Egaleo, Athens 122 10, Greece

Archontoula Chatzilazarou
Department of Oenology and Beverage Technology, Faculty of Food Technology and Nutrition, Technological Educational Institute of Athens, 12 Ag. Spyridonos St., Egaleo, Athens 122 10, Greece

Evangelos Katsoyannos
Department of Food Technology, Faculty of Food Technology and Nutrition, Technological Educational Institute of Athens, 12 Ag. Spyridonos St., Egaleo, Athens 122 10, Greece